Materials Metrology and Standards for Structural Performance

Materials Metrology
and Standards for
Structural Performance

Edited by

B. F. Dyson, M. S. Loveday

and

M. G. Gee

Division of Materials Metrology
National Physical Laboratory
Teddington, UK

CHAPMAN & HALL

London · Glasgow · Weinheim · New York · Tokyo · Melbourne · Madras

Published by Chapman & Hall, 2–6 Boundary Row, London SE1 8HN, UK

Chapman & Hall, 2–6 Boundary Row, London SE1 8HN, UK

Blackie Academic & Professional, Wester Cleddens Road, Bishopbriggs, Glasgow G64 2NZ, UK

Chapman & Hall GmbH, Pappelallee 3, 69469 Weinheim, Germany

Chapman & Hall Inc., One Penn Plaza, 41st Floor, New York NY 10119, USA

Chapman & Hall Japan, ITP-Japan, Kyowa Building, 3F, 2-2-1 Hirakawacho, Chiyoda-ku, Tokyo 102, Japan

Chapman & Hall Australia, Thomas Nelson Australia, 102 Dodds Street, South Melbourne, Victoria 3205, Australia

Chapman & Hall India, R. Seshadri, 32 Second Main Road, CIT East, Madras 600 035, India

First edition 1995

© 1995 Chapman & Hall

© 1995 Crown copyright: Chapters 1, 5 (Appendix), 7–10, 12

Typeset in 10/12 Palatino by Pure Tech Corp., Pondicherry, India

Printed in Great Britain by the University Press, Cambridge

ISBN 0 412 58270 8

A catalogue record for this book is available from the British Library

Library of Congress Catalog Card Number: 94-71197

∞ Printed on acid-free text paper, manufactured in accordance with ANSI/NISO Z39.48-1992 (Permanence of Paper).

Contents

Contributors

J. Albright, MTS Systems Corporation, Eden Prairie, MN, USA

P. M. Cooper, National Physical Laboratory, Teddington, UK

M. J. Crowder, University of Surrey, UK

H. Czichos, Federal Institute for Materials Research and Testing, Berlin, Germany

G. D. Dean, National Physical Laboratory, Teddington, UK

M. J. Dixon, Cambridge Consultants Limited, Cambridge, UK

B. F. Dyson, National Physical Laboratory, Teddington, UK

P. E. Francis, National Physical Laboratory, Teddington, UK

M. G. Gee, National Physical Laboratory, Teddington, UK

P. J. Goodhew, University of Liverpool, UK

W. T. K. Henderson, NAMAS Executive, NPL, Teddington, UK

C. E. Hinton, Instron Limited, High Wycombe, UK

M. K. Hossain, National Physical Laboratory, Teddington, UK

F. A. Kandil, National Physical Laboratory, Teddington, UK

R. D. Lohr, Instron Limited, High Wycombe, UK

M. S. Loveday, National Physical Laboratory, Teddington, UK

I. McEnteggart, Instron Limited, High Wycombe, UK

R. Morrell, National Physical Laboratory, Teddington, UK

B. E. Read, National Physical Laboratory, Teddington, UK

B. Roebuck, National Physical Laboratory, Teddington, UK

S. R. J. Saunders, National Physical Laboratory, Teddington, UK

I. R. Sced, National Physical Laboratory, Teddington, UK

G. B. Thomas, NAMAS Executive, NPL, Teddington, UK

D. J. Walters, Consultant, Bristol, UK

L. C. Wolstenholme, City University, London, UK

Preface

Materials metrology is the measurement science used for determining materials property data. An essential element is the symbiosis between the understanding of materials behaviour and the development of suitable measurement techniques which, through the provision of standards, enable design engineers and plant operators to acquire materials data of appropriate precision. This book is concerned only with those aspects of materials metrology and standards that relate to the design and performance in service of structures and consumer products. It does not consider their important role in the processing of materials.

The editors are grateful for the commitment and patience of the experts who contributed the various chapters. In addition, help from staff in the Division of Materials Metrology, National Physical Laboratory, in assisting with the task of refereeing the chapters is gratefully acknowledged.

The production of this book was carried out as part of the Materials Measurement Programme of underpinning research financed by the United Kingdom Department of Trade and Industry.

Brian F. Dyson
Malcolm S. Loveday
Mark G. Gee
Division of Materials Metrology
National Physical Laboratory
Teddington, TW11 0LW UK

CHAPTER 1

Materials metrology and standards: an introduction

B. F. Dyson, M. S. Loveday and M. G. Gee

1.1 MATERIALS ASPECTS OF STRUCTURAL DESIGN

Knowledge concerning the behaviour of materials has always been vital for the success of manufactured products, but never more so than at the present time. The consumer revolution of the late twentieth century has resulted in a demand for products that combine functionality at an acceptable price with quality and aesthetics: products that range from humble toasters and irons to high-tech tennis racquets and golf clubs.

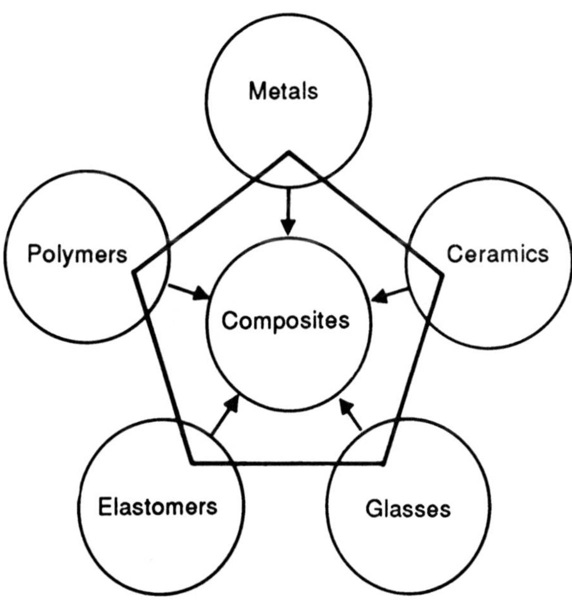

Fig. 1.1 The menu of engineering materials (Ashby, 1989a, p. 17).

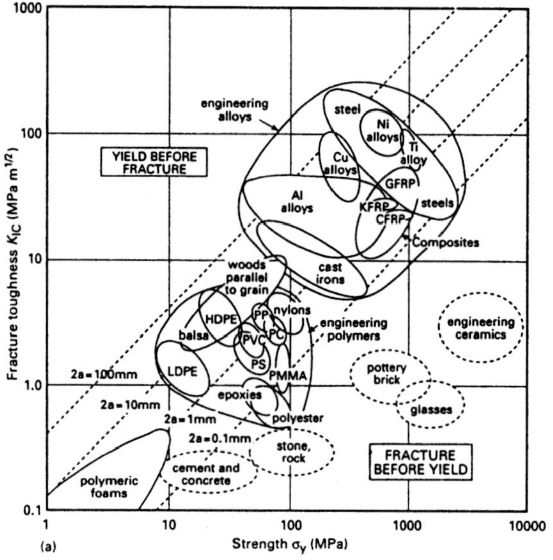

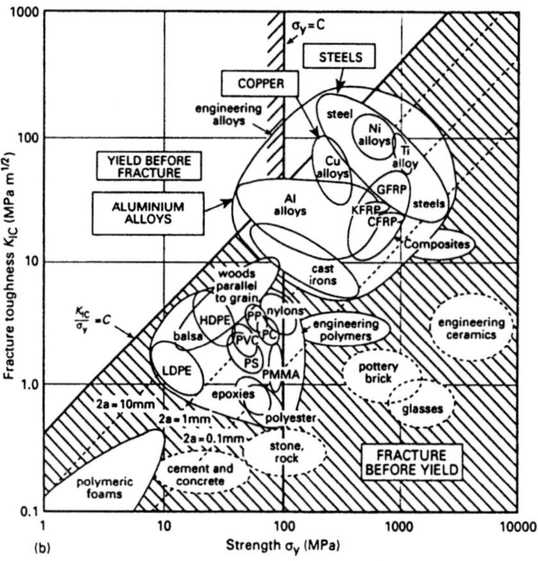

Fig. 1.2 (a) Materials selection chart: fracture toughness plotted against strength. (b) Materials for pressure vessels. Steel, copper alloys and aluminium alloys best satisfy the 'yield before break' criterion. In addition, a high yield strength allows a high working pressure. The materials in the remaining triangle are the best choice (Ashby, 1989a, p. 25).

Competition between international manufacturing companies increases remorselessly and key elements for increasing market share are access to knowledge and its speedy implementation. Vast changes have occurred during the last few decades with respect to the types of materials available for use in manufacturing consumer products. Engineers can now utilize any combination of materials from a menu of six broad classes – metals, polymers, elastomers, ceramics, glasses and composites (Fig. 1.1) – giving a staggering 50 000 materials from which to choose (Ashby, 1989a; 1989b; Waterman and Ashby, 1991). In order to help the design engineer narrow the choice of materials to a manageable few at the conceptual stage of design, Ashby (1989a; 1989b) has proposed a series of materials selection charts, an example being shown in Fig. 1.2. The axes of the charts have been chosen to display the performance-limiting thermal and short-term mechanical properties of materials: modulus, strength, toughness, density and thermal properties (conductivity, diffusivity, expansion etc.). The logarithmic scales allow performance-limiting combinations of properties (merit indices) to be examined and computed. For example, in Fig. 1.2 the merit index K_{IC}/σ_y (fracture toughness/yield stress) ensures the yield before break of a pressure vessel. Ashby (1991) has recently extended the concept to consider the influence of component geometry (shape) in order to optimize combinations of material and shape to give the best value of merit index for a given loading mode (torsion, bend etc.). The charts are now available for use on a PC. The important point as far as measurement of properties is concerned is that data for initial materials selection need not have a high degree of precision, since logarithmic coordinates are used. However, once a given class of material has been identified and a specific material chosen, the designer will invariably require more precise data to be provided by the materials supplier.

Another key market for materials lies in transport systems – cars, trains, ships and aircraft – and large capital-intensive engineering structures such as bridges, power stations and petrochemical plant. Not only are these products required to last longer but public safety becomes an important issue, since functional failure of certain critical components may (and often does) lead to loss of life. The materials property base now extends beyond the range of short-term properties considered above and the level of precision required of the data often exceeds that given by the materials supplier – if the data are supplied at all. In-house testing may then be undertaken on the specific batch of material used to manufacture the critical component. However, it is unfortunate that little if any guidance is given by designers as to the level of precision required of the data: this is a fertile area for collaboration between designers and materials metrologists.

Bridges, power stations and petrochemical plant are designed by code to last for a prescribed time and it has been common practice to extend their lifetimes for obvious financial reasons. A whole new industry has been created during the last 20 years that is aimed at predicting the remaining lifetime of these structures, particularly power stations. The cause of loss of function is usually material fracture but can be component distortion due to excessive (local) deformation or corrosion; sometimes it is a synergistic combination of all three. To meet the exacting predictive requirements, not only has the material properties base been expanded to include creep, low-cycle fatigue (with a variety of cycle types) and sometimes wear – with and without environmental interaction – but the level of precision required of the data is greater than historically demanded in design.

Storage and implementation of materials knowledge is being revolutionized by the ubiquity of personal computers. This presents an exciting new challenge to the engineer at a time when the design process itself has been radically transformed by computer-aided design and manufacture. The need for lifetime prediction of critical components is likely to increase even more in the foreseeable future as safety and environmental issues continue to exert their constraints. Metallic materials at high temperatures and composite materials at all temperatures suffer global damage mechanisms rather than the localized single-crack damage associated with fatigue and brittle fracture in metals and ceramics. This has led to an upsurge in computer-based modelling using state variable approaches to global damage which has and will continue to result in new experimental property measurements being developed.

The need (perceived or otherwise) for extended and more precise databases, in order to improve the efficiency of usage of materials, is driving the considerable effort that is being devoted world-wide to the development of new and improved testing procedures. The cost of producing the sophisticated data being demanded for advanced applications is high and thus the emphasis on international harmonization and accreditation of procedures to help remove potential barriers to trade is to be doubly welcomed, since it also facilitates collaborative research.

1.2 TOWARDS HARMONIZED MEASUREMENT TECHNIQUES

The role of accreditation organizations such as the National Measurement Accreditation Service (NAMAS) in the UK (Chapter 10) has become pivotal in the world-wide drive for improved quality and reliability of test results; a review of various national accreditation agencies has been given in Leemput (1992). The topic of quality management and assurance in testing laboratories has been recognized to be of major importance and was the subject of the first Eurolab symposium held in Strasbourg in 1992 (Eurolab, 1992a; 1992b). Matters such as the mutual recognition of test results were addressed (Forstén, 1992) and in this context the need for a wider portfolio of certified reference materials (CRMs) was recognized. These can be used for proficiency testing, for accreditation audits and for bench mark testing in collaborative research activities, as Hossain and Sced (Chapter 9) discuss in some detail. In the field of mechanical testing, the Community Bureau of Reference (BCR) is able to supply CRMs for Charpy impact testing (Marchandise, 1992) and for creep testing (Gould and Loveday, 1992). In addition, work is in hand to produce a CRM for room temperature tensile testing (Loveday, 1992).

Documented testing procedures, codes of practice and written standards provide guidelines for accrediting testing laboratories and considerable progress has been made to harmonize testing standards within the Single European Market (Roche and Loveday, 1992).

Collectively, the above aspects of a measurement infrastructure, together with the primary standards maintained at national standards laboratories such as the National Physical Laboratory (NPL), UK, the National Institute of Science and Technology (NIST), USA, and the Physikalisch-Technische Bundesanstalt (PTB), Germany, form a **national measurement system** (NMS). This topic has been reviewed by Hossain and Sced in Chapter 9: details are given of the hierarchical series of calibrations which provide traceability from the test house, workshop or research laboratory to the appropriate national primary standards for the material properties under investigation. It is important to assess the precision of the measurement at each calibration stage and this has now been recognized by official bodies: accreditation agencies and documented standards now require that measured results are accompanied by an estimate of the uncertainty of measurement. A compilation of the terms used for this purpose is given in Appendix 1.A.

International collaboration in the materials field has been fostered by a number of organizations, each of whom has acted to focus and coordinate pre-competitive research or to produce codes of practice as precursors to standards. Outline details of some of these organizations are given in Appendix 1.B and further information is given elsewhere, e.g. for COST in Marriott (1991) and for VAMAS in Hossain (1992).

1.3 STANDARDS

Documentary standards, known commonly as 'standards', are defined in BS 0 : 1981; they are distinguished from measurement standards (known in French as *étalons*), which are more commonly known as the primary or fundamental standards of mass, length, time etc. of the type maintained at establishments such as NPL. Documentary standards (known in French as *normes*) for some materials properties such as the measurement of tensile or creep behaviour have been in existence as national standards for many years. Such standards often form the mainstay of quality assurance schemes and provide a framework for traceability to the national measurement system.

Comparisons of design data become meaningful only when those data have been acquired under similarly controlled testing conditions, which are specified in agreed national or international standards. In the United Kingdom, standards are produced by national committees of experts under the auspices of the British Standards Institution (BSI), and one of the major changes that has taken place over the last decade is that the phraseology used in the standards is now couched in mandatory terms following the guidelines laid down in BS 0 : 1981 rather than in terms of recommendations or suggestions as previously used.

It is important that standards specify the tolerances in the testing conditions such that material property data of 'appropriate accuracy' can be measured. Appropriate accuracy depends upon the purpose for which the data are required: a level of precision sufficient for product release may be inadequate for determining reliable data for design or life prediction of high-technology plant, which often operate under severe thermal, mechanical or environmental conditions. Relatively simple product release tests such as hardness, impact strength, tensile testing, or even the 100 hour high-temperature stress rupture test do not provide the design engineer or plant operator with the necessary data to safely design or predict the remanent life of components.

It should be of concern that in the drive towards harmonization of standards across the Single European Market, many of the new European Standards have wider testing tolerances than existed in the individual national standards on which they were based. This will inevitably lead to wider scatter in the material databases at a time when design engineers require more sophisticated and precise data to take full advantage of advanced finite element modelling or other computer-based design techniques. It is also ironic that the widening of the testing tolerances has come during a period that has seen considerable advances in control technology of testing machines (see Chapter 3), so enabling data to be acquired with far less scatter than hitherto possible!

Standards produced under the guidance of the International Standards Organization (ISO) invariably have less stringent specifications for the testing conditions, since they tend to encompass the different specifications in the various national standards. The national delegations serving on the ISO committee were happy to accept this state of affairs since work carried out in their own countries to their own national standards would then automatically comply with the ISO Standards. However, with the introduction of the Single Market it was agreed to adopt European Standards which would supersede the national standards of the individual member states. Where possible, it was also agreed to adopt ISO Standards as the working drafts for the European Standards, generally without further tightening of the testing tolerances accepted in the ISO Standards.

An example of the less stringent requirements in testing tolerances can be seen in the field of tensile testing of metallic materials at elevated temperatures. If equivalent errors from all sources are considered then the permitted tolerances between 600 and 800 °C in the ISO and EN (European) Standards are ± 7 °C; in comparison, the tolerance in the superseded British Standard (BS 3688) was only ± 5 °C. Further detailed comparisons of various testing standards have been given in Roche and Loveday (1992).

The importance of the testing standards may be appreciated by considering Fig. 1.3. The schematic illustration shows the relationship of the testing and calibration standards (developed by materials metrologists) to the design codes (used by design and plant engineers) and to the primary standards (maintained at national standards laboratories). Before some components – e.g. pressure vessels – can be put into service, there is a legal requirement for a certificate to be issued by the appropriate accredited authority to ensure that its design is in accordance with the procedures laid down in the design codes. Typically, the design codes (e.g. BS 5500) specify materials property data that must be measured in accordance with the appropriate testing standard, e.g. room temperature tensile properties such as the 0.2% proof strength ($R_{\mathrm{P0.2}}$) or the tensile strength (R_{M}) have to comply with EN 10002/1. In turn, the tensile machine used to determine the tensile properties has to be calibrated following the procedure laid down in EN 10002/2 using a force-proving device verified in accordance with EN 10002/3 using a deadweight machine. The latter has to have traceability to the national primary standard, the kilogram, which is maintained at the national standards laboratory. To ensure that the traceability chain is not broken, the laboratories undertaking the testing and calibrations should be audited by an independent accreditation agency such as NAMAS.

Thus the legal requirements for the design of critical components are based upon materials property data acquired in accordance with testing and calibration standards which have traceability to the national primary standards as illustrated in Fig. 1.4.

1.4 MATERIALS METROLOGY

Measurements of structural properties of materials such as yield, ultimate strength and toughness are normally carried out using mechanical testing machines. These machines should be designed, chosen and used in such a way that the required measurements can be carried out easily, effectively and with the necessary precision and traceability of measurement. In Chapter 2, McEnteggart and Lohr discuss how these aims can be achieved within the context of the different types of mechanical testing machine that are currently available.

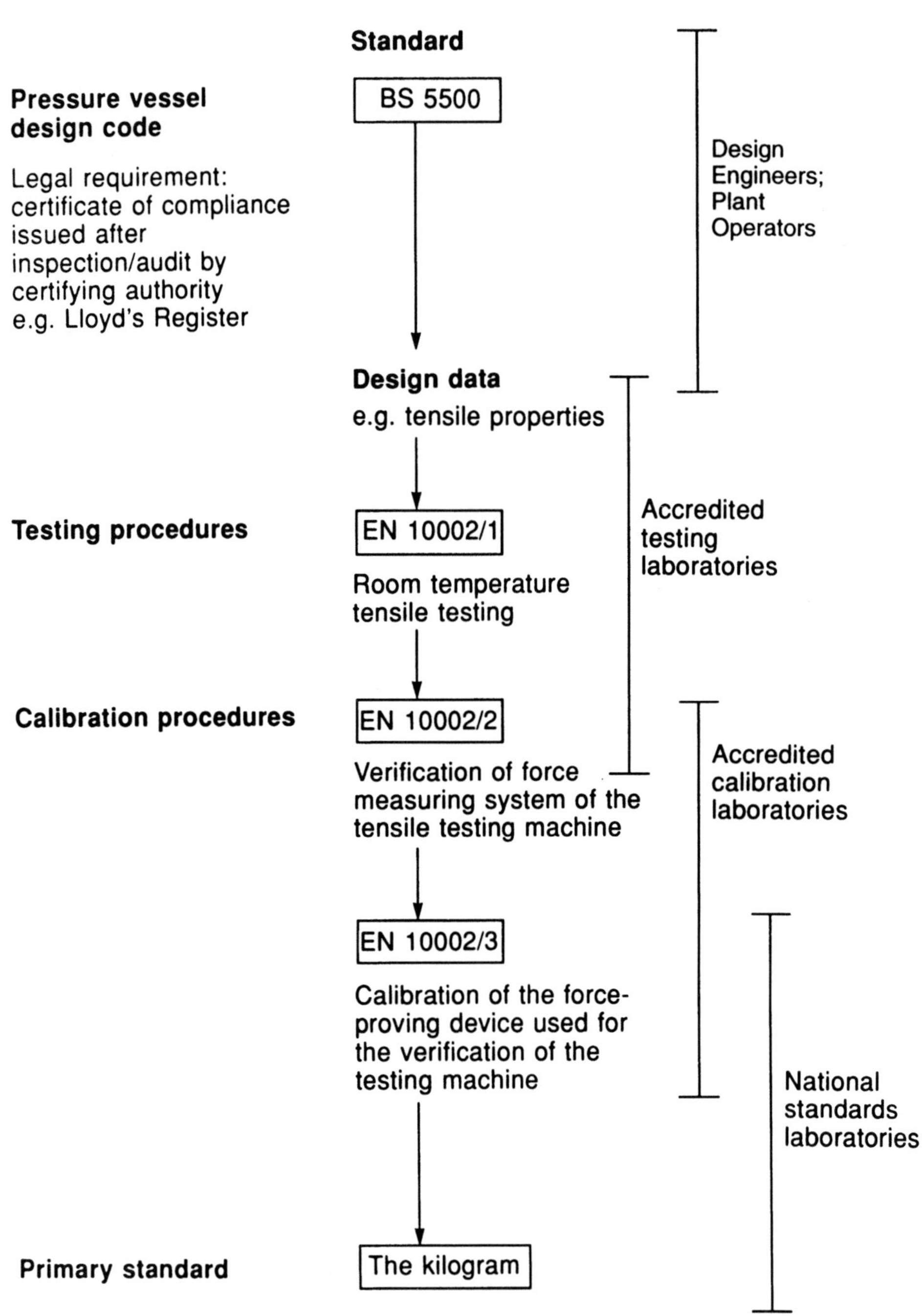

Fig. 1.3 An example of the traceability route for a product via its design code through the testing and calibration standards to a primary standard.

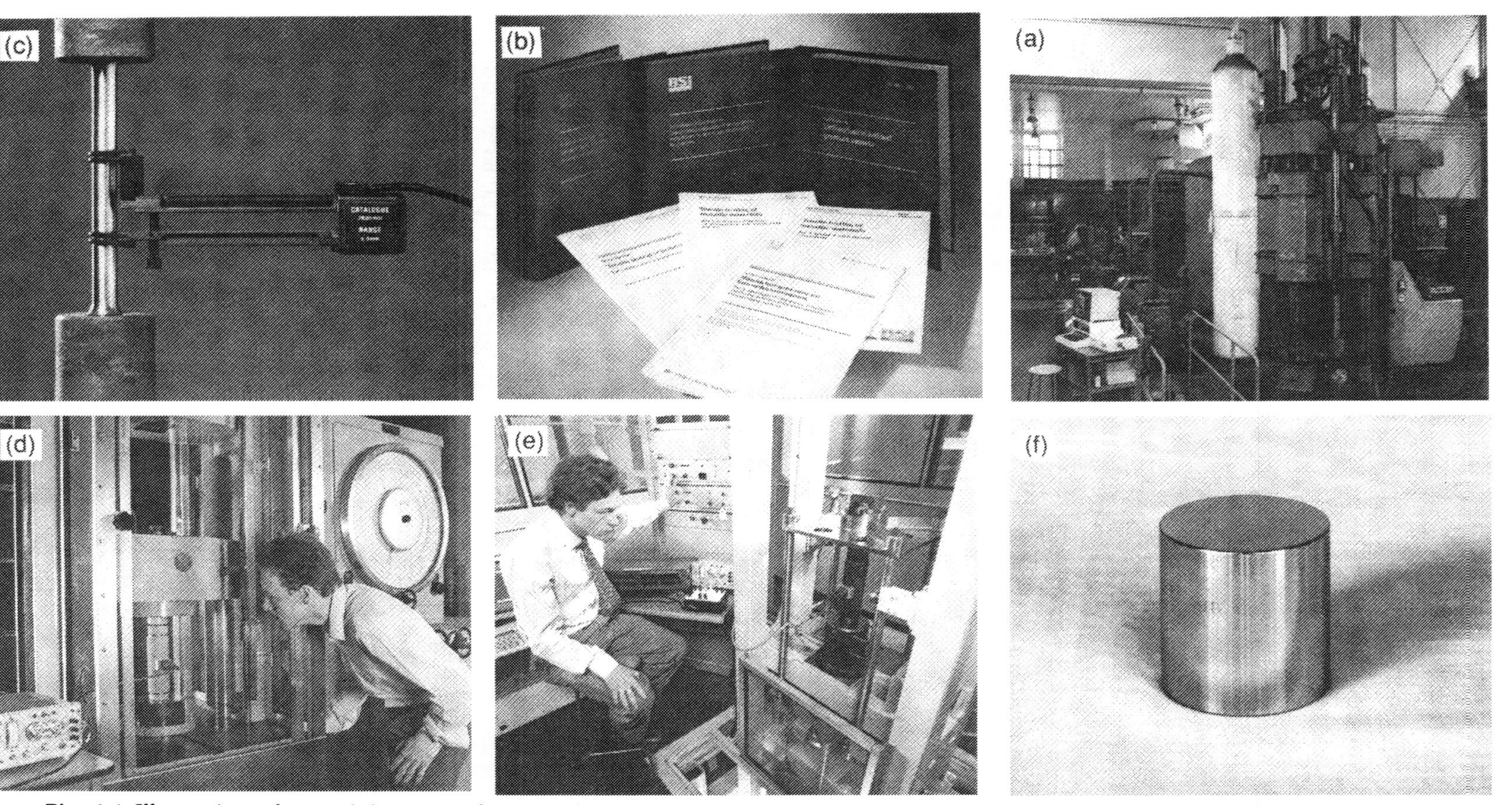

Fig. 1.4 Illustration of traceability route from product to primary standard. (a) Hydraulic pressure vessel controlling 700 tonne extrusion press. (b) Design codes, testing and calibration standards. (c) Room temperature tensile testing. (d) Calibration of testing machine using a force-proving device. (e) Verification of force-proving device in deadweight machine. (f) UK primary kilogram.

Alignment of the loading train in mechanical testing systems is often not considered with the care that is required. Misalignment can have an adverse effect on many test results, giving large scatter in modulus data and a decrease in fatigue life under cyclic loading. One interesting approach is the development of self-aligning grips that minimize the need for precise load train alignment. Techniques have also been devised that incorporate sensors for the measurement of bending within the testpiece: mechanical adjustments are used to eliminate the bending that is generated as the testpiece is mounted in the test machine.

The development of digital electronics and powerful personal computers/workstations is having a considerable effect on the means of controlling modern mechanical testing machines and for the capture and analysis of data. Digital controllers, which are increasingly being used in newer generations of testing machines, have many advantages such as insensitivity to noise, flexibility, the possibility for adaptive control, and simple interfacing to computers for data acquisition and analysis. Modern materials testing systems are now usually equipped with a computer that is used for the overall control of the test, data logging, data storage, and data analysis. Integration of computers into test systems can give flexibility and ease of use, but these very factors may also lead to the danger of too much reliance being placed on the computer so that errors are introduced by improper control of the test or choice of operating conditions. Software validation is an issue of considerable concern.

There is increasing interest in the use of materials under high rates of application of load or strain. Servohydraulic testing machines are used for this type of task. In Chapter 3, Hinton describes how the tuning of the machine controller can be optimized to reduce errors in the control feedback loop. In particular, consideration is given to the limitations of the well-known proportional + integral + derivative (PID) controller, and to strategies which lead to improved control such as active suppression of the effects of resonance in the actuator and load frame, and the use of adaptive control of load and frequency.

Until recently, the conventional static procedure was used for calibrating load cells in dynamic cyclic and high-rate tests. There is considerable evidence to suggest that this static procedure is not adequate under dynamic conditions. Dixon (Chapter 4) describes the progress that has been made towards the development of a reliable, traceable and cost-effective procedure for dynamic force measurement and calibration. This new procedure uses compensation of the inertial loading generated within the test system by acceleration of components such as the testpiece grips. The compensation is achieved by a correction signal that is derived from measurements of acceleration of a well-defined inertial mass which is firmly attached to the loading train. The new procedure has the potential to form a reliable basis for traceable load measurements under dynamic conditions.

Accurate measurements of strain within a testpiece need to be made whenever knowledge about the deformation of materials under applied stresses is required. Materials are tested under a wide range of imposed stress states, and over a wide range of temperatures from room temperature to above 1500 °C (for ceramics). In many ways, the measurement of strain in creep tests can be particularly severe, since small displacements are measured over periods that can extend to many thousands of hours. Traceable, accurate measurements of strain need to be made under all of these conditions. In Chapter 5, Walters describes the use of contact strain measurement methods including extensometry, emphasizing the modifications necessary to make good measurements at high temperature. Particular issues that are raised include the design of ambient temperature extensometers with foil resistance gauges, the use of diametral extensometry, the measurement of biaxial and triaxial strain, and the measurement of strain in notched testpieces. The chapter is supplemented by an appendix by Loveday that discusses the impact of the new ISO and EN Standards for the verification of extensometers used in uniaxial testing. Non-contacting strain measurement is not reviewed in this book, but the subject has recently been considered in detail in McEnteggart (1992).

Dynamic testing requires the measurement of strain as well as load, but there is only limited knowledge of the dynamic characteristics of strain measurement extensometers, and there are no widely accepted standards for the dynamic calibration of extensometers. In Chapter 6, Albright describes potential sources of dynamic errors such as the resonances that are induced in the mechanical parts of the extensometer. The extensometer response is sensitive to the quality of the applied load waveform, and is test machine dependent. The framework that is necessary for an acceptable dynamic strain measurement standard is described in some detail.

The subject of load train misalignment and its effect on scatter in test results is revisited in the discussion by Kandil and Dyson of the errors introduced by misalignment of testpieces within the loading train in low-cycle fatigue (LCF) experiments (Chapter 7). The importance of this source of scatter was brought to the fore by an analysis of a large international LCF interlaboratory comparison which demonstrated that whilst the repeatability of results within a single laboratory was quite small, the reproducibility of results from one laboratory to another was unacceptable. The analysis by Kandil and Dyson showed that misalignment accounts for the majority of the scatter in data, and was consistent with the observed poor reproducibility in the interlaboratory exercise.

One of the most important properties of a material is its stiffness modulus. This determines the magnitude of the deformation of a structure when subjected to loading under elastic conditions and, together with the density, the dynamic response of structures. Chapter 8 by Dean, Loveday, Cooper, Read and Roebuck describes how measurements of modulus can be made by calibrated measurements of these responses.

Analysis of the linear part of a stress–strain curve can be used to obtain values for modulus with an accuracy as good as ±1%, when care is taken. This is contrasted with acoustic resonance and ultrasonic wave techniques which are particularly suited to materials having a very high modulus and a low failure strain. Ultrasonic techniques also offer an accurate and convenient method for measuring all the elastic constants of materials that are anisotropic. The main part of the chapter is followed by an appendix by Morrell which defines the limitations to accuracy when using different equations for calculating elastic properties from resonance techniques for ceramics.

The measurement of the degradation of materials is the subject of Chapters 11 and 12. Czichos (Chapter 11) describes a systematic framework for wear metrology which subdivides the parameters controlling wear into the three groups. These are the structural parameters, defining the components (materials, lubricant and environment) of the test and their physical, chemical and technological properties; the operational parameters, including the loading kinematic and temperature conditions and their duration; and the interaction parameters, which characterize the action of the operating parameters on the structural components of the tribological system. The chapter is concluded with a discussion of the impact of computers on test procedures, and the use of a systematic methodology for wear testing.

Francis and Saunders (Chapter 12) outline the factors governing the precision of corrosion measurements. In aqueous corrosion, the measurement of potential and temperature, and the control and characterization of the test environment, are critical. For high-temperature corrosion, continued protection depends upon the diffusion barrier characteristic of the corrosion product and its chemical stability. The mechanical properties of the reaction product are also important since any mismatch in thermal expansion coefficient induces stresses during growth of the oxide. Thus, in the section on high-temperature corrosion measurement, the problems associated with gas composition and the determination of the mechanical properties of the protective layers are considered.

Quantitative descriptions of the microstructure of materials are beginning to be used routinely to characterize engineering materials, with applications in cleanness assessment, inclusion counting, phase analysis, grain size determination, volume fraction and size and distribution of particles and fibres in composites, and porosity in ceramics. In Chapter 13, Goodhew assesses the extent to which metrological reproducibility and traceability can be achieved in these fields, and points out some of the areas where procedures need to be refined. The chapter includes a useful appendix listing standards relevant to microstructural metrology.

Statistical procedures are increasingly being used in the design of experiments in general and in the analysis of test results: they give a proper basis for the quantification of the uncertainty associated with measurements. In Chapter 14, Wolstenholme and Crowder assess the statistical methods that are likely to be appropriate in the context of materials metrology. The importance of exploratory data analysis based on graphical methods is emphasized, and common measures to describe the spread and position of data such as the mean and standard deviation are outlined. Other more sophisticated techniques such as analysis of variation (ANOVA) are then developed in more detail.

Interlaboratory comparison studies are a necessary part of the standardization process for measurements. In Chapter 15, Crowder and Wolstenholme consider the statistical issues concerned with the design of interlaboratory comparisons, and the analysis of the results forthcoming from such studies. It is concluded that the best practice in the design of interlaboratory exercises will only be achieved when proper account is taken of the requirements that arise from the statistical design and analysis of such exercises.

In this book, space dictated that only certain aspects of materials metrology could be covered. The large and important area of high-temperature mechanical property measurements were not included because they have been treated extensively over the past decade (Loveday, Day and Dyson, 1982; Sumner and Livesey, 1985; Gooch and How, 1986; Dyson, Lohr and Morrell, 1989; Loveday and Gibbons, 1992). Physical properties have also been omitted.

APPENDIX 1.A GLOSSARY OF TERMS FOR THE EXPRESSION OF UNCERTAINTY IN MEASUREMENT

Term	*Definition*	*Notes*
Accuracy (VIM, 1990)	The degree to which the result of a measurement agrees with the true value of a measurand. Accuracy combines random and systematic errors and thus should be expressed in terms of precision and trueness (bias).	Accuracy is a qualitative concept. The term 'precision' alone should not be used for 'accuracy'. ISO Standards now require trueness (or bias) to be stated rather than accuracy.
Bias (ASTM, 1988)	Persistent or systematic error that remains constant over a series of replicated measurements.	Bias can be estimated by the difference between a measured average and an accepted standard or reference value.
Certified reference material (CRM) (Seah, 1992)	A reference material whose property value(s) is certified by technically valid procedures accompanied by, or traceable to, a certificate or other documentation which is issued by a certifying body, e.g. BCR or ASTM.	The material (or substance) must be stable and homogeneous and have properties sufficiently well established for use for calibration of apparatus, or assessment of a measurement method (ISO, 1992, Section 2.1).

Term	*Definition*	*Notes*
Confidence level	The probability that the true value lies within the quoted range of uncertainty. For most purposes a value of 95% probability is acceptable, i.e. the reading lies within $\pm 2\sigma$, where σ is the standard deviation of the measured results.	
Error of measurement (*see* random error and systematic error) (VIM, 1990)	The result of a measurement minus the true value of the measurand.	'Absolute error', which has a sign, should not be confused with 'absolute value of an error', which is the modulus of an error.
Measurand (VIM, 1990)	A quantity subjected to measurement.	As appropriate, this may be the quantity to be measured, or the measured quantity.
National standard (Seah, 1992)	A standard recognized by an official national decision as the basis for fixing the value, in a country, of all other standards of the quantity concerned.	
Precision (*see* repeatability and re-producibility) (BS 5497 : Part 1 : 1987; ISO 5725)	A general term for the closeness of agreement between replicate test results.	A concept to describe the **dispersion** of repeated measurements with respect to a measure of location of central tendency, usually the mean (ASTM, 1988).
Random error (VIM, 1990)	A component of the error of measurement that, in the course of a number of measurements, varies in an unpredictable way.	It is not possible to correct for random error; however, its expectation is zero. The sample variance, calculated from the variations of repeated observations of the measure and, serves to characterize the random error. The standard deviation of the mean of the observations is used to estimate the component of the uncertainty of measurement arising from the random variations.

Term	*Definition*	*Notes*
Repeatability (ISO 5725)	Repeatability r is the value below which the absolute difference between two single test results obtained with the same method on identical test materials, under the same conditions (same operator, same apparatus, same laboratory and within a short interval of time), may be expected to lie with a specified probability (in the absence of other indicators, the probability is 95%).	
Reproducibility (ISO 5725)	Reproducibility R is the value below which the absolute difference between two single test results obtained with the same method on identical test materials, under different conditions (different operators, different apparatus, different laboratories, different times etc.) may be expected to lie with 95% probability (unless a different probability level is specified).	
Range (ASTM, 1988)	The largest minus the smallest of a set of numbers.	
Scatter	A generic term used to describe the range or spread of test results.	
Standard deviation	A statistical measure of the dispersion of a set of values: symbol σ.	
Standard error	A statistical measure of the dispersion of a set of values expressed as the ratio of the standard deviation to the square root of the number of values.	
Systematic error (*see* bias) (VIM, 1990)	A component of the error of measurement that, in the course of repeated measurement of the same measurand, remains constant or varies in a predictable way.	Systematic error, and its cause, may be known or unknown.

Term	*Definition*	*Notes*
Traceability (Seah, 1992)	The property of a result of a measurement whereby it can be related to appropriate standards, generally international or national standards, through an unbroken chain of comparisons.	
Transfer standard (Seah, 1992)	A standard used as an intermediary to compare standards, material measures or measuring instruments.	
Trueness (ISO 5725)	Closeness of agreement between the average value of a large number of test results and the true or accepted reference value.	
True value (Thomas, 1991)	The value of a quantity which would be obtained if the uncertainty associated with its measurement could be reduced to zero.	The value of a measurand that is completely defined (VIM, 1990, Section 1.18) The result obtained by a perfect measurement. True value is an idealized concept.
Uncertainty	A statement of the limits of the range within which a measured value of a measurand is expected to lie with respect to the true value.	A generic term characterizing the inability of a measurement process to measure the true value. The term has been used to encompass both precision and accuracy (ASTM, 1988).

See also BS 5532 : Parts 1 and 3 and ISO (1991).

APPENDIX 1.B   ORGANIZATIONS CONCERNED WITH PRE-COMPETITIVE RESEARCH OR CODES OF TESTING PRACTICE.

Acronym	*Organization/location*	*Activity*
BCR	Community Bureau of Reference, CEC DG XII, Brussels, Belgium	Industrial and materials technologies, measurement and testing, reference materials
COST	European Cooperation in Scientific and Technical Research, CEC DG XII G1, Brussels, Belgium	19 countries; over 40 programmes including telecommunications, transport, materials, agriculture, metrology, food technology

ESIS (formerly EGF)	European Structural Integrity Society, c/o Laboratory for Metal Science and Engineering, Attention Professor A. Bakker, PO Box 5025, 2600 GA Delft, The Netherlands	24 nations; fosters research into the prevention of failure of engineering materials, components and structures
Eurolab	Eurolab Secretariat Laboratoire National d'Essais, 75015 Paris, France	17 countries (EC and EFTA); forum for testing and analytical laboratories for scientific and technical cooperation
HTMTC	High Temperature Mechanical Testing Committee, NPL, Teddington, UK	Improvement of HT testing techniques, codes of practice, conferences and publications
VAMAS	Versailles Project on Advanced Materials and Standards, c/o Dr H. Rook, Materials Science and Engineering Laboratory, NIST, Gaithersburg, Maryland 20899, USA	International pre-competitive collaborative research

REFERENCES

Ashby, M. F. (1989a) Materials selection in conceptual design, in *Materials and Engineering Design* (eds B. F. Dyson and D. R. Hayhurst), Institute of Metals, London, Part I, Chapter 2, pp. 13–25.

Ashby, M. F. (1989b) On the engineering properties of materials. *Acta Metallurgica*, **37**, 1273–93.

Ashby, M. F. (1991) Materials and shape. *Acta Metallurgica et Materialia*, **39**, 1025–39.

ASTM (1988) Proposed terminology relating to precision and bias statements in test method standards, in *Standards on Precision and Bias for Various Applications*, 3rd edn, C26, ASTM, pp. 473–480.

BS 5497 : Part 1 : 1987 Precision of Test Methods: Guide for the Determination of Repeatability and Reproducibility for a Standard Test Method. Equivalent to ISO 5725.

BS 5532 : Part 1 : 1973 Statistical Terminology. Part 1: Glossary of Terms Relating to Probability and General Terms Relating to Statistics. Equivalent to ISO 3534/1.

BS 5532 : Part 3 : 1986 Statistical Terminology. Part 3: Glossary of Terms Relating to the Design of Experiments. Equivalent to ISO 3534/3.

Dyson, B. F., Lohr, R. D. and Morrell, R. (1989) *Mechanical Testing of Engineering Ceramics at High Temperatures*, Elsevier Applied Science.

Eurolab (1992a) *Quality Management and Assurance in Testing Laboratories: Proceedings of the 1st Eurolab Symposium*, January 1992, Strasbourg, France.

Eurolab (1992b) Workshop and Seminar on Measurement Uncertainty in Testing, December 1992, Barcelona, Spain.

Forstén, J. (1992) Mutual recognition of test results, in *Proceedings of the 1st Eurolab Symposium*, vol. 2, pp. 52–62.

Gooch, D. J. and How, I. M. (1986) *Techniques for Multi-Axial Creep Testing*, Elsevier Applied Science.

Gould, D. and Loveday, M. S. (1992) A reference material for creep testing, in *Harmonisation of Testing Practice for High Temperature Materials* (eds M. S. Loveday and T. B. Gibbons), Elsevier Applied Science, London, Chapter 6, pp. 85–109.

Hossain, M. K. (1992) Current status and future trends, Isprom '91; and VAMAS: the next phase. *VAMAS Bulletin*, no. 15, 2–9, National Physical Laboratory, UK.

ISO 5725 : 1986 Accuracy (Trueness and Precision) of Measurement Methods and Results. Part 1: General Principles and Definitions.

ISO(1991) *Guide to the Expression of Uncertainty in Measurement*, 5th draft, ISO/TAG4/WG3, circulated by BSI, document 91/89017.

ISO (1992) *Terms and Definitions Used in Connection with Reference Materials*, Guide 30.

Leemput, P. J. van de (1992) Laboratory accreditation: standards, reference materials and proficiency testing, in *Harmonisation of Testing Practice for High Temperature Materials* (eds M. S. Loveday and T. B. Gibbons), Elsevier Applied Science, London, Chapter 4, pp. 53–65.

Loveday, M. S. (1992) Towards a tensile reference material, in *Harmonisation of Testing Practice for High Temperature Materials* (eds M. S. Loveday and T. B. Gibbons), Elsevier Applied Science, London, Chapter 7, pp. 111–53.

Loveday, M. S., Day, M. F. and Dyson, B. F. (1982) *Measurement of High Temperature Mechanical Properties of Materials*, HMSO, London.

Loveday, M. S. and Gibbons, T. B. (eds) (1992) *Harmonisation of Testing Practice for High Temperature Materials*, Elsevier Applied Science, London.

Marchandise, H. (1992) The calibration and standardisation of the impact toughness test, in Chapter 5, pp. 67–83, *Harmonisation of Testing Practice for High Temperature Materials* (eds M. S. Loveday and T. B. Gibbons), Elsevier Applied Science, London, Chapter 5, pp. 67–83.

Marriott, J. B. (1991) Collaborative materials research in power engineering: an overview of the COST initiative. *Materials at High Temperatures*, **9** (3), 122–6.

McEnteggart, I. (1992) Contacting and non-contacting extensometry for ultra high temperature testing, in *Proceeding of the International Symposium on Ultra High Temperature Mechanical Testing* (eds R. D. Lohr *et al.*). To be published 1994 by Woodhead Publications.

Roche, R. and Loveday, M. S. (1992) Harmonisation and improvements in European Standards, in *Harmonisation of Testing Practice for High Temperature Materials* (eds M. S. Loveday and T. B. Gibbons), Elsevier Applied Science, London, Chapter 4, pp. 33–52.

Seah, M. P. (1992) Interlaboratory studies. *VAMAS Bulletin*, no. 15, 10–14.

Sumner, G. and Livesey, V. B. (1985) *Techniques for High Temperature Fatigue Testing*, Elsevier Applied Science.

Thomas, G. B. (1991) *Reporting Uncertainty into the Measurement of Mechanical Properties*, 1st draft, MMTC (91)3, Mechanical and Metallurgical Testing Committee, NAMAS, NPL.

VIM (1990) *International Vocabulary of Basic and General Terms in Metrology*, draft revision, International Bureau of Weights and Measures (BIPM).

Waterman, N. A. and Ashby, M. F. (eds) (1991) *Elsevier Materials Selector*, vol. 1, Elsevier Applied Science.

CHAPTER 2

Mechanical testing machine criteria

I. McEnteggart and R. D. Lohr

2.1 INTRODUCTION

This chapter considers the important characteristics of the main types of universal testing machines. These machines are used in both research and quality control to provide engineering property data on a wide range of materials, such as metals, polymers and ceramics and their respective composites.

The relative importance of the various machine characteristics for different types of testing are assessed. The effects of the machine performance and mode of operation on the outcome of typical types of test including tensile, creep and fatigue (high and low cycle) are considered. The relative merits of analogue and digital control are outlined and consideration is given to applications software.

2.2 THE MAIN TYPES OF UNIVERSAL TESTING MACHINE

There are three main types of 'universal' materials testing machine in common use today.

2.2.1 The electromechanical testing machine

In the electromechanical testing machine, rotary motion from a servo-motor is transmitted to a moving crosshead via a belt or gear reduction and a pair of lead screws (Fig. 2.1). The machine is servo controlled via feedback from an incremental encoder located close to the motor.

Electromechanical testing machines are used for static and quasi-static testing. These machines are available in force capacities from below 5 kN to above 600 kN, and speed ranges of $1\,\mu\text{m min}^{-1}$ to $2000\,\text{mm min}^{-1}$ can be provided.

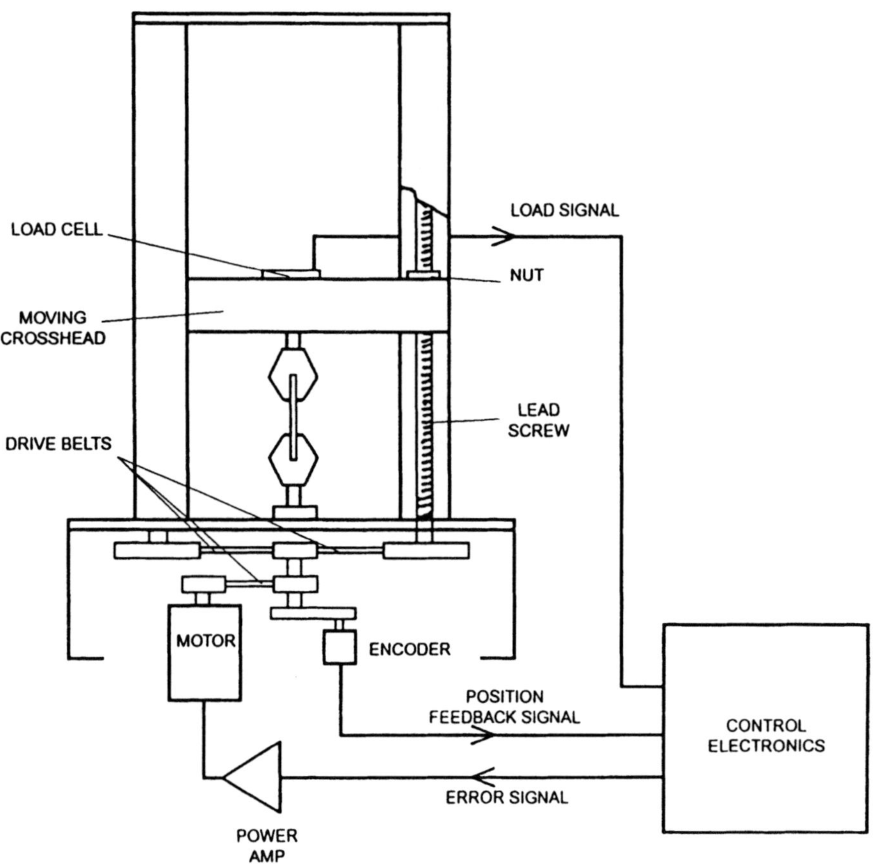

Fig. 2.1 Electromechanical testing machine.

2.2.2 The servohydraulic testing machine

In the servohydraulic testing machine, the drive is provided directly by the piston of a double-acting hydraulic actuator controlled via a servovalve (Fig. 2.2). Feedback from a linear position sensor (often a linear variable differential transformer (LVDT)) mounted in the actuator is used to achieve position control. Load and strain control operation are also provided.

Servohydraulic machines are capable of operation over a wide frequency range (DC to 500 Hz) and hence satisfy the requirements of a very wide range of applications. These machines are available to cover force capacities from below 10 kN to above 1 MN.

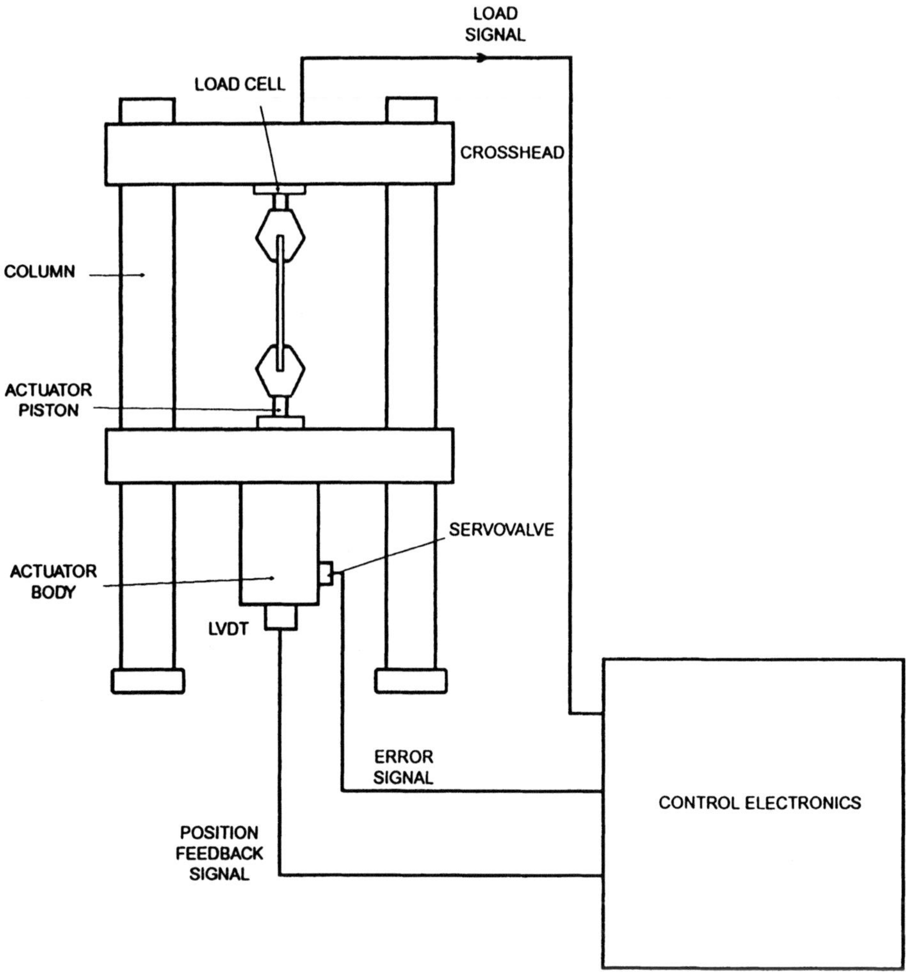

Fig. 2.2 Servohydraulic testing machine (hydraulic connections not shown).

2.2.3 The electric actuator machine

The electric actuator testing machine has a similar mechanical layout to a servohydraulic machine: however, the hydraulic actuator is replaced by a single recirculating ball screw driven from a servomotor via a reduction gearbox. As in the servohydraulic machine, position control is provided via a linear position sensor. Load and strain control are also provided.

Electric actuator machines have a typical maximum operating frequency of 1 Hz, and are capable of operating at very slow rates ($1 \, \mu\text{m h}^{-1}$ in strain control). They are well suited to creep/fatigue testing where long dwell periods are required during which the servovalve in a servohydraulic machine may 'silt up'.

2.3 TESTING FRAME

2.3.1 Force capacity

The testing machine frame must have a force capacity (fatigue rated for dynamic applications) matched to the capability of the drive system. This is rarely a problem as most frames are designed to achieve a high stiffness.

2.3.2 Daylight

The testing frame should provide a workspace with sufficient horizontal and vertical daylight to accommodate the testpiece, load string and other accessories. In some electromechanical machines, workspaces may be provided above and below the moving crosshead or a side testing space may be fitted. Ease of operator access to the working area is an important consideration.

In servohydraulic machines, achieving easy access to the working area may mean positioning the actuator above the top crosshead or sinking the machine into a pit. Locating the actuator in the crosshead allows the fitting of T-slot tables, fluid baths (for conducting tests in corrosive media), autoclaves etc. to the machine platen.

2.3.3 Stiffness

A key characteristic of a testing frame is stiffness. In general axial, lateral and torsional stiffnesses need to be considered.

In most routine tensile testing, axial stiffness is the most important parameter. A high value of frame stiffness will minimize frame deflections and hence the elastic energy stored in the frame during a test. Frame deflections decrease the accuracy of measurements based on crosshead or actuator movement. Excessive elastic energy stored in the frame will influence the results of tests conducted in position or load control and lead to high-energy testpiece failures at reduced elongation values (Newman and Sigvaldason, 1965). It can also lead to premature failure of machine components due to repeated shock loading.

For an electromechanical machine, stiffness may be specified for the frame only or for the frame plus the drive. The frame-only stiffness reflects the stiffness of the structural components of the frame, i.e. the crosshead, table and drive screws. The drive stiffness includes the compliance of the drive train components outside the control loop, e.g. belts, pulleys and gears. A typical 100 kN capacity electromechanical testing machine (Instron 4505) has an axial frame plus drive stiffness of 300 kN mm^{-1}.

Servohydraulic machine stiffness is essentially determined by the stiffness of the frame. The compliance of the servohydraulic actuator does not influence the stiffness as it is inside the control loop.

Tests involving compressive loading, including low-cycle fatigue, require a frame with high lateral stiffness able to react to the side loads generated by non-axial loading and hence minimize testpiece bending. For high lateral stiffness the frame design should incorporate structural guidance columns with high second moment of area in both lateral directions. Servohydraulic frames include structural columns to support the top crosshead. Guidance columns are not a feature of all electromechanical frames: many frames rely on the drive screws to guide the crosshead. Typical values of electromechanical machine lateral stiffness (load applied to the side of the moving crosshead) are 44 kN mm^{-1} for a 100 kN frame with guidance (Instron 4505 using 100 mm diameter guidance columns).

The lateral stiffness of a servohydraulic machine is determined by the frame and the actuator. In applications requiring the highest lateral stiffness, actuators with hydrostatic bearings are to be preferred over those with plain bearings.

Some tensile tests, e.g. tensile testing of wire and rope, can generate torsional loads. However, the largest torques will be generated in tension–torsion machines used for example in biaxial low-cycle fatigue and the testing of components such as drive shafts. Clearly such tests require frames with good torsional stiffness, which can be achieved by incorporating large-diameter guidance columns and, where applicable, by going to four columns.

2.4 CONTROL MODES

2.4.1 General

All modern materials testing machines, both electromechanical and servohydraulic, employ closed-loop control systems (Chapter 3). Many electromechanical machines offer position control only. Most servohydraulic machines and some electromechanical machines provide facilities for load and strain control.

The type of control mode used in a test can dramatically influence the results. Figure 2.3 shows the (typical) effect of different control modes on the result of a tensile test on annealed mild steel. The difference between position and strain control is greatest for tests on high-modulus materials, such as metals and composites, where the testpiece stiffness is significant with respect to the testing machine and load string stiffness.

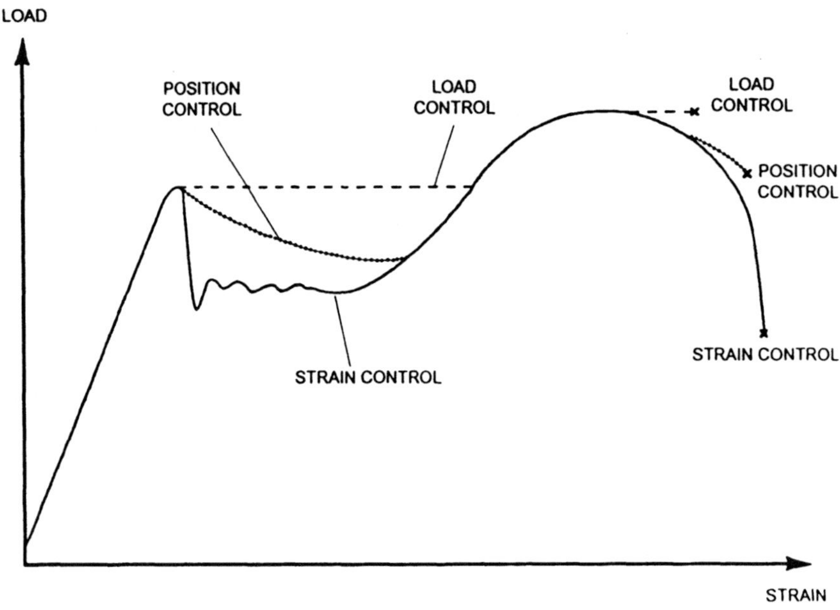

Fig. 2.3 Effect of control mode on result of tensile test on annealed mild steel.

For low-modulus materials, such as elastomers and textiles, where the testpiece stiffness is very low in comparison with that of the testing machine, the differences between position and strain control will be reduced. Conducting a tensile test in load control results in stable control during the elastic region. However, once the testpiece yields the control loop goes essentially open loop, leading to rapid extension through the yield elongation phase. Control is re-established in the work hardening phase (though with reducing gain) up to the tensile strength (R_m), when a high-energy break can be expected.

2.4.2 Position control

Position control is suitable for monotonic and some simple forms of cyclic testing, by reversing direction under the control of machine limits or a computer. Position control is invariably used for positioning the crosshead or actuator and for loading and unloading testpieces.

Position sensing on electromechanical machines is usually via a rotary optical encoder. Control performance can be very good indeed, with resolutions of better than 0.1 μm and steady state errors within 0.1% of set speed over a wide speed range.

Position sensing on servohydraulic machines is frequently based on LVDT. LVDTs have excellent resolution; however, a non-linearity of

± 0.5% of full travel is typical, providing a limit to the position control performance.

2.4.3 Load and strain control

Position control cannot provide accurate control of testpiece load or strain because of the presence of extensions taking place outside the testpiece gauge length in the grips, load string, frame and drive.

The limitations of position control can be overcome by using closed-loop control of load or strain when the feedback signal for the control is derived from either the load cell or an extensometer. Load control is usually used to conduct tests in the elastic region, e.g. high-cycle fatigue, when changes in testpiece compliance due to crack growth can be expected. Strain control is needed in situations where the testpiece is deforming plastically, e.g. low-cycle fatigue.

A complicating factor when using load or strain control is that the testpiece becomes part of the control loop. There is a need to optimize the control loop settings, a process known as loop shaping, in order to achieve good control performance with each type of testpiece. Furthermore if the stiffness of the testpiece changes during a test (e.g. elastic to plastic transition) the control loop response can change dramatically.

2.4.4 Outer-loop control

In some cases a basic machine with position control may be provided with the ability to control the rate of increase of load or strain in a monotonic test using outer-loop control. In this form of control, a computer monitors the load or strain feedback signal and modifies the speed of the machine during the test to maintain a constant rate of loading or extension.

The dynamic accuracy of an outer-loop control system is inferior to the equivalent direct control loop. This is due to the additional time delays in communicating with and processing data in the computer.

Outer-loop control may also be used in other situations such as amplitude control in cyclic testing and stress intensity control in crack growth testing. In these applications the testpiece characteristics change slowly and the outer loop can react sufficiently rapidly to maintain good control.

2.5 RESPONSE CHARACTERISTICS

2.5.1 Electromechanical machine response

As noted in section 2.2, the dynamic performance of the electromechanical testing machine is limited. However, this type of machine is often used for cyclic testing of extensible materials such as rubbers, textiles and foams and thus measures of dynamic performance are relevant. The basic measure of dynamic performance is the acceleration time, i.e. the time required to reach a given speed from rest. Other measures are the turnaround and stop times, at or from a given speed. Turnaround and stop performance may also be specified in terms of position overshoot, i.e. the extra distance the crosshead would travel if commanded to reverse or stop at a given position. Dynamic performance is usually measured under no-load conditions and for speeds greater than 10 mm min^{-1}. The maximum dynamic performance obtained from an electromechanical machine is essentially a function of the available motor torque and the polar moment of inertia of the motor and the first drive pulley (or gear).

2.5.2 Servohydraulic machine response

The dynamic response of a servohydraulic machine is usually specified in terms of a set of performance curves. These curves plot the maximum

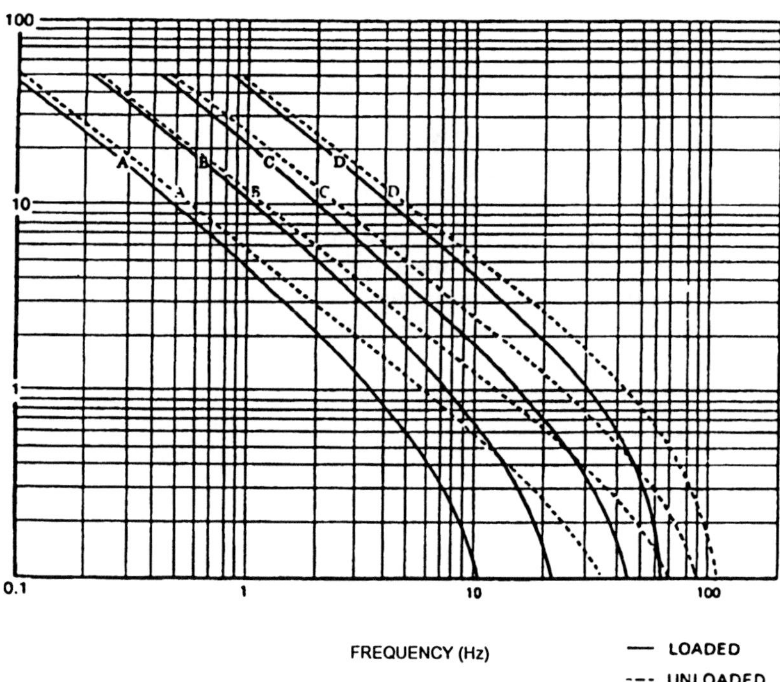

Fig. 2.4 Typical performance curves for 100 kN servohydraulic actuator with ± 50 mm stroke: servovalve flow (A) 10 (B) 20 (C) 40 (D) 80 l min^{-1}.

available position amplitude against frequency for loaded and unloaded operation. Figure 2.4 shows a typical set of performance curves.

The actual performance of a servohydraulic system is determined by a number of factors, which include the working pressure, the actuator piston area, the mass of the actuator piston and grip, the actuator stroke, the flow ratings of the servovalve and the hydraulic power pack and the volume of the hydraulic accumulators.

2.6 DIGITAL VERSUS ANALOGUE CONTROL

The increasing use of digital control electronics within testing machines means that it is relevant to ask how these systems compare with analogue systems and which parameters are important in a digital controller.

2.6.1 Outline of analogue and digital controllers

A block diagram of a servohydraulic testing machine with analogue control is shown in Fig. 2.5. In this type of machine the various physical parameters are represented by analogue voltages. Transducer conditioning

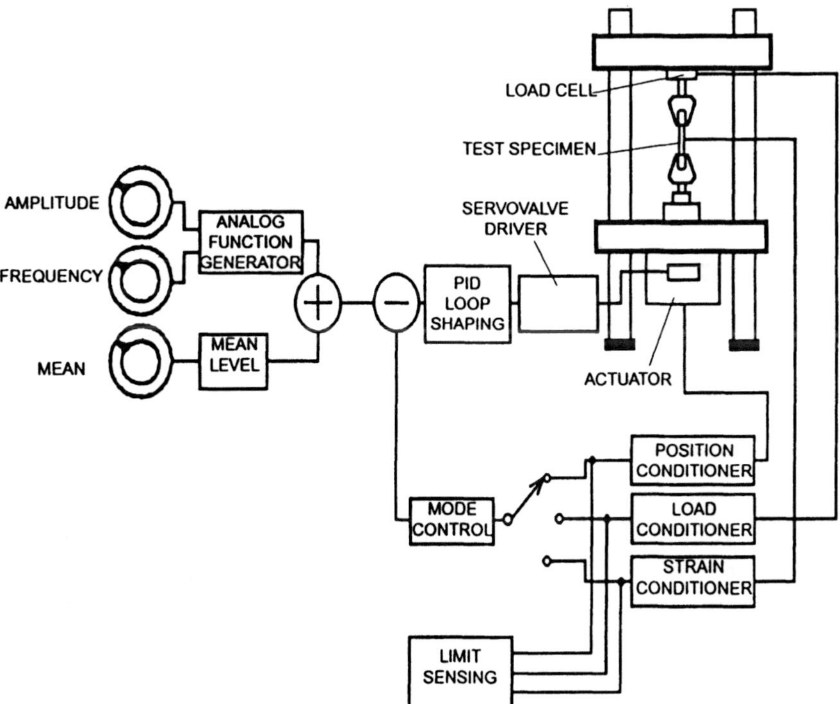

Fig. 2.5 Analogue control loop.

electronics convert the low-level analogue signals from the various transducers into calibrated high-level analogue voltages. These voltages can then be routed to digital display devices for monitoring purposes. One transducer can be selected to provide the control loop feedback signal. The control loop demand signal, the sum of a mean level and a cyclic function generator, is an analogue voltage. The demand and feedback signals are subtracted in a summing junction and the resulting difference or error signal is amplified and used to drive the servovalve which controls the actuator.

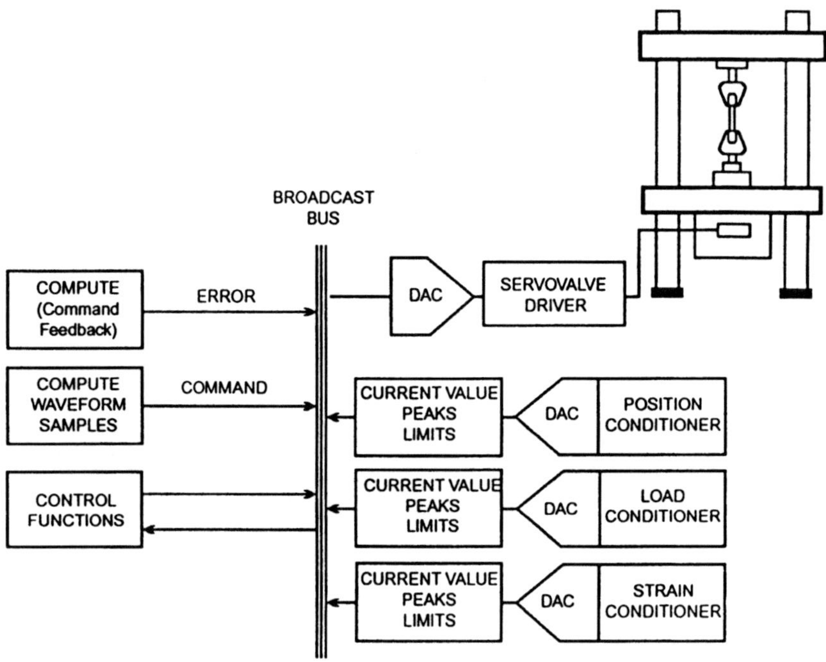

Fig. 2.6 Digital control loop.

A block diagram of an equivalent servohydraulic testing machine with digital control is shown in Fig. 2.6 (Nicolson, 1989). Transducer signals, after initial analogue conditioning, are converted into a stream of digital values at the system sampling rate. The demand signal is simply a sequence of numbers updated at the same sampling rate. The summing of the demand and feedback signals to generate the error signal and the subsequent processing of the error signal are entirely digital processes. The servovalve requires an analogue drive and this is provided via a digital to analogue converter.

2.6.2 Advantages of digital controllers

Unlike an analogue voltage, the digital numbers used to represent the physical parameters within the digital controller have a finite amplitude resolution. Also unlike the analogue voltages, the digital numbers do not vary continuously in time but change at the loop update rate.

At first sight it may seem that the resolution achieved by the digital controller will be worse than that of the analogue controller. In a good digital controller design this is not the case since the resolution of the digital numbers is less than the inherent system noise. This can be achieved by the use of a high-resolution analogue to digital converter (ADC) and techniques such as dynamic ranging in which the gain of the electronics prior to the ADC is continually adjusted to make best use of the ADC resolution.

The additional control loop delays introduced by the sampling process mean that the ultimate frequency response of a digital controller is lower than that of an analogue controller. However, with current digital controllers there are no significant disadvantages below 100 Hz.

Compared with an analogue controller, the digital controller is very stable. The digital numbers are not subject to zero and span drifts. Processes such as accurate mode changing, which require very good matching between control channels, are very easy to achieve in a digital controller.

The digital controller is very flexible and can easily support a number of advanced features such as load protect, which limits the maximum applied force during testpiece loading to a small fraction of capacity, and automatic loop shaping, which provides initial control loop gain setting in accordance with the testpiece stiffness. The implementation of fully self-adaptive control, which continually modifies the control loop parameters during the course of a test to account for slow and fast (cyclic) changes in testpiece stiffness, is achievable with the digital controller.

Interfacing an analogue controller to a computer has to be accomplished by using some form of dedicated interface containing computer addressable demand signal generation and data acquisition hardware. Linking a digital controller to a computer is a more straightforward task: the computer can communicate directly with the digital demand generator and can collect the digital data. Also it is possible for the computer to access all of the internal functions of the digital controller, e.g. transducer balance and calibration functions, giving great flexibility.

2.7 ALIGNMENT CRITERIA

Load-string misalignment introduces testpiece bending and has an adverse effect on many test results. In tensile tests, misalignment can result in large scatter in modulus data and low measured values of yield stress and ultimate tensile strength. In cyclic testing, misalignment can lead to a decrease in fatigue life.

Alignment is influenced by both the frame and the load string components. Frame alignment is described in terms of the errors in parallelism (angularity) and concentricity between the fixed and moving crossheads (electromechanical) or actuator piston face and crosshead (servo-hydraulic). For a typical electromechanical machine the specifications usually include runout over the total travel and maximum variation over shorter travels. For servohydraulic machines the alignment is often specified at a fixed crosshead height.

In order to help maintain good alignment of the load string, methods of positively locating the load string components to the frame may be provided, e.g. the use of locating rings. However, for the most demanding applications, e.g. low-cycle fatigue and ceramics testing, in which testpiece bending must be very low, adjustable alignment fixtures should be installed.

A recent development in load strings for the tensile testing of ceramics and other brittle materials has been the Instron Super-grip. This is a self-aligning tensile grip using a low-friction flexible coupling, based on a number of small hydraulic pistons (Liu, Pih and Voorhes, 1989). The use of such grips minimizes the need for precise load string alignment.

2.8 LOAD CELLS

The majority of load cells designed for use in materials testing machines use strain gauges to convert the deflections of elastic members into electrical signals. Load cells are available for use in both static or quasi-static testing and for dynamic operation. Full load capacities range from less than 10 N to over 5 MN.

In general, modern load cell designs together with their conditioning and measurement systems are capable of easily meeting the requirements of the various international standards (e.g. BS 1610 : Part 1; ASTM E4; ISO 376 and EN 10002/2). Furthermore, the ability to accurately measure force to within $\pm 0.5\%$ of a reading can be maintained down to a small fraction of the full rating (e.g. 1/250th). This large 'turndown' can significantly reduce the need to exchange cells to match the requirements of individual tests.

In addition to accuracy, other load cell parameters are important. The cell should be insensitive to extraneous forces and moments. Since the load cell in a materials testing machine is part of the load string, it should have high axial and lateral stiffness and the fixing method should not adversely affect the load string alignment. Finally the temperature coefficients of zero and span should be low (typically 20 ppm $°C^{-1}$ or better) to ensure stable behaviour in varying ambient conditions.

2.9 COMPUTERS AND SOFTWARE

2.9.1 General

Modern materials testing systems, whether intended for simple tensile testing or complex multiaxial fatigue, are usually equipped with a computer. Indeed some of today's testing machines are so intimately integrated with a computer that they will not function without one. The functions which can be provided by a computer are test control, data logging and storage, data analysis, report generation and data management. The functions which are actually performed are determined by the specific software package in use on the computer. Software packages for materials testing can be either general purpose or application specific.

2.9.2 Test control

Today's software packages provide a very comprehensive range of test control facilities. In addition to the ability to set up speeds, waveforms and control modes, complex functions such as automatic drive to zero the load and balance of extensometers prior to the start of the test can be implemented. Outer-loop control of parameters such as stress intensity in crack growth tests can also be provided via software.

In automated and semi-automated testing systems the software has to coordinate the operation of the testing machine and the other equipment, e.g. robots, testpiece measuring devices and automatic extensometers.

2.9.3 Data logging

Data logging is the process by which calibrated data from transducers are passed to and stored in the computer. The rate at which data are transferred is important: if the rate is too slow, important information can be lost. It is important to realize that very rapid changes in load can take place in a slow-speed tensile test such as at upper/lower yield and break. The speed of data transfer may be limited by either the controller or the computer. For some applications, for example single-shot high-rate testing, 'burst mode' may enable much higher data rates for a short period than the normal maximum continuous rate.

For consistent results it is important that the values from the various transducer channels which make up each data point are all measured at the same time and as a result are not subject to data skew.

2.9.4 Data analysis

A general purpose software package will provide a variety of different algorithms for analysing the test data in accordance with relevant product standards. In some cases several different algorithms will be available for the calculation of the same test parameter, e.g. modulus of elasticity. This provides the user of such software with a high degree of flexibility; however, care must be taken in the selection of the appropriate algorithm. Well-conceived software should provide the user with the ability for comprehensive re-analysis of data from old tests.

2.9.5 Report generation and data management

Finally application software needs to be capable of producing a variety of report formats to suit user needs. Increasing emphasis is being placed on the ability to link the output of application software to a variety of other packages such as databases, statistical process control (SPC) software and laboratory management systems.

2.10 CONCLUSIONS

The range of materials testing systems available to the materials scientist is wide and varied. The main difference between the two basic types of machine, i.e. electromechanical and servohydraulic, is in their power and frequency response characteristics. However the capital costs of servohydraulic machines, below about 500 kN, are higher than those of an equivalent capacity electromechanical machine. The installation and especially running costs (electrical power and cooling) are also higher for servohydraulic machines.

The latest generation of digital control electronics provides a wide range of advanced control and data collection options whilst maintaining stable drift-free operation without the need for frequent recalibration. The use of computers and applications software provides the operator with the ability to set up test methods to control the operation of the test machine as well as the storage and analysis of the data.

Advances in digital electronics have not, of course, altered the need for accurate load string alignment and the choice of the correct control mode. The mechanical characteristics of the frame, load string and drive system are still of paramount importance in ensuring the integrity of test data.

ACKNOWLEDGEMENT

The author would like to express his thanks to G. R. Murfitt, Instron Ltd.

REFERENCES

Liu, K. C., Pih, H. and Voorhes, D. W. (1989) Uniaxial tensile strain measurement for ceramic testing at elevated temperatures: requirements, problems and solutions, in *Mechanical Testing of Engineering Ceramics at High Temperatures* (eds B. F. Dyson, R. D. Lohr and R. Morrell), Elsevier Applied Science.

Newman, K. and Sigvaldason, O. T. (1965) Testing machine and specimen characteristics and their effect on the mode of deformation, failure, and strength of materials, in *Proceedings of the Institution of Mechanical Engineers SEE Symposium on Developments in Materials Testing Machine Design*, Institution of Mechanical Engineers.

Nicolson, A. M. (1989) A quantum leap into direct digital control, in *Conference Proceedings, Ideas in Science and Electronics Exposition and Symposium*, May 1989, Albuquerque, New Mexico.

CHAPTER 3

Dynamic control methods

C. E. Hinton

3.1 INTRODUCTION

This chapter describes methods used to control servohydraulic materials testing machines. Machines of this type are used when high-speed load reversal is required such as during fatigue testing. They are able to apply high-frequency alternating loads superimposed upon a mean tensile or compressive load – a task for which electromechanical machines are not well suited.

Figure 3.1 shows the main mechanical components of a servohydraulic materials testing machine. The crosshead can be moved and clamped at any position up and down the columns to accommodate testpieces of different lengths. The upper and lower grips are normally hydraulically operated although other gripping mechanisms are sometimes used. These grip the ends of the testpiece firmly to prevent slipping. The columns are fixed to the base platen which also carries the hydraulic actuator (some configurations have the actuator mounted in the crosshead and the load cell mounted on the platen). The actuator piston position is usually monitored and controlled using an internal linear variable differential transformer (LVDT) as the feedback transducer, and the force applied to the testpiece is measured by the load cell mounted beneath the crosshead. In many tests, an extensometer is attached to the testpiece to provide a more accurate measure of extension than can be obtained from the LVDT which is sensitive to load frame deflections. The measurement resolution of an extensometer is also superior because it usually has a smaller range than the LVDT.

The capability of machines of this type is specified by the manufacturer in terms of a maximum actuator displacement versus frequency graph. This is the so-called **performance envelope** of the machine. But for the user to get the best out of a machine within this envelope, it is necessary to know more than just its maximum capability. An appreciation of how the machine is controlled and how to optimize the controller is also required. Machine users often mistakenly believe that load or strain excursions entered on the controller front panel will be exactly replicated on the testpiece. In fact

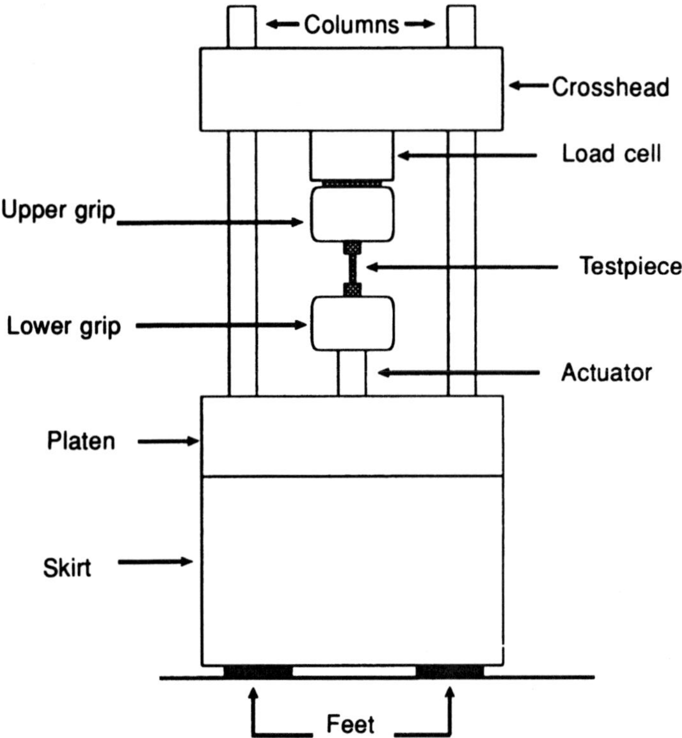

Fig. 3.1 Servohydraulic materials testing machine.

this is never the case, and optimizing the control loop is all about reducing the error between what is desired and what is actually achieved to an acceptably small level. Sadly, the quality of some tests has been called into question because the importance of correct control loop tuning was not understood, resulting in significant applied load and strain errors!

Proportional + integral + derivative (PID) control is widely used on servohydraulic machines. In view of the importance of correct control loop tuning, this chapter provides a tutorial introduction to this type of control, outlines the practical effect that each term has upon a materials test and explains how a PID controller should be set up to give optimum performance. Although PID control is well suited for dynamic testing, there are circumstances when it requires assistance from other methods. The most commonly encountered limitations are described including the need for active suppression of the effects of actuator and load frame resonances, the use of outer-loop amplitude control, demand hold and the adaptive control of amplitude and frequency.

Finally, the chapter concludes with an overview of future control strategies that are likely to appear for dynamic control. The development of fast digital processing has made feasible control algorithms which are more sophisticated than PID.

3.2 THE CONTROL LOOP

Figure 3.2 is a simplified block diagram of a typical servohydraulic materials testing machine control loop. Three control modes are provided: actuator position control, testpiece load control and testpiece extension (strain) control. As long as appropriate transducers are fitted, the position, load and extension signals are available simultaneously. On high-quality machines care is taken to match the signal conditioning electronics which process each transducer output so that all three signals stay in phase with each other.

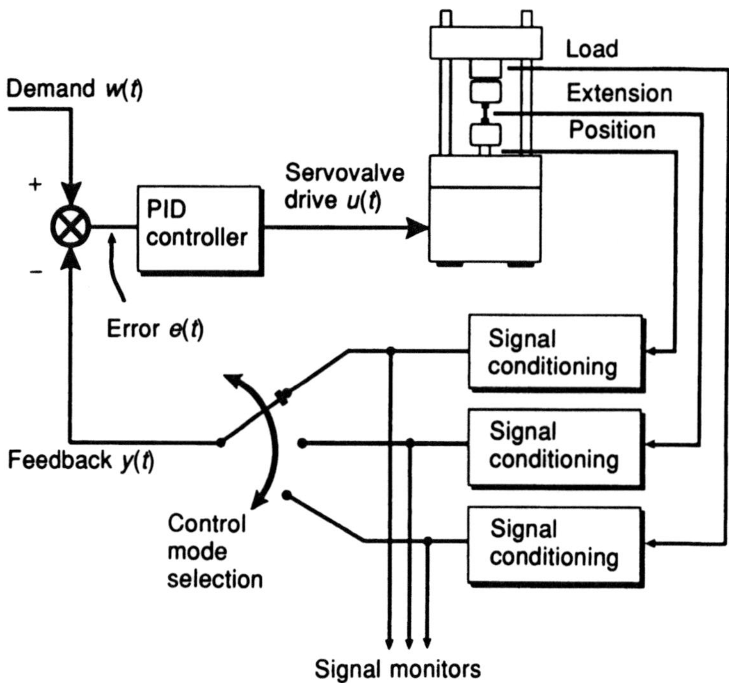

Fig. 3.2 Position, load and strain control.

The mode selector is used to select which conditioned transducer signal is to be the controlled variable $y(t)$. As some tests require the control mode to be changed during the course of the test, most machines have some form of bumpless mode transfer mechanism.

The function of the PID controller is to produce an appropriate control signal $u(t)$ to drive the testing machine actuator in a direction which minimizes the servo-error $e(t)$:

$$e(t) = w(t) - y(t) \tag{3.1}$$

where $w(t)$ is the demand signal which is usually derived from a function generator or a computer. The output of the controller goes directly to the servovalve which on servohydraulic machines acts as the power amplifier.

Figure 3.2 makes no distinction between analogue controllers and digital ones. The signals in early control loops were all analogue; then in the early 1980s a number of manufacturers produced control loops consisting of hybrid electronics in which the three-term controller was still all-analogue but its terms could be digitally addressed. Mode selection and demand generation could be digitally addressed, and monitor signals – with the inclusion of appropriate analogue to digital converters (ADCs) – could be collected digitally. The impetus for this move away from totally analogue electronics was the need for computer control and more sophisticated processing of test results. The subsequent development of fast digital signal processors (DSPs) in the late 1980s heralded another leap – the replacement of the analogue PID controller with a totally digital equivalent. In these direct digital controllers the error signal $e(t)$ is represented as a series of discrete digital samples as opposed to the continuous signal of its analogue predecessor. Figure 3.2 can be modified to this form of control by adding an ADC after each signal conditioner and a digital to analogue converter (DAC) after the controller, with the DSP executing the PID algorithm.

These advances have as yet made little difference to the type of control used. PID control is still employed regardless of whether the electronics are analogue, hybrid or digital, although direct digital control has the potential to execute more complex algorithms. We look at likely future developments in section 3.8, but for the moment we stay with PID control which is featured on almost all past and present servohydraulic machines.

3.3 PID CONTROL

Figure 3.3 shows a canonical arrangement of a PID controller. Implementation variants abound (Astrom and Hagglund, 1988) each with their own special features, but the essential operating principles of all PID controllers are the same. Provided the sampling frequency is high enough, direct digital versions behave just like their analogue counterparts.

The PID controller has three adjustable terms: loop gain K_p, integral gain K_i and derivative gain K_d. There is a set of these three controls for each mode because different settings are required in position, load and strain control. On analogue controllers the terms are adjusted using potentiometers. On digital and some hybrid controllers the values are entered using a front panel numeric key pad or via increment and decrement buttons, or they can be computer addressed. It is important to be able to read the set values so that previous test conditions can be repeated, and to this end it is good practice to record the settings along with the test data.

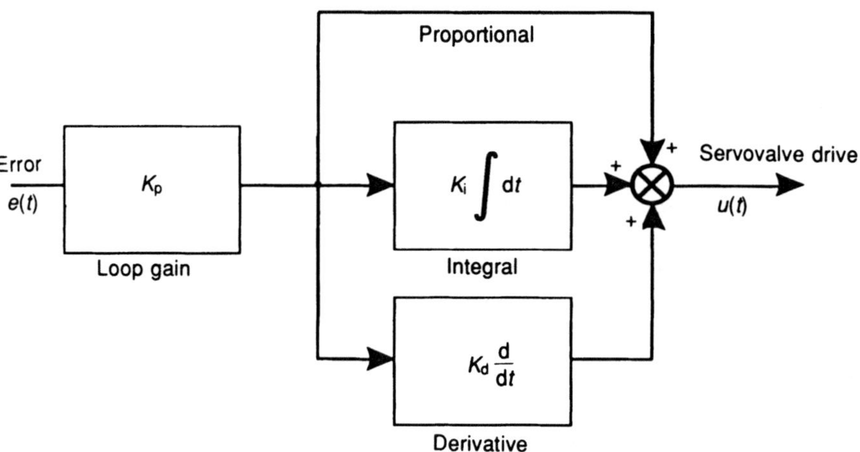

Fig. 3.3 PID controller.

Referring to Fig. 3.3, the servo-error signal is first passed through an amplification stage. The amount of amplification is controlled by the loop gain setting. (This is so called because it alters the gain of the control loop. Total loop gain is the sum of the gains of all the components in the control loop and is therefore not the same as the magnitude of K_p.) This amplified error signal is then passed along three parallel paths: the proportional path, the integral path and the derivative path. The proportional path has unity gain. The integral path produces an output proportional to the time integral of error (i.e. an output proportional to the area beneath the error versus time profile). The derivative path produces an output proportional to the rate of change of error (i.e. to the slope of the error versus time profile). The outputs of all three paths are finally summed together to form the servovalve drive signal $u(t)$:

$$u(t) = K_p \left\{ e(t) + K_i \int e(t)\mathrm{d}t + K_d \frac{\mathrm{d}e(t)}{\mathrm{d}t} \right\} \tag{3.2}$$

If $e(t)$ and $u(t)$ have the same units (usually volts), then K_p is dimensionless, K_i has units of seconds^{-1} and K_d has units of seconds.

To help understand how each term of the PID controller affects performance, a number of response traces are presented in Figs 3.4 to 3.7. These are simulated responses calculated using a model for a servohydraulic machine which has been shown to correspond well with real measurements (Hinton and Clarke, 1989). They are typical of those obtained in position, load and strain control. The time axis conveys the sort of response time expected on servohydraulic machines.

3.3.1 Loop gain K_p

This term simply amplifies or attenuates the servo-error signal $e(t)$. On good machines it has a range of at least -100 dB to $+100$ dB (10^{-5} to 10^5). This wide range is needed to accommodate the various actuators, servovalves and transducers used on materials testing machines and the diverse nature of test applications. Note that on digital controllers, adjustment of the loop gain term can be made to have a linear feel in terms of its effect on machine performance. On analogue controllers, where potentiometers are used to alter the gain, it is difficult to obtain this even sensitivity over such a large gain range and care often needs to be exercised near the extremes of the setting range.

The influence of loop gain is illustrated by the response $y(t)$ of the machine to a step change in the demand signal $w(t)$. Figure 3.4 shows three responses each corresponding to a different loop gain setting. For all three traces the integral and derivative gains are zero. The top trace is the step response obtained when the loop gain value is low. This setting has been given the notional value g for the purpose of comparison, its actual value is unimportant. The low-gain response is rather sluggish. Increasing the gain makes the response crisper, as illustrated by the centre trace, and also brings the final value closer to the step demand. Too high a value, though, results in oscillatory behaviour – depicted in the bottom trace – with a large and lightly damped overshoot. Increasing the loop gain much beyond the $6g$ setting caused the response to become unstable.

Without exception, it is always desirable to have the loop gain set to the highest possible value commensurate with not producing unwanted

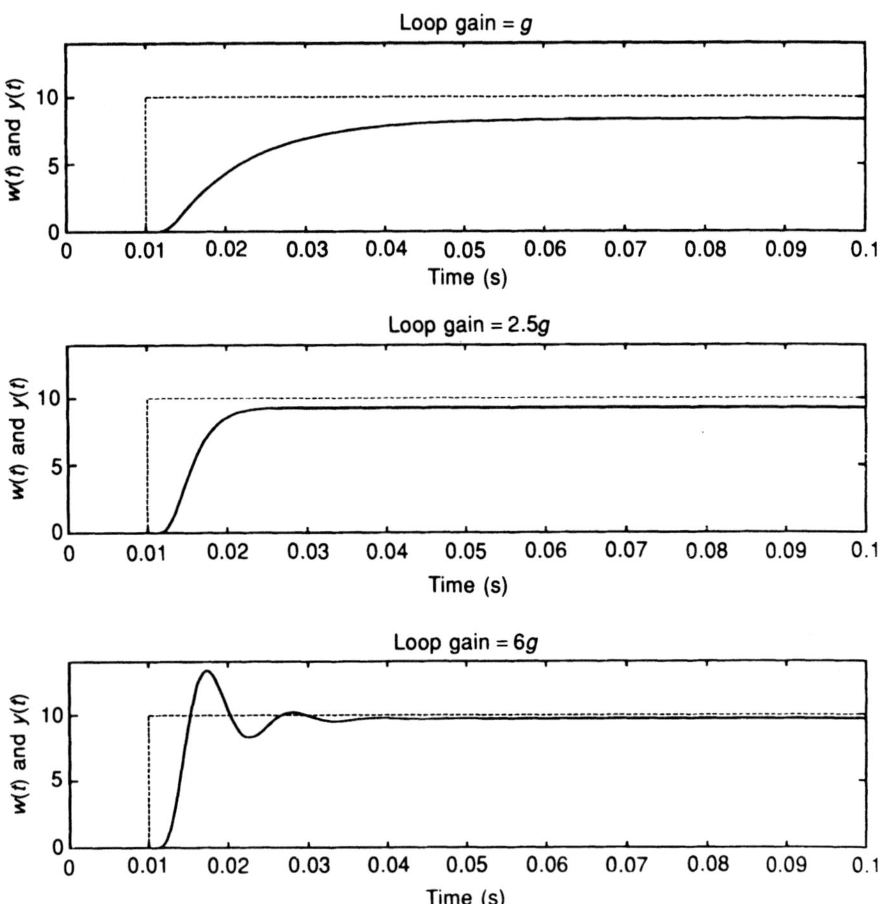

Fig. 3.4 Typical step responses showing the effect of increasing the loop gain K_p from low setting g through to $6g$. For all three traces the integral-gain K_i and the derivative-gain K_d are zero.

overshoot. Maximizing loop gain must be the aim in any control loop optimization procedure. This is because the testing machine is much more able to closely follow demand profiles and achieve accurate loading if the loop gain is high.

Figure 3.4 shows that the controlled variable $y(t)$ – after transients have died away – is always less than the demanded value. The difference is called the **static steady state error** or **steady state offset** and is a consequence of the fact that the servovalve must be open for the actuator to apply load to the testpiece. Since in proportional control the valve opening is proportional to the servo-error $e(t)$, the servo-error must be non-zero. As loop gain is increased the static steady state error becomes less but it can never be made zero as this would require infinite loop gain which would render the system unstable. The problem of static steady state error is the very reason the integral term is included within the PID controller. How the integrator removes this error is discussed next.

3.3.2 Integral term K_i

Because the integrator produces an output proportional to the time integral of error, any finite error will cause the integrator output to keep on increasing, driving the actuator harder and harder and forcing the error to zero. Once the error has reached zero, the integrator output stops increasing but holds the value necessary to remove the error. Some PID controller designs remember the output value needed to drive the static error to zero even when the integrator is switched off. This feature can be useful because it allows the integrator to be switched off without upsetting mean-level control.

Integral action is essential in materials testing. Without it, static loads cannot be applied accurately. As well as removing the static steady state error associated with having to open the servovalve for the actuator to apply load, the integrator also eradicates other loop offsets. An important offset is the **valve null offset** which occurs when a non-zero drive signal $u(t)$ is needed to close the servovalve. Although mechanical adjustment is provided to minimize valve null offset, it is difficult to set it precisely to zero. (Even though integral action removes valve null offset, it is still important to mechanically null the servovalve to prevent the actuator jumping when the integrator is turned off.) The null offset also changes with oil temperature and time (null offset drift).

The integrator essentially affects the low-frequency or static performance of the testing machine. It has little influence at high frequency. Figure 3.5 shows step responses obtained with different integral gain settings. The loop gain for all three traces corresponds to the centre trace of Fig. 3.4, i.e. $K_p = 2.5g$. Altering the integral gain setting affects the speed with which the static steady state error is removed. The top trace of Fig. 3.5 is for a low integral gain setting – given the notional value i. The static steady state error is removed very slowly which, if accuracy is to be preserved, would severely limit the frequency at which different loads could be demanded. In the centre trace ($K_i = 1.5i$) the static steady state error is removed more rapidly. Too much integral gain, though, produces the low-frequency overshoot shown in the bottom trace ($K_i = 3i$) with a slow pull-in to the demand signal. The centre trace is the best response for this system. It offers the fastest removal of the static steady state error without overshoot.

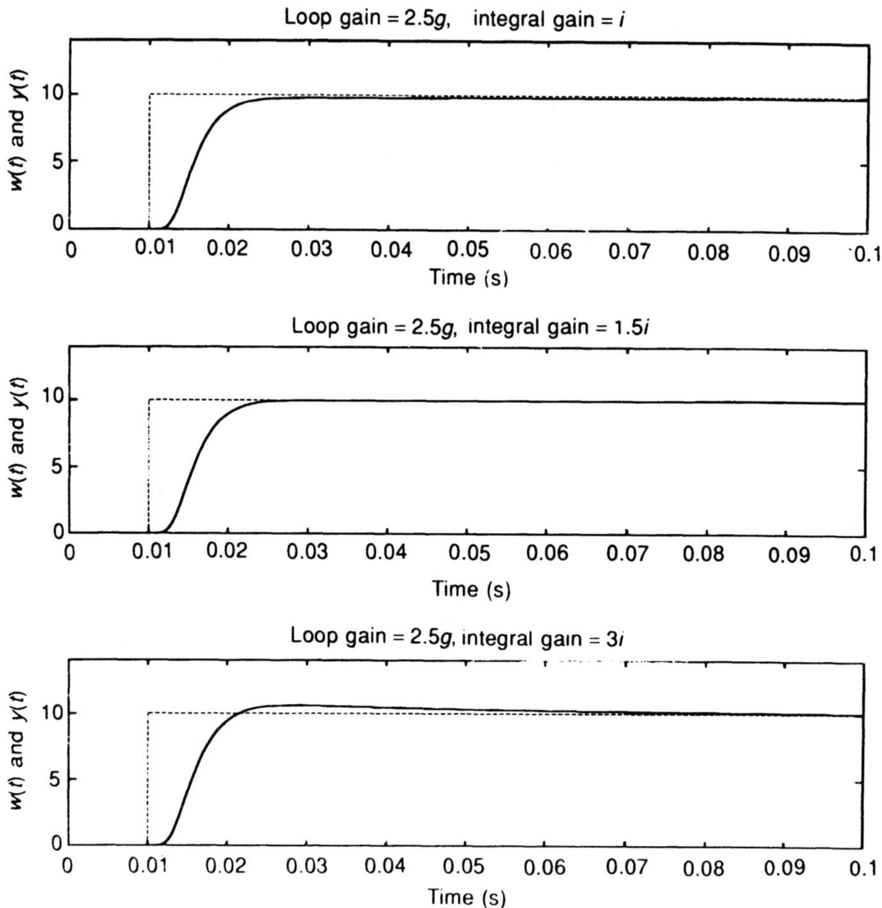

Fig. 3.5 Typical step responses showing the effect of increasing the integral gain K_i from low setting i through to $3i$. For all three traces the loop-gain K_p is $2.5g$ and the derivative-gain K_d is zero.

3.3.3 Derivative term K_d

Derivative action introduces damping into the control loop. This allows a higher loop gain to be set with the attendant benefits of faster response and better control accuracy. Derivative action introduces phase lead, cancelling some of the phase lag produced by the loop components. An unfortunate side effect is that it also causes the gain to increase at high frequency which if excessive makes the loop unstable. There is therefore an optimum derivative gain setting which offers some phase advance benefit but without too much increase in high-frequency gain.

Another problem is that increasing the gain at high frequency exaggerates signal noise. For this reason most PID controllers limit the gain of the derivative path by incorporating a series low-pass filter which cancels or spoils derivative action at high frequency. Advice is given by Clarke (1984) regarding the frequency at which spoiling should start. Its level is usually fixed or related to the derivative setting. It is rarely accessible as an operator control.

The effect that derivative control has upon machine response is illustrated in Fig. 3.6. The top trace is a repeat of the high-gain trace of Fig. 3.4 but with the integral gain set to $1.5i$ so that the static steady state error is removed. The derivative gain for this trace is zero. Its lightly damped response is a good starting point from which to demonstrate the damping properties of derivative action.

The centre trace shows what happens when the derivative gain K_d is increased to the notional value d. Improved damping is reflected by the smaller overshoot. The setting $K_d = d$ was found to be the optimum for this system. Increasing the derivative gain further caused the overshoot to get worse. The bottom trace is the response obtained when $K_d = 2.5d$. Damping is less than for the centre trace and the frequency of oscillation is higher. This worsening condition is a consequence of increased gain of the derivative path at high frequency.

Returning to the optimum centre trace setting: even though the loop gain must be turned down to remove the overshoot, we are still left with a higher loop gain than was possible with just proportional + integral (PI) control. The resulting response is the bottom trace of Fig. 3.7, which can be compared with the optimum responses obtained using proportional control and PI control. The top two traces of Fig. 3.7 are copied from Figs 3.4 and 3.5. The problem with proportional control is that the final value of the controlled variable is always less than the demand. Integral action removes this static steady state error, but the damping offered by derivative control – because it enables the loop gain to be set higher – gives PID control the most responsive performance.

3.4 SETTING UP THE PID CONTROLLER

The importance of optimizing the PID controller to suit the test conditions cannot be overemphasized. This procedure is referred to differently by different people: American users and users of machines of American origin call it 'loop shaping'. Others – particularly if they have connections with the control community – refer to it as controller 'tuning'. For still others the term *'loop gain'* has come to mean the combined

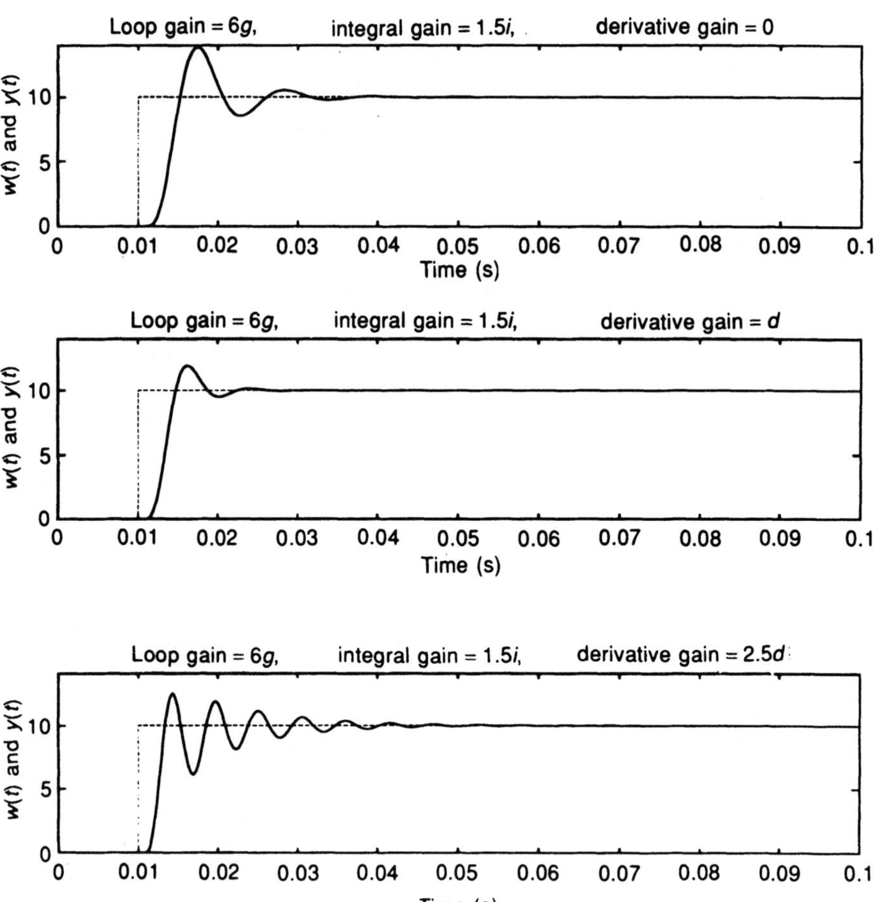

Fig. 3.6 Typical step responses showing the effect of increasing the derivative-gain K_d. For all three traces the loop gain K_P is $6g$ and the integral-gain K_i is $1.5i$.

effect of loop gain, integral gain and derivative gain. They call optimizing the controller 'setting the loop gain': this easily causes confusion because it implies that only the loop gain has to be adjusted. The term **tuning** is the best description of the optimization process: the controller is being tuned or matched to suit the parameters of the testing machine and the test conditions.

The dynamics of a servohydraulic materials testing machine are affected by the following parameters:

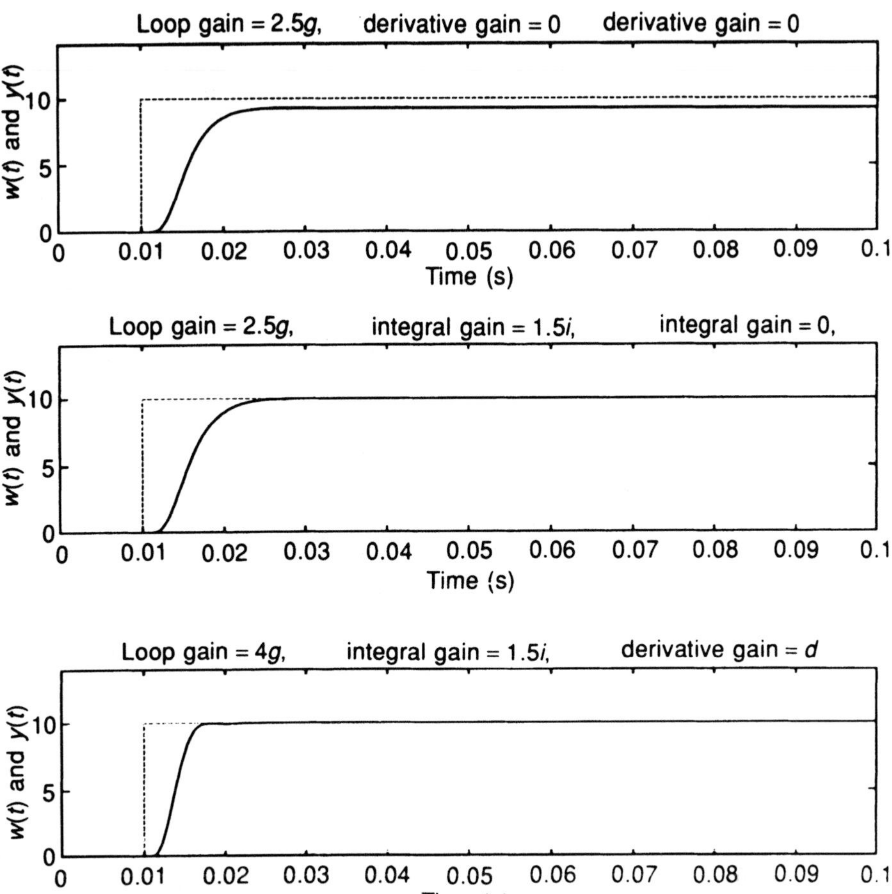

Fig. 3.7 Optimum step responses for (top) proportional control only, (centre) PI control and (bottom) PID control.

- load capability and stroke of the actuator;
- servovalve flow rating and type;
- load frame stiffness;
- grip mass (especially that attached to the piston);
- testpiece stiffness and damping (and mass if significant);
- feedback transducer sensitivity;
- mean position of the actuator;
- mean load.

Different combinations of these parameters call for different settings of loop gain K_p, integral gain K_i and derivative gain K_d and different settings are needed for each control mode. Setting up the controller is not a once-and-for-all operation. It should be checked every time any of the above parameters is altered. A most likely change is a change of testpiece stiffness and it is very important to retune when changing from one testpiece type to another.

There are many bad tuning practices. Most are the result of ignorance about how the PID controller works and the significance of each of the three terms. It is for this reason that the previous section went into such detail. A deplorable practice (unfortunately too often encountered) is to increase the loop gain until the machine goes unstable and then lower the gain until stability is restored. This leaves the machine in an uncertain state without any indication of response time or the magnitude of any overshoot. Even more deplorable is the practice of not tuning the controller at all!

The procedure set out below has been found to work well on servo-hydraulic machines. It is based upon the working principle of the PID controller and aims to maximize the loop gain setting without producing overshoot. It should be conducted either with the first testpiece of a test series installed or using a testpiece of the same type. The procedure monitors response to a square-wave demand signal. Square waves are ideal for tuning because they excite high-frequency modes. The method compares the demand and feedback signals. This can be done using an oscilloscope (digital or analogue storage types are best) or using computer displayed waveforms.

The amplitude of the demand waveform during tuning is important. If it is too small, response will be dominated by friction effects and false results will be obtained. If it is too large, the testpiece may sustain premature damage. Usually an amplitude equivalent to 2% of the range of the feedback transducer is satisfactory. If this produces excessive load when operating in position or strain control, limit the demand amplitude to produce loads which are equivalent to 2% of the load cell range. Amplitude sensitivity can be checked after completing the tuning procedure by ensuring that the response is still good for a larger amplitude.

The frequency of the square-wave demand should be set to 5 Hz. This gives the machine enough time to respond and provides sufficient time axis resolution on oscilloscope displays.

3.4.1 Tuning procedure

1. Set the loop gain to a low value, i.e. a setting which does not give obvious oscillatory behaviour.
2. Set the derivative gain term to zero.

3   If the integrator design is such that it remembers the last offset value, set the integral gain to zero. If the integrator does not remember the last offset value it may be necessary to set a finite but low integral gain in order to hold the mean-level setting.

4. Gradually increase the loop gain until the output waveform just overshoots its final values at the top and bottom of the square wave (do not exceed 10% overshoot). Note that without integral action these final values will be less than the demand signal.

5. Gradually increase the derivative gain setting until the overshoot is at its minimum. If the overshoot disappears, return to step 4 and further increase the loop gain.

6. When the the derivative setting that produces the smallest overshoot has been determined, adjust the loop gain until the output waveform just does not exhibit overshoot.

7. Monitor both the demand signal and the feedback signal and gradually increase the integral gain so that the feedback signal approaches the demand signal as rapidly as possible but without overshoot.

8. Record settings.

3.5 DITHER

Most machines have a dither control which can be adjusted by the operator. This governs the amplitude of a high-frequency signal applied to the servovalve to overcome the effects of silting and stiction. Clearances around the final-stage spool of the servovalve are very small and, during any long periods of inactivity, small silt particles in the hydraulic oil build up and prevent the spool from moving. A large valve drive signal $u(t)$ is then required to break out the spool. Dither stops this happening by keeping the valve spool active. Dither also helps overcome the bad effects of sliding friction within the actuator. By providing a drive signal which just causes the actuator to break out of its friction bound, any superimposed desired signal will command immediate response from the actuator.

The dither amplitude should be set so that it is just inaudible but not so large that it appears on the load signal. It is essential for tests involving long hold periods.

3.6 INTEGRAL WINDUP

We have seen how the integrator within the PID controller removes the static steady state error and the valve null offset error. Things can go wrong if for some reason the actuator is prevented from moving! The integrator output will then increase, driving the servovalve harder and harder in an attempt to move the reluctant actuator, and the servovalve will eventually end up hard over. It is sometimes difficult to recover from this situation. It can take an appreciable time, once the actuator is released, for the integrator to unwind. For this reason, it is advisable to turn the integrator off when deliberately stalling the actuator to pre-load the load string.

The actuator can also be immobilized if the supply pressure fails. Most systems incorporate low-pressure detection, one function of which is to reset and switch off the integrator to prevent it winding up while the pressure is off. Also many PID controller designs incorporate an anti-windup mechanism to stop integration while the error signal is saturated. This prevents windup while the actuator is slewing and limits the overshoot that would otherwise occur when attempting to operate outside the actuator's performance envelope.

3.7 WHEN PID CONTROL NEEDS ASSISTANCE FROM OTHER METHODS

3.7.1 Coping with actuator and frame resonances

Some actuators, notably those with low stall force ($< \pm 20$ kN) and long stroke, tend to have a lightly damped resonance within the operating frequency range of the machine. Machines fitted with these actuators are difficult to operate using just PID control. In fact, derivative action actually makes the resonance worse and it is best to operate with just PI control. Even so, it is difficult to increase the loop gain sufficiently to achieve good waveform following.

The resonance is associated with the mass of the piston and its attached grip bouncing on the oil trapped within the cylinder volumes on either side of the piston. Without a testpiece installed, the undamped natural resonant frequency is given by

$$\omega_n = \sqrt{\frac{K_A}{M}} \tag{3.3}$$

where K_A is the stiffness of the actuator and M is the piston and attached grip mass. Thus the frequency is lower and more of a problem when the actuator is soft and the mass is large. This is why low-stall-force, long-stroke actuators which are softer than higher-stall-force, shorter-stroke versions are most vulnerable. Also the advisability of minimizing grip mass is obvious.

Actuator resonance is a well-known problem which has been studied in other fields. Many solutions have been proposed: piston bleed, the use of underlapped servovalves, pressure feedback etc. Two methods favoured for materials testing are phase lag compensation and acceleration feedback.

Phase lag compensation is appropriate when the frequency of actuator resonance is greater than 100 Hz. It consists of a low-pass filter placed in series with the controller. To appreciate how it works requires an understanding of stability analysis which is outside the scope of this text. However, the fact that it does work was demonstrated when direct digital control was first introduced in place of analogue control. Direct digital systems have an inherent phase lag associated with their finite sampling period. It was soon noticed that the problem of actuator resonance was less severe on systems controlled digitally. Although not good as far as other performance objectives are concerned, sample delay has been very effective in damping actuator resonance. Future developments of digital control will undoubtedly result in higher sampling frequencies and the unfortunate reappearance of the actuator resonance problem. Specific phase lag compensation will then have to be included within the controller.

A disadvantage of phase lag compensation is that it tends to reduce machine bandwidth. This is not a serious problem if the frequency of resonance is above 100 Hz but, for lower resonant frequencies, acceleration damping gives better performance. (Bandwidth is defined as the frequency range over which the closed-loop gain stays above − 3 dB (70%) of its zero-frequency value. For example we may expect a servohydraulic machine to have a bandwidth of 100 Hz which means that it has a flat frequency response up to nearly 100 Hz. Bandwidth is a measure of machine responsiveness and can be degraded by a poorly tuned controller.)

Acceleration damping is effective at any frequency but is less convenient and more expensive than simple phase lag compensation because it requires attachment of an accelerometer to the actuator piston and a separate conditioning circuit. Figure 3.8 is a block diagram of a typical arrangement. The signal from the accelerometer attached to the actuator piston is summed with the control signal $u(t)$ and provides damping by opening the servovalve in a direction which relieves the differential pressure across the piston associated with the resonance. To be effective, the valve spool must move in phase with the acceleration signal. The phase advance circuit is essential to ensure this happens. It compensates for the phase lag of the servovalve's own control loop.

As well as actuator resonance, resonance of the load frame can badly affect performance. This problem affects load control – especially when testing soft testpieces. Mass attached to the load cell moving in sympathy with any frame resonance generates unwanted inertia force signals superimposed upon the testpiece load signal. Not only is the monitored load different from that actually carried by the testpiece but the detected inertia force reduces stability. The solution when testing low-stiffness testpieces – such as elastomers or long testpieces in an extra-height frame – is to employ a dynamically compensated load cell.

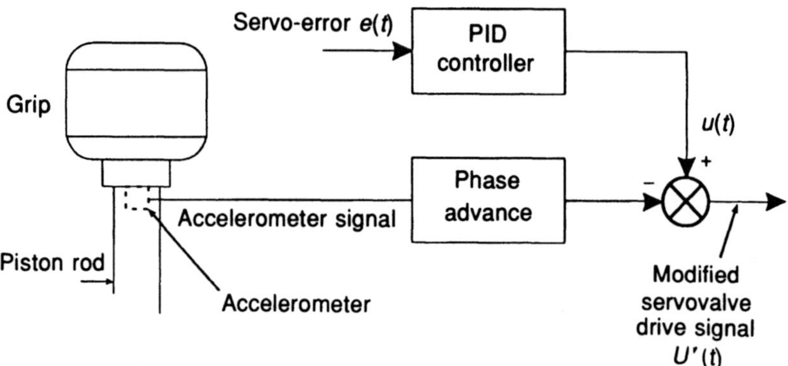

Fig. 3.8 Acceleration feedback for damping actuator resonances.

3.7.2 Outer-loop control

Materials tests which employ swept-frequency periodic waveforms often specify constant output amplitude. This is difficult to achieve accurately with PID control alone. Accurate peak levels for sine and triangular waveforms can be obtained using the outer-loop control scheme scheme shown in Fig. 3.9. The normal control loop with its PID controller is shown as the inner loop within the dashed box. Output amplitude is kept constant by using the outer loop to automatically adjust the demand signal to the inner loop. The amplitude controller is often another PID controller which, instead of processing the difference between demand and feedback, has as its input the difference between the actual amplitude and the desired amplitude. Outer-loop amplitude control like this is also useful during constant-frequency fatigue testing because it automatically compensates for the loss of amplitude that would otherwise occur as stiffness reduces due to crack propagation.

A useful extension of outer-loop amplitude control is **mixed-mode control**. This controls the amplitude of the dynamic periodic waveform in a different mode from that used to maintain the mean level. Mixed-mode control is the only way a viscous damping component such as a shock absorber can be tested in load control. The load developed by such a device is proportional to the rate of extension; it is not able to withstand static load. Load control cannot therefore be used in the conventional way because the actuator would simply drift off to one end of its stroke. Instead in mixed-mode control the dynamic component can be regulated in load control while the mean level is maintained in position control.

Amplitude control is fine for periodic waveforms but random waveforms call for different approaches. One method uses **demand hold** to hold the demand signal at each level until the feedback catches up to

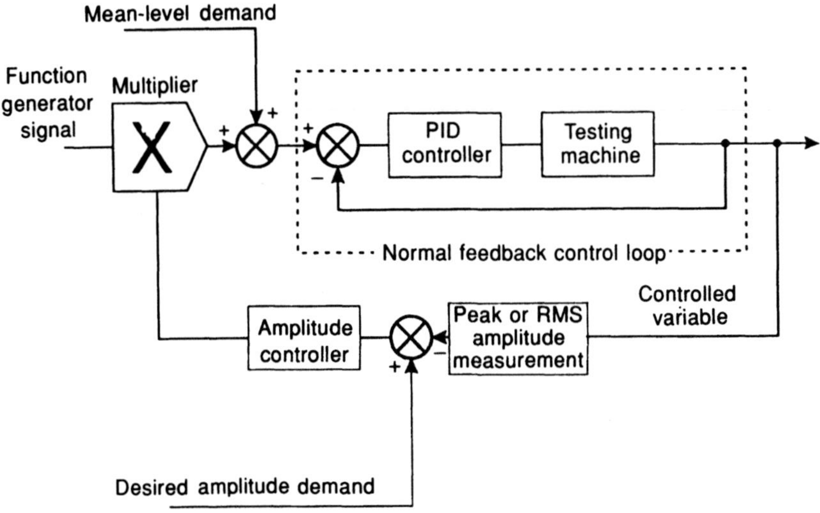

Fig. 3.9 Outer-loop amplitude control.

within some pre-defined tolerance. The function generator then sends out the next transition. Demand hold is particularly useful if some of the excursions within a sequence make demands of the machine outside its performance envelope. When this happens the actuator is given time to slew to the next turning point. A disadvantage of the method is that it unnecessarily slows down the frequency of transitions that lie within the performance envelope. This can seriously lengthen the time needed to complete a fatigue test. Under closed-loop control the feedback signal lags behind the demand signal. The function generator therefore always has to wait at every transition for the feedback to catch up even when the correct output would have been reached without waiting. Demand hold is also unable to protect against overshoot.

A better approach for random waveforms is to use adaptive control of amplitude and frequency (Edwards, 1986). This method monitors the accuracy of each transition and gradually builds a correction matrix whose elements are used to modify the demand signal the next time the sequence is run through. Both the amplitude and the frequency of the demand signal are altered so that not only is accuracy obtained but the test time is minimized.

Attention has been focused recently on the accuracy of applied random loading by the publication of the second issue of the Fatigue Research Advisory Group (FRAG) requirements for control of fatigue testing (Cook, 1990). This calls for the accuracy of testing to be recorded in reports and classifies test results according to the accuracy obtained.

Various round-robin collaborative test programmes have also discussed random loading accuracy. Appendix D of AGARD report 732 (Newman and Edwards, 1988), which records the accuracy of the FALSTAFF (1976) sequence in eight laboratories, states that some had specific problems that could have caused significant deviations in fatigue life.

3.8 THE FUTURE

The advent of direct digital control using the fast arithmetic of DSPs has opened up the possibility of more complex algorithms. Two immediate needs are:

- an automated method of setting up the controller terms (auto-tuning);
- a means of keeping the controller at its optimum setting for tests in which the characteristics of the testpiece alter significantly enough to affect performance (adaptive control).

Satisfying the first of these simplifies machine operation and should make testing more reliable because the human factor – manual tuning – is avoided. Satisfying the second will ensure that performance is maintained throughout the duration of the test.

Adaptive control is needed because during most material tests the characteristics of the testpiece alter as the test proceeds. Usually stiffness changes either due to the onset of plasticity or as the result of crack propagation. This can adversely affect performance in ways which are different in load and strain control. In load control, a reduction in testpiece stiffness tends to lower the control loop gain with an attendant increase in the error between the demanded and actual applied loads. In strain control the reverse happens: the gain increases as the testpiece becomes less stiff. This gain increase can lead to instability and is sometimes observed during low-cycle fatigue (LCF) testing as indicated in the real experimental result of Fig. 3.10. This example shows part of an LCF test cycle. The control loop performed well while the testpiece was essentially elastic but exhibited *limit-cycle* instability past the material's yield point.

Some machines are more sensitive to stiffness change than others. Generally it is better to conduct strain control tests using a machine which has a stiff actuator and frame. For load control, the opposite is true: sensitivity to stiffness is less if the actuator and frame are soft.

If maintaining waveform shape is not important and the demand is periodic, outer-loop amplitude control can be useful. It is able to keep the amplitude constant even in the face of stiffness change. For random waveforms, using an error matrix to modify the amplitude and frequency of the demand signal is only satisfactory if the stiffness changes slowly. Adaptive control on the other hand can cope with rapid stiffness

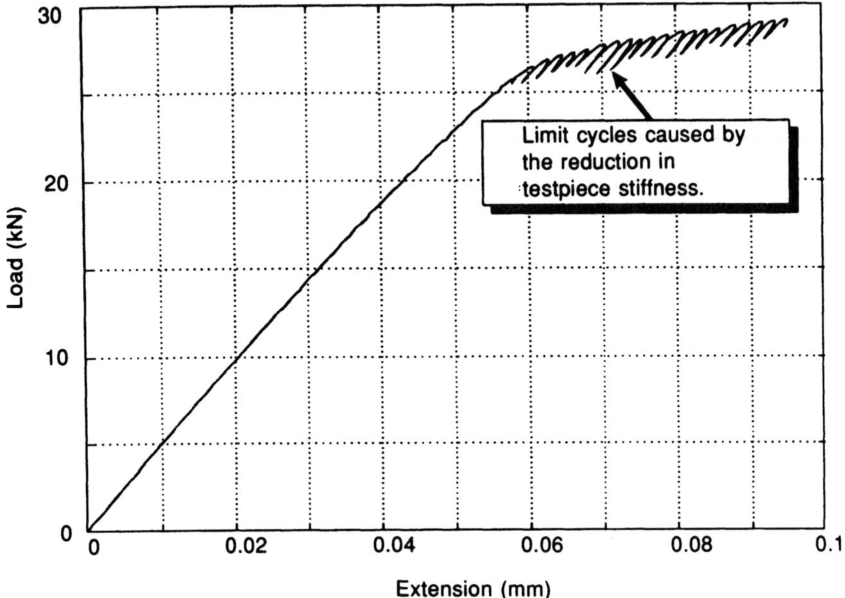

Fig. 3.10 Instability in strain control.

changes and should be able to preserve waveform fidelity for any type of waveform shape.

There is a rich literature on adaptive control. A comprehensive account is provided by Astrom and Wittenmark (1989). There have been many cited applications in areas as diverse as aerospace, process plant and robotics. Most modern adaptive control schemes assume little prior knowledge about the device being controlled and attempt to glean information about it from its response to probing signals. Normal operating signals can be used providing they are sufficiently dynamically rich. Unfortunately, many materials tests such as LCF employ signals which have an essentially low-frequency spectrum and the use of unwanted probing could jeopardize the test. For this reason, materials testing requires an alternative approach and is the subject of current research.

REFERENCES

Astrom, K. J. and Hagglund, T. (1988) *Automatic Tuning of PID Controllers,* Instrument Society of America.
Astrom, K. J. and Wittenmark, B. (1989) *Adaptive Control,* Addison-Wesley.
Clarke, D. W. (1984) PID algorithms and their computer implementation. *Transactions of the Institute of Measurement and Control,* **6** (6).
Cook, R. (1990) *FRAG Requirements for Control of Fatigue Testing,* 2nd issue, RAE.

Edwards, P. R. (1986) Control and monitoring of servo-hydraulic fatigue machines using a computer network with adaptive control of amplitude and frequency, in *Measurement and Fatigue – EIS '86*, EMAS, pp. 3–17.
FALSTAFF (1976) *Description of a Fighter Aircraft Loading Standard for Fatigue Evaluation (FALSTAFF)*, common report of F. & W. Emmen, LBF, NLR, IAGBG.
Hinton, C. E. and Clarke, D. W. (1989) Mathematical modelling of servo-hydraulic materials-testing machines, in *Conference Proceeding Mechatronics '89: Mechatronics in Products and Manufacturing*, Lancaster University.
Newman, J. R. and Edwards, P. R. (1988) *Short-Crack Growth Behaviour in Aluminium Alloy: an AGARD Cooperative Test Programme*, AGARD report 732.

CHAPTER 4

Dynamic force measurement

M. J. Dixon

4.1 INTRODUCTION

A formal procedure for the static calibration of load cells for use in materials testing machines has been available for many years: British Standard 1610 : 1985 provides a mechanism, traceable to the primary force standards, for the verification of the static force indicated by a load cell. The procedure requires the load cell to be calibrated against a transfer standard which has itself been calibrated in one of the deadweight force standard machines at the National Physical Laboratory (NPL). The uncertainty is known for each step of the calibration chain and hence a figure for the overall accuracy of the load cell may be calculated.

Load cells, however, when used in materials testing machines are not subjected to purely static loads. Servohydraulic machines and electro-mechanical shakers are designed to apply cyclic loads of varying amplitude to the testpiece at frequencies up to several hundred hertz. At present, these machines use load cells calibrated statically in the manner described above. There is, however, considerable evidence to suggest that the static calibration may be insufficient when the system is operated dynamically.

In 1986 a project was set up at the NPL with the aim of producing a new British Standard for dynamic force calibration of testing machines with a traceable and cost-effective procedure. This chapter summarizes the work of the project to date and indicates the likely form of the new standard.

4.2 THE EXISTING DRAFT BRITISH STANDARD

A draft British Standard, giving a procedure for the dynamic force calibration of testing machines, was produced in the early 1970s (BSI, 1971). Some years later a successor to this draft was issued, with the same basic principle but with greater technical refinement in that it recognized the use of modern peak monitors and digital voltmeters (DVMs) (BSI, 1986). This has remained a draft in the UK, but standards based on the same procedure have been released in the USA (ASTM : E467 : 1976 (1982)) and in Europe (ISO 4965 : 1979). None of the three documents, however, has been adopted by users of testing machines, because the procedure is too time consuming and expensive.

There are three possible sources of error in dynamic force measurement in a testing machine. Firstly, the strain gauges on the load cell may fail to reproduce the strain in the load cell element correctly at high frequency. Secondly, there is the possibility of an inertia error being generated by the mass between the load cell and the testpiece. This mass is usually made up by the gripping system of the testing machine and the testpiece. Thirdly, bandwidth limitations of the instrumentation and readout system may prevent them from recording the dynamic force amplitudes correctly.

The existing draft standard ignored the first source of error and assumed that the strain gauges would reproduce the dynamic strain correctly, irrespective of type or method of attachment. The inertia error was accounted for by individual calibrations for each type of testpiece to be used. Calibration of the instrumentation was also required, but by a rather impractical method.

For each type of testpiece, strain gauges were to be applied to form a simple load cell. The resulting device was known as a reference testpiece; it was calibrated statically to establish traceability and mounted in the testing machine to be calibrated, and its dynamic output was then compared with that indicated by the testing machine load cell. This quickly became an expensive proposition where a user had a large number of different testpieces to test.

4.3 THE DYNAMIC STANDARD FORCE TRANSDUCER

The first requirement of the project was to produce a dynamic force standard at the NPL, traceable to the existing primary standards, against which all other measurements could then be made. This was achieved by the construction of a dynamic force standard transducer which could then be used for measurement in a materials testing machine (Dixon, 1989; 1990a). The transducer employed simultaneous measurement of the dynamic strain across an elastic element by two independent techniques. The possibility of the same systematic error occurring in both methods of measurement was minimized by selecting two techniques as dissimilar from each other as possible and by separate calibrations of each stage of the instrumentation and measurement system.

The outputs from both systems were calibrated statically in a dead-weight force standard machine at the NPL and the transducer was then operated dynamically in a servohydraulic testing machine. The difference between the two force outputs over the frequency range of interest was then a measure of the overall uncertainty of dynamic force measurement.

4.3.1 The transducer

The two methods of strain measurement chosen were the electrical resistance strain gauge and a capacitance gauge, designed and built at the NPL. Two transducers were built, the first with a steel element made from En 24T, and the second, incorporating several design improvements, with an aluminium element made from HE 15 TF (Duralumin). This second device is shown in Fig. 4.1.

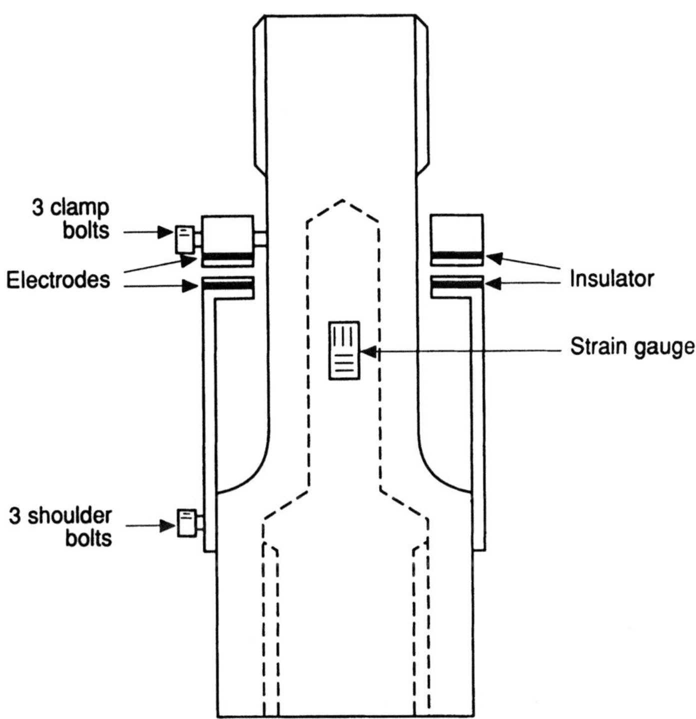

Fig. 4.1 The aluminium standard transducer.

The transducer elements were designed for a maximum static load of 25 kN in both tension and compression, corresponding to a strain level of 350 microstrain. A hollow column configuration was used with an internal thread at one end and an external thread at the other. This allowed the attachment of one capacitance electrode to a low-strain area and the other to a high-strain area, thereby making the conditions of strain measurement as different as possible from those of the resistance strain gauge transducer.

Each strain gauge bridge was formed of four 90 degree rosettes, equi-spaced around the element and aligned so that one gauge measured the axial strain and the other the transverse or Poisson's strain. The strain gauge types and adhesives are detailed in Appendix 4.B.

The capacitance gauge operated by changing the air gap between two parallel, flat, ring electrodes. The electrodes were made from brass shim 0.025 mm thick, attached to a polystyrene sheet insulator 0.05 mm thick using a commercial contact adhesive. One of these electrodes was attached with the same adhesive to an annulus at one end of an aluminium tube. The other end of this tube fitted over the outside diameter of the transducer element in a low-strain region and was attached to it by three shoulder bolts, equispaced around the circumference.

The other electrode, with its insulator, was attached with contact adhesive to an aluminium ring. This ring was designed to be as light and stiff as possible and was attached to the other end of the element by three radial bolts. These bolts had conical ends which fitted into small holes on the element surface. The interface was therefore formed between the edge of the hole and an annulus of the conical bolt tip, resulting in a very positive location. The electrodes were screened from electrical interference by an earthed brass shield which could be slid into place after assembly.

4.3.2 The instrumentation system

A block diagram of the complete instrumentation and measurement system is shown in Fig. 4.2. Excitation for the strain gauge bridge came from an adjustable DC power supply. The required excitation level was calculated to be approximately 6 volts and was set precisely during the first static calibration to achieve a convenient span. The output from the bridge was amplified by a commercial instrumentation amplifier, the AD 521, with the gain fixed at 1000. The output from this was passed through an eight-pole low-pass Butterworth filter, cornered at 500 Hz. This removed high-frequency noise and provided anti-aliasing protection during the subsequent digitization of the signal.

The instrumentation for the capacitance gauge consisted of a commercial AC bridge, the ASL 1055, with some modifications. These were necessary because the bandwidth of the instrument was limited by the final two-pole low-pass filter, required to remove the high-frequency components from the demodulated signal. This filter was found to apply significant attenuation (15%) to a 100 Hz signal and was therefore replaced by a similar filter to that constructed for the strain gauge instrumentation.

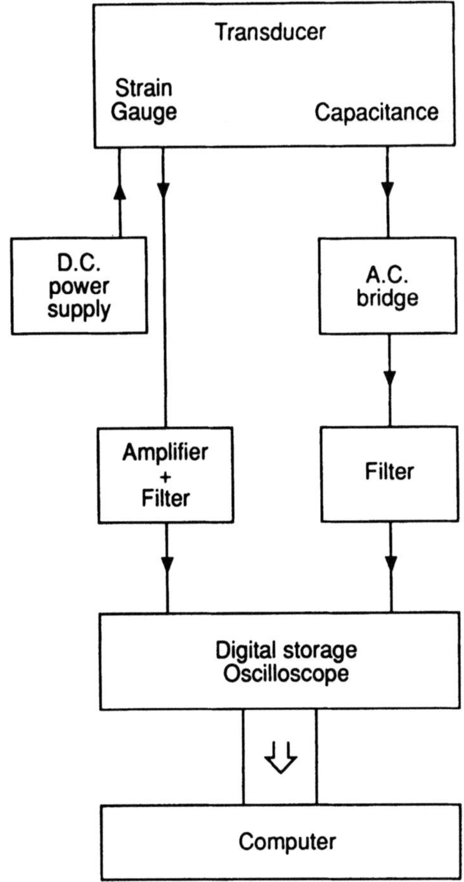

Fig. 4.2 The instrumentation and measurement system.

Data from both channels were recorded on a Nicolet 3091 digital storage oscilloscope (DSO). This instrument sampled both channels simultaneously with 12 bit resolution (one part in 4096) and stored 4096 points for each channel. At each test frequency, the sampling rate was set to record approximately 12–15 cycles of data. The sampling rate varied, therefore, from 10 kHz for an applied force at 100 Hz down to 0.1 Hz for an applied force at 0.001 Hz. The data were then sent, via an IEEE bus, to an HP 300 series computer.

A cross-correlation algorithm, written at the NPL, was then used for measurement of the dynamic amplitudes. Both signals were correlated against reference sine and cosine waveforms of the test frequency, generated in software. The general cross-correlation function is as follows (see Appendix 4.A for notation):

$$C_{ab}(\tau) = \frac{1}{t_1} \int_0^{t_1} V_a(t)\, V_b(t-\tau)\, d\tau \qquad (4.1)$$

τ was set to zero and the waveform to be measured, $V_a(t)$, was correlated against $V_b(t)$, where $V_b(t)$ took the form of $\sin(\omega t)$ and $\cos(\omega t)$. The two cross-correlations became

$$V_{\sin} = \int V_a(t)\, \sin(\omega t)\, dt \qquad (4.2)$$

$$V_{\cos} = \int V_a(t)\, \cos(\omega t)\, dt \qquad (4.3)$$

and the signal amplitude was given by

$$A = \omega/\pi \, \sqrt{\left(V_{\sin}^2 + V_{\cos}^2\right)} \qquad (4.4)$$

$V_{\sin}$ and $V_{\cos}$ may be thought of as the first-term Fourier coefficients of the time series with fundamental frequency $\omega/2\pi$.

4.3.3 Calibration of the instrumentation system

The measurement system consisting of the Nicolet DSO, the computer and the analysis software was first calibrated against a standard source from the Division of Electrical Science of the NPL. This source was known to be accurate in amplitude to within $\pm 0.01\%$.

The frequency range of 10 to 100 Hz was swept in 10 Hz steps at a variety of different voltage amplitudes, thereby allowing several voltage ranges of the DSO to be calibrated. From the amplitudes recorded, the absolute accuracy of measurement was calculated for each channel. The calibration of the strain gauge and capacitance instrumentation is detailed in Appendix 4.C.

4.3.4 Calibration of the standard transducer

Both methods of strain measurement were calibrated statically in the 50 kN deadweight force standard machine at the NPL. Three runs were taken, in both tension and compression, the device being removed, rotated through 180 degrees and replaced before the third run in each mode. Ten points were taken, the lowest at 2 kN and the highest at 20 kN.

Both methods of strain measurement showed a repeatability of better than 0.1% of reading and hysteresis of better than 0.2% of reading. The transducer was then installed in a servohydraulic testing machine and dynamic comparisons were made between the two methods of strain measurement. Dynamic amplitudes in the range ± 2.5 to ± 15 kN were used, corresponding to strain ranges of ± 35 to ± 210 microstrain. For each test a mean level of load was also applied, such that the device was never operated through the zero-load region.

Some typical test results from the aluminium transducer are shown in Fig. 4.3. The difference in strain (and hence force) measurement between the capacitance gauge and the strain gauge bridge was in all tests less than ± 0.5%. At frequencies above 1 Hz, no difference was detected between the outputs from the three types of strain gauge tested (to within the repeatability of the tests), indicating that they are all equally suitable for dynamic measurements. Below 1 Hz, the output from the EA bridge was reduced by approximately 0.5% relative to the MA and TK bridges, owing to creep in the gauge backing material.

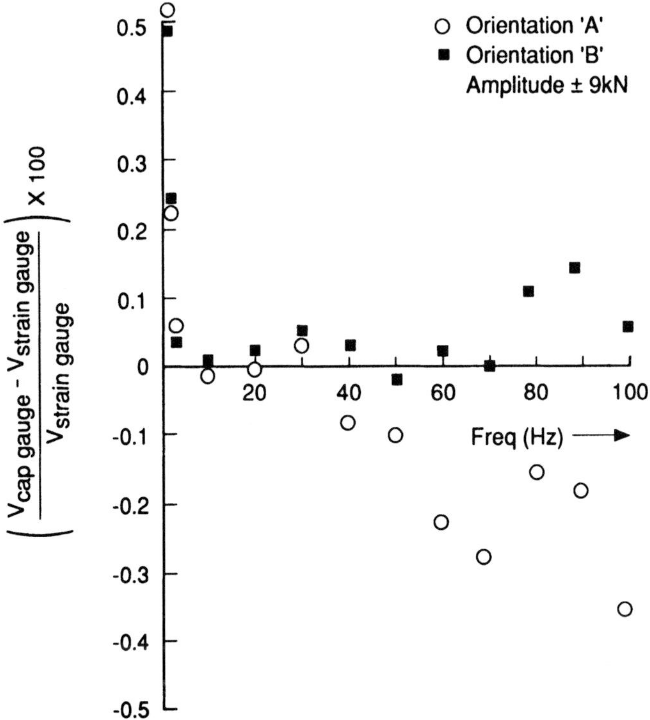

Fig. 4.3 Test results from the aluminium standard transducer.

A detailed error analysis showed that the degree of correlation between the two methods of force measurement was within the uncertainty limits for the experiment. A normal distribution with a large number of degrees of freedom was assumed in the computation of random uncertainties, and the overall figure was arrived at by adding the individual figures in quadrature. Systematic uncertainties were combined by addition. Using 95% confidence limits (2σ) for the random uncertainty, a figure of ± 0.4% was calculated for the overall uncertainty of dynamic force measurement.

 Dynamic force measurement

4.4 EVALUATION OF THE EXISTING DRAFT STANDARD

Reference testpieces were then produced to evaluate the existing draft standard. As part of the procedure, these testpieces were calibrated statically against a deadweight force standard but they were also calibrated dynamically against the dynamic force standard to verify the performance of the strain gauge installations.

Three reference testpieces showed no measurable difference from the standard load cell, but a fourth testpiece with EA-06-125-TM-120 gauges showed a progressive attenuation with frequency, to a maximum of – 1.5% at 100 Hz. The error was found to be linear with strain amplitude up to the maximum tested, 350 microstrain. This experiment was repeated several times over a period of months and after recalibration of the instrumentation, but the attenuation remained constant. Further experiments were therefore undertaken to establish the reason (Dixon, 1991b).

The conclusion eventually reached was that the original strain gauges or adhesive were defective in some way. Future installations were always checked against the force standard.

The reference specimens were then used to calibrate a testing machine at the NPL to the procedure in the draft standard.

4.4.1 Conclusions reached about the existing draft standard

Inertia errors of up to 3% were recorded with a stiff testpiece (1400 kN mm^{-1}) (Dixon, 1988) and up to 40% with a soft testpiece (4 kN mm^{-1}). Weibull (1961) mentioned errors of up to 30% of the applied load and Geers (1982) measured errors of 10% at 55 Hz, 59% at 100 Hz and greater than 100% at higher frequencies.

It was concluded from the evaluation that, although the reference testpiece method was technically sound, more sophisticated instrumentation was needed in the calibration equipment than the peak monitors specified in the standard. The method for calibrating the instrumentation was also felt to be unworkable because of the low signal levels required. It was confirmed that for a user with several types of testpiece the procedure was unworkable, because of the time required both to construct and calibrate the reference testpieces and to test them in the machine.

4.5 THE DEVELOPMENT OF AN ACCELEROMETER COMPENSATION SYSTEM

An accelerometer compensation system was then developed for a testing machine to test the viability of this method as an alternative way of measuring the inertia error (Dixon, 1991a).

4.5.1 Theory

The significant masses and stiffnesses in the testing machine are shown in Fig. 4.4. The load frame is assumed to be rigid compared with the stiffness of the load cell and specimen. Thus, equating forces (for notation see Appendix 4.A):

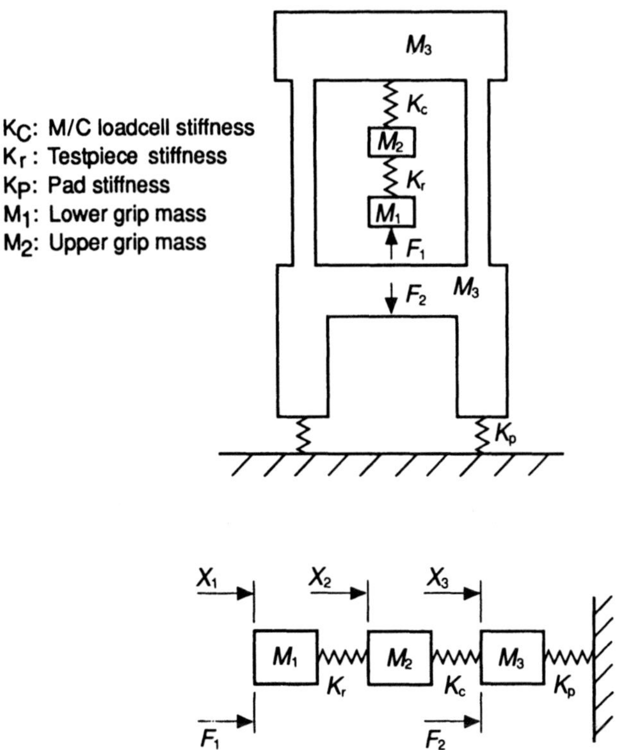

Fig. 4.4 Mass spring model of a testing machine.

$$F_1 = M_1 \ddot{x}_1 + K_r (x_1 - x_2) \tag{4.5}$$

$$K_r (x_1 - x_2) = M_2 \ddot{x}_2 + K_c (x_2 - x_3) \tag{4.6}$$

$$K_c (x_2 - x_3) = M_3 \ddot{x}_3 + K_p x_3 - F_2 \tag{4.7}$$

$$F_1 = -F_2 \tag{4.8}$$

$$K_c (x_3 - x_2) = F_c \quad \text{(force recorded by load cell)} \tag{4.9}$$

$$K_r (x_2 - x_1) = F_r \quad \text{(force seen by specimen)} \tag{4.10}$$

Substituting equations 4.9 and 4.10 into equation 4.6,

$$F_c = M_2 \ddot{x}_2 + F_r \qquad (4.11)$$

The inertia error is thus created only by the acceleration of M_2, the upper grip.

The error may be calculated by measurement of M_2 and x_2, or more conveniently $\ddot{x}_2$, using an accelerometer mounted on the mass. Alternatively, a proportion of the output from the accelerometer may be subtracted from the load cell signal, removing the error. This is commonly known as load cell compensation (McConnell and Park, 1981; Milz, 1978; Collier, Care and Hilyard, 1986; Darrouj and Faulkner, 1989).

Upon rearrangement of the above equations:

$$K_p x_3 = - M_1 \ddot{x}_1 - M_2 \ddot{x}_2 - M_3 \ddot{x}_3 \qquad (4.12)$$

The resultant of the inertia forces in the system is thus equal to the dynamic reaction from the mounting pads.

For calibration of a compensation system, the inertia force can in principle be generated by cycling the actuator without a testpiece. From equation 4.11, as F_r is now zero,

$$F_c = M_2 \ddot{x}_2 \qquad (4.13)$$

The gain of the compensation controls the proportion of the accelerometer signal that is subtracted. This gain is adjusted until F_c remains as close to zero as possible over the entire frequency range. The relative magnitudes of M_1, M_2 and M_3 usually mean that only small values of F_c can be generated and difficulty may be experienced in setting the compensation precisely, owing to the noise level in the system. The calibration is valid for all testpiece stiffnesses but must be repeated if M_2 is changed.

The acceleration signal must also be in phase with the load cell signal at the point of subtraction. The error in subtraction of the inertia force that will result from a small phase difference between the two signals may be calculated as follows. If the gain control has been set correctly then, in magnitude, $M_2 \ddot{x}_2 = F_c$ when the system is operated without a testpiece. If the angle between the two signals is θ then

$$F_c \sin(\theta/2) = q/2 \qquad (4.14)$$

where q is the magnitude of the error. For small θ,

$$q/F_c = \sin\theta \qquad (4.15)$$

For example if θ is 5 degrees, $q/F_c = 0.09$ or 9%. In this case, therefore, only 91% of the inertia error will be removed.

4.5.2 Construction of the compensation system

A system was therefore built to compensate the load cell of the testing machine. A Bruel and Kjaer type 4338 accelerometer was mounted, axial

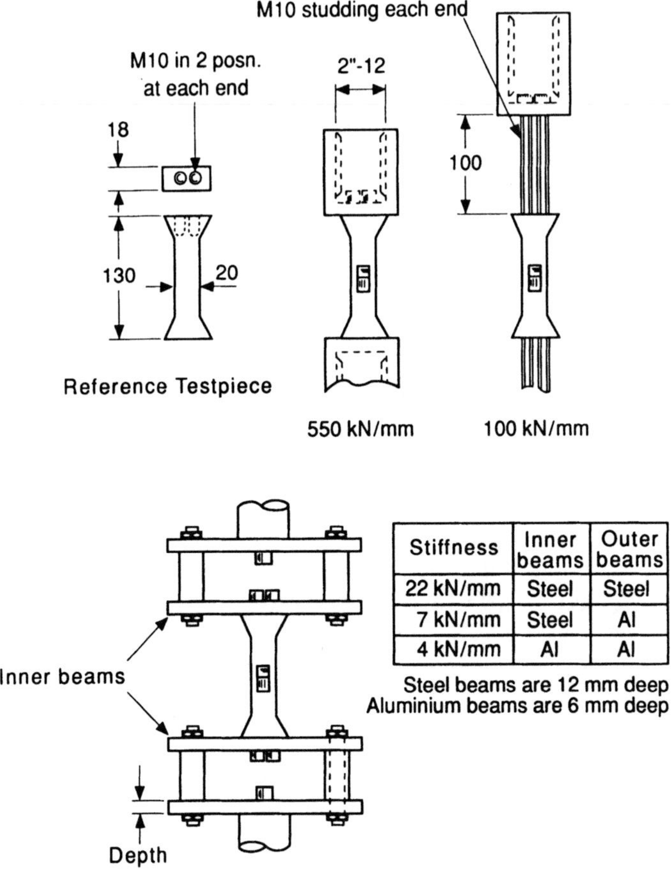

Stiffness	Inner beams	Outer beams
22 kN/mm	Steel	Steel
7 kN/mm	Steel	Al
4 kN/mm	Al	Al

Fig. 4.5 The reference testpiece stiffness fixtures.

with the load string, on the stud adaptor to the load cell and the cable was led out through drillings in the stud. Square steel plates, with steel blocks bolted to the edges, were attached to the lower flange of the load cell and to the end of the actuator rod, to simulate a pair of grips, and the reference testpieces were mounted between these. The mass of each simulated grip was approximately 100 kg.

Two reference testpieces were used, with stiffnesses of 1400 kN mm^{-1} and 550 kN mm^{-1}. A range of adaptors (studs and beams) was also used to reduce the stiffness of the second reference testpiece, to a minimum of 4 kN mm^{-1}. These were attached, in pairs, to the ends of the reference testpiece, as shown in Fig. 4.5. Each configuration was thus symmetrical, and the adaptors were of negligible mass when compared with the simulated grip mass. The dynamic characteristics of each configuration were therefore assumed to be similar to those of a single-element test-piece of the same stiffness.

The output from the accelerometer was amplified using a Bruel and Kjaer type 2624 charge amplifier and then applied to the compensation circuit. This enabled a proportion of the accelerometer signal to be subtracted from the load cell signal, the amount depending on the mass to be compensated for. The two signals were first passed through similar filters to remove noise. The accelerometer signal was then passed through an adjustable phase shift network allowing the phase difference between the two signals to be minimized. The accelerometer signal was then passed through a gain stage, which controlled the amount of compensation, and the output was subtracted from the load cell signal.

The strain gauge instrumentation and calibrated measurement system described in section 4.3 were used throughout the experiments.

4.5.3 Calibration of the compensation system

The compensation system was calibrated initially by cycling the actuator with the masses attached but without a testpiece. The phase of the accelerometer signal was adjusted until it was in phase with the load cell signal, as viewed on a DSO, over the frequency range of 10 to 100 Hz. The gain control of the accelerometer compensation circuit was then adjusted until the output from the load cell remained at a minimum over the frequency range.

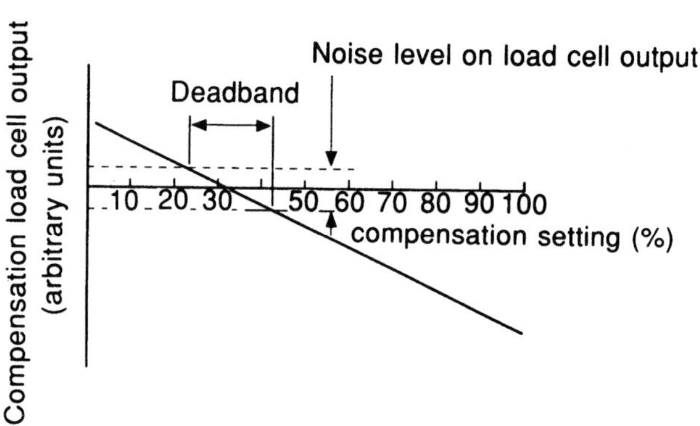

Fig. 4.6 The deadband in the compensation setting.

With zero compensation, the load cell output indicated the force moving the grip mass M_2 in reaction to the actuator. As the amount of compensation was increased, this signal reduced to a minimum, and then increased again as the accelerometer signal became dominant (too much compensation). The noise level on the load cell output, however,

Force amplitude (kN)	Testpiece stiffness (kN / mm)	Symbol
2	1400	■
2	550	△
1	100	x
0.5	22	●
0.25	7	▲
0.25	4	○

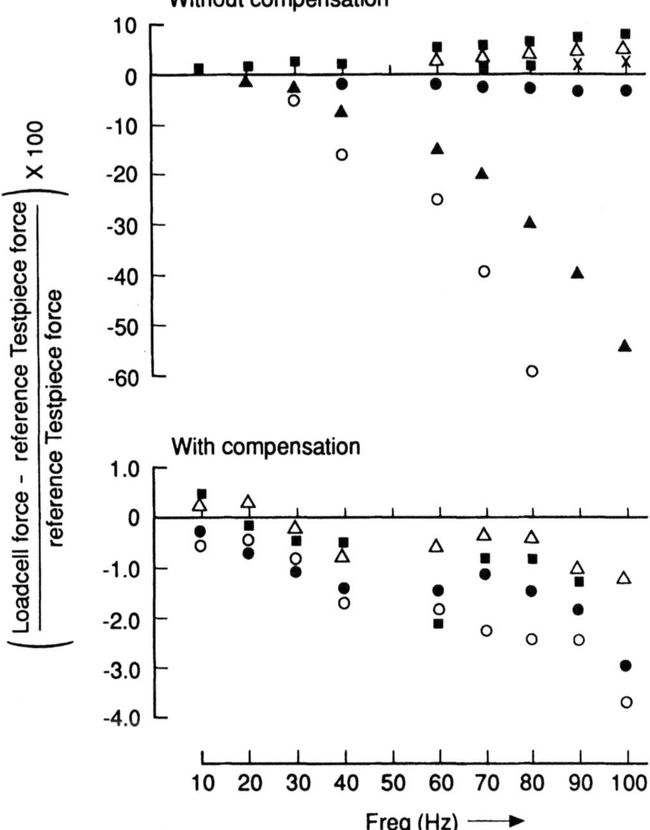

Fig. 4.7 The effect of compensation.

caused a deadband within which the load cell output did not visibly change, as illustrated in Fig 4.6. For the purposes of initial tests, the compensation was set in the middle of this deadband.

The outputs from the six reference testpieces were compared with the load cell output in turn, with and without compensation, over the frequency range of 10 to 100 Hz. The results in Fig. 4.7 show that compensation removed the majority of the inertia error but that an error of up to 4% was still present for the least stiff reference testpiece. When presented graphically, many of the points overlapped at low frequencies where the error was small, and some have hence been omitted for clarity. Tests were not conducted at 50 Hz as some mains interference was present on the load cell signal.

It was also confirmed that, in accordance with the theory, the magnitude of the inertia error without compensation was not affected by the mass between the testpiece and the actuator (the lower grip), or by the compliance of the mountings between the testing machine and the floor.

The compensation was then reset using the least stiff reference testpiece, which had produced the largest error, by adjusting the setting until the error remained at a minimum over the entire frequency range. The adjustment required was within the deadband and reduced the error to less than $\pm 2\%$ for the entire range of testpiece stiffnesses.

The experiments were then repeated using two accelerometers mounted opposite each other on the inertia mass. Their outputs were averaged in the calibration equipment and fed into the circuit as before. To within the uncertainty of the experiment, there was no difference between these results and those obtained with a single accelerometer mounted on axis, as long as the two accelerometers were at 180 degrees to each other and hence cancelled out any transverse motion.

4.5.4 Discussion

The experimental results from the uncompensated load cell may first be explained from the equations of motion.

The motion of the frame in space x_3 is in reaction to the motion of the lower grip mass x_1. When the testpiece is stiff, the motion of the lower grip will be small and the reactive motion of the frame will be even smaller (as $M_3 \gg M_1$). The frame motion may therefore be neglected. From equation 4.11, the machine load cell will record the load on the testpiece, plus the force required to move the upper grip. The motion of the upper grip is in phase with the lower grip, and hence with the excitation force. The error is therefore positive and the testing machine load cell over-reads.

As the testpiece stiffness is decreased, the motion of the lower grip becomes larger. Hence the motion of the frame in reaction to it becomes significant. This frame motion is in anti-phase with the lower grip and the excitation. Since the testpiece is now much less stiff than the machine load cell, the motion of the upper grip will now closely resemble that of the frame in both magnitude and phase. The motion of the upper grip, shown by equation 4.11 to be the only cause of the inertia error, is now in anti-phase with the excitation. The error is therefore negative, and the machine load cell under-reads.

The magnitude of the frame motion and its phase relative to the excitation will also be influenced by the method of attachment to the floor (through K_p) and the level of damping in the mountings.

The results show that, for this system, the frame motion becomes equal and opposite to the upper grip motion (from the load cell deflection) at a testpiece stiffness between 100 and 22 kN mm^{-1}. Theoretically, it is therefore possible to have zero error at some stiffness within this range. Below 22 kN mm^{-1} the frame motion is dominant and a large negative error results, inversely proportional to the testpiece stiffness. The magnitude of the resultant error, as plotted in Fig. 4.7, is non-linear at high stiffnesses because of the opposing error from the load cell deflection.

The compensated results show that cycling the actuator without a testpiece is not an accurate way of calibrating the compensation. Better results were achieved by comparison with a reference testpiece of low stiffness. This reduced the error to less than ± 2% over the frequency and stiffness range tested.

4.6 THE NEW DRAFT STANDARD

The main objections to the existing draft standard are that it is too time consuming and expensive to implement and that the measurement of amplitudes by detection of the peaks is insufficient for calibration purposes. A new standard has therefore been developed (Dixon, 1991c) using accelerometers to measure the inertia error and the cross-correlation method for measurement of the amplitudes.

4.6.1 A summary of the new draft standard

The new standard uses only two strain gauged reference testpieces, the two varying from each other in stiffness by a factor of 100. The stiffer is known as the type 1 reference testpiece and the other as the type 2 reference testpiece. This establishes an envelope of stiffness and frequency within which the machine is calibrated. For calibration outside this stiffness range, the actual testpieces may be used in conjunction with the accelerometers to establish the inertia error.

Two accelerometers, mounted opposite each other on the inertia mass, are used to measure its acceleration. They may be mounted via magnetic mounts or wax, or by studs into the grip mass. Their outputs are averaged in the calibration equipment and then used to measure the inertia error.

For machines without load cell compensation, the inertia mass is established at the time of the first calibration by weighing the detachable parts and by calculation of the end mass of the load cell. The end mass is defined as that part of the load cell element which contributes to the inertia force. This total mass is assumed to be invariant with time. Other load string combinations are weighed when calibration is required. Two sets of measurements, as detailed in sections 4.6.1(a) and 4.6.1(b), are performed in the calibration procedure each year.

For machines with load cell compensation, the procedure outlined in section 4.6.1(c) is performed each year to verify the compensation system.

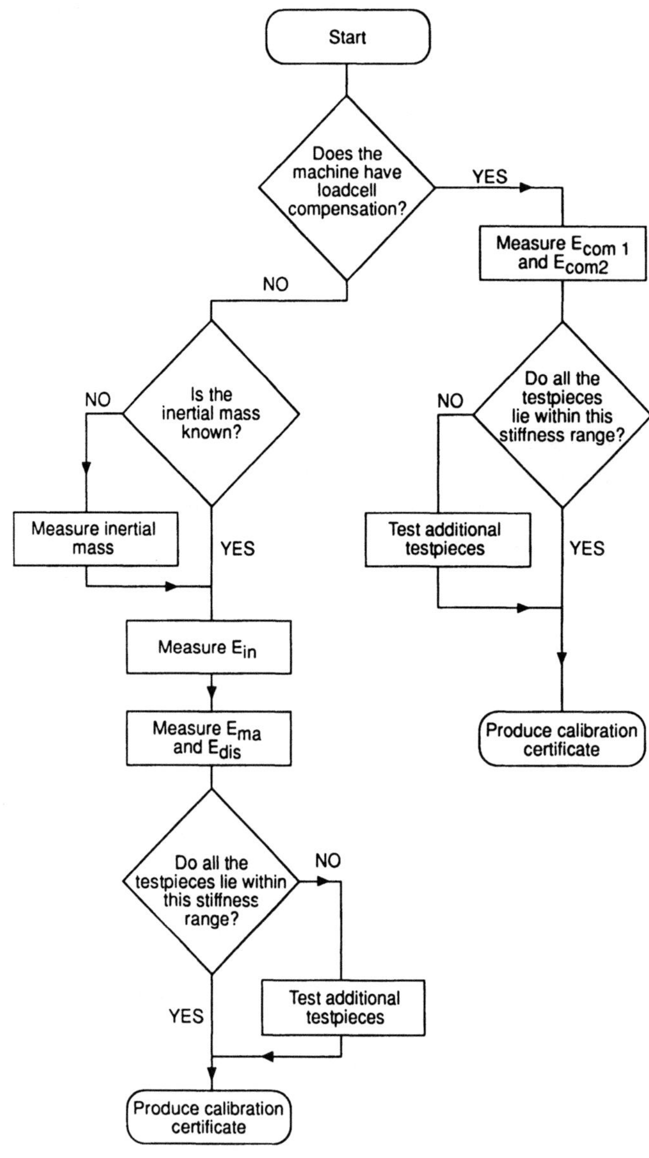

Fig. 4.8 The calibration procedure.

The calibration equipment uses the cross-correlation method for measurement of the signal amplitudes from the reference testpieces, the machine load cell and the accelerometers. The output from the machine's peak monitor is also recorded.

Traceability of the new standard may be achieved by calibration of the reference testpieces against the dynamic standard transducers at the NPL (described in section 4.3) and by calibration of the accelerometers against a national standard. Traceable electrical standards are also available at the NPL for calibration of the instrumentation.

A flow diagram of the new calibration procedure is shown in Fig. 4.8.

(a) Measurement of the instrumentation bandwidth

Two accelerometers are attached to the inertia mass and the average of their outputs is measured for a given force amplitude over a range of frequencies. The stiff reference testpiece is mounted between the grips and its output recorded. The inertia error is then calculated as Ma_1.

The inertia error plus the output from the reference testpiece F_{r1} equals the force recorded by the testing machine load cell. The difference between this sum and the force amplitude indicated by the testing machine instrumentation is the instrumentation error. Calibration errors for the machine instrumentation q_{in} may therefore be defined at different frequencies:

$$q_{in} = \left(\frac{F_i}{F_{r1} + M_2 a_1} \right) - 1 \qquad (4.16)$$

The frequency at which q_{in} first exceeds ± 0.01 is the instrumentation limiting frequency f_i.

(b) Determination of the stiffness and frequency envelope

For each inertia mass, the limits of testpiece stiffness and frequency, within which the inertia error is within a predefined limit, are determined, and a calibration error is calculated at each frequency tested. This procedure is carried out for both the type 1 and the type 2 reference testpieces.

$$q_{ma} = \frac{M_2 a}{F_c} \qquad (4.17)$$

The magnitudes of F_c at the harmonic frequencies $2f_0$, $3f_0$ etc. up to $9f_0$ are also measured and expressed as a distortion factor q_{dis}, calculated as

$$q_{dis\,(h)} = \frac{F_{c\,(h)}}{F_{c\,(f0)}} \qquad (4.18)$$

where h runs from 2 to 9.

The frequency at which either q_{ma} or q_{dis} first exceeds ± 0.04 for the type 1 reference testpiece is the type 1 testpiece limiting frequency f_1. Similarily, for tests with the type 2 reference testpiece, f_2 is derived.

(c) *Calibration of a machine with load cell compensation*

Load cell compensation systems normally measure the acceleration of the inertia mass and subtract a proportion of this signal from the load cell signal. The proportion required varies only with the inertia mass and, once set, should be constant for all values of testpiece stiffness and frequency. The calibration determines the error for the low- and high-stiffness reference testpieces:

$$q_{com1} = \left(\frac{F_i}{F_{r1}}\right) - 1 \tag{4.19}$$

$$q_{com2} = \left(\frac{F_i}{F_{r2}}\right) - 1 \tag{4.20}$$

The frequency at which either q_{com1} or q_{com2} first exceeds ± 0.04 is the compensated limiting frequency f_c. The machine is therefore calibrated for use up to this frequency with testpieces of stiffnesses between that of the types 1 and 2 reference testpieces.

(d) *Presentation of results*

The calibration therefore defines an envelope of stiffness and frequency within which the total error, made up of the instrumentation and inertia errors, is known to be less than $\pm 4\%$. If some of the testpieces for which calibration is required lie outside this stiffness range, then they may be tested individually and recorded as individual points, additional to the envelope. A typical calibration envelope is shown in Fig. 4.9.

4.7 CONCLUSIONS

The existing draft standard for dynamic force measurement has been evaluated at the NPL. Errors of up to 40% were recorded and the procedure was confirmed to be very time consuming, taking about half a day for each reference testpiece.

A traceable dynamic force standard was then established at the NPL and the principle of accelerometer compensation was verified on a testing machine. It was found possible to compensate for errors of at least up to 60% of the applied load. A new draft standard was then developed for calibration using a combination of reference testpieces and accelerometers

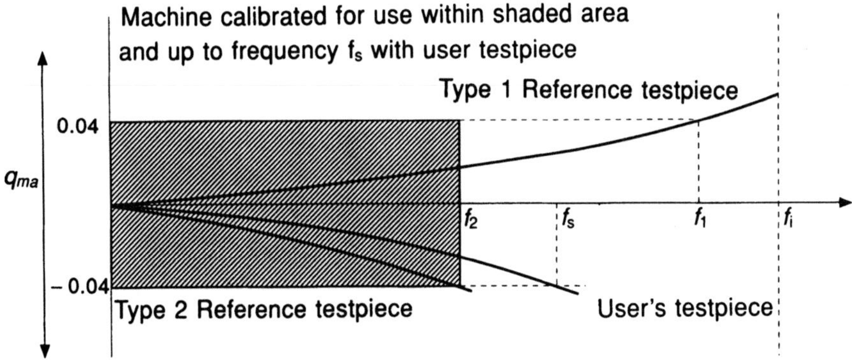

Fig. 4.9 A typical calibration envelope.

mounted on the inertia mass. This is much quicker and easier to implement than the draft that it replaces and can be made traceable to the
relevant primary standards.

ACKNOWLEDGEMENTS

The work described in this chapter was carried out at the NPL under the
Research Associate Scheme. The author, then an employee of Instron
Ltd, would like to express his thanks to Dr R. D. Lohr and to Mr R. F.
Jenkins for advice, help and encouragement.

APPENDIX 4.A   NOTATION

A	signal amplitude
C	calibration factor
C_{ab}	correlation coefficient
F_i	force sensed by the load cell and measured by the testing machine instrumentation
F_r	force sensed by the reference testpiece and measured by the reference measurement system
F_c	force sensed by the machine load cell and measured by the reference measurement system
F_1	force applied by actuator to lower grip
F_2	reaction force of frame to actuator
K_c	machine load cell stiffness

                    Dynamic force measurement

K_r testpiece stiffness
K_p pad stiffness
M_1 lower grip mass
M_2 upper grip mass
M_3 frame mass
$V_a(t)$ waveform to be measured
$V_b(t)$ reference waveform

a acceleration of the inertia mass
f_0 carrier frequency
f testpiece limiting frequency
f_i instrumentation limiting frequency
f_c compensated limiting frequency
q error in compensation of load cell
q_{in} instrumentation error
q_{ma} inertia error
q_{dis} distortion error
q_{com} compensation error
t time
x_1 displacement of lower grip
x_2 displacement of upper grip
x_3 displacement of frame

θ phase angle between force and acceleration signals
τ delay time
ω frequency

APPENDIX 4.B   TYPES OF STRAIN GAUGE USED IN THE STANDARD TRANSDUCERS

Three types of strain gauge, all supplied by Micro-Measurements, were used in the transducers. Each of the transducers' flexure elements had one bridge of high-quality transducer gauges, type TK-06-125-VB-350 for the steel element and type MA-13-125-VB-350 for the aluminium element. The aluminium element also had a second bridge of type EA-13-125-TM-120 gauges, equispaced between the first bridge. The object of this was to evaluate any differences in dynamic performance between the three types. The TK and MA gauges have thin, rigid, epoxy-based backing materials and are rather brittle in nature, whereas the EA gauges have a much thicker, more flexible polyimide backing. This makes them more rugged, easier to apply and hence more appropriate for stress analysis work. It was not known, however, whether the thicker backing would prevent them from recording the strain correctly when used at frequencies up to 100 Hz.

The ED series gauges, recommended for dynamic measurements, were not used. The principal advantage of these iso-elastic foil gauges is a gauge factor approximately 50% higher than that of constantan and Karma foils, thereby improving the signal to noise ratio in a dynamic measurement environment. They do not, however, have any self-temperature compensation mechanism and the resultant apparent strain makes them unsuitable for a transducer which is to be calibrated statically as well as dynamically.

All three bridges were attached with M-Bond 610 adhesive from Micro-Measurements. This is a high-performance epoxy adhesive with a low solids content, resulting in a very thin, hard, glue line.

Note The symbols TK, EA, MA and the name Karma refer to Micro-Measurements products.

APPENDIX 4.C   CALIBRATION OF THE STRAIN GAUGE AND CAPACITANCE INSTRUMENTATION

4.C.1 The strain gauge instrumentation

The simplest method of dynamic calibration for this type of instrumentation is to apply a sinusoidal signal of known magnitude, from a calibrated voltage source, to the input terminals. The frequency of the calibration signal is then varied and the output voltage from the instrumentation is measured at each frequency.

A calibration factor C, defined as output/input, may be established for the instrumentation, and calibration results recorded as C_{f1}, C_{f2}, C_{f3} etc. for different frequencies. Note that it is not the absolute value of C that is important, as this is usually included in the static calibration of the transducer, but rather the variation between the values of C at different frequencies.

The disadvantage of this method is that a sinusoidal input signal of similar magnitude to that produced by the transducer is required. For a strain gauge bridge this will be in the range of 1 to 10 millivolts. This is lower than the minimum voltage available from most voltage sources and requires a high degree of stability, at least a decade better than the accuracy required from the calibration. Therefore, for a calibration accuracy of 0.1% and an input voltage level of 10 mV, a source stable to 10 μV would be required. It is also difficult to inject such a voltage from an external supply into a circuit without introducing noise and mains pick-up.

An alternative solution is to use a voltage source of similar magnitude to the output from the instrumentation, but input to the circuit via a potential divider which drops the voltage by approximately the same ratio that it is amplified by the instrumentation (Fig. 4.C.1) (Dixon, 1990b).

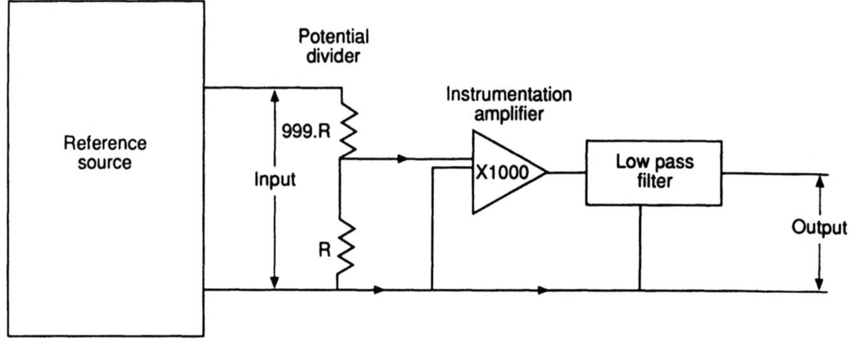

Fig. 4.C.1 The calibration set-up for the strain gauge instrumentation.

The precise value of the voltage division is not important, as the results may be normalized against the lowest frequency point (at perhaps 0.1 or 1 Hz). The only critical requirement for the voltage divider is that the division should be constant and this may be ensured by the use of high-quality resistors. The instrumentation may therefore be calibrated to the same degree of accuracy using a voltage source of higher output level and lower stability.

4.C.2 The capacitance instrumentation

A diagram of the capacitance gauge instrumentation is shown in Fig. 4.C.2. A sinusoidal waveform of fixed frequency f_0 is produced in the carrier generator and applied to the bridge at point A. A square wave of logic amplitude and the same frequency is also derived from this sinusoid. In the normal operating mode, the effect of the transducer operating (at frequency f for instance) is to amplitude modulate the bridge output signal. It now consists of a carrier at f_0 (caused by the mean level of signal) and two sidebands at $f_0 - f$ and $f_0 + f$ (the carrier frequency minus/plus the frequency of mechanical excitation).

After amplification the signal enters the phase-sensitive detector. This may be thought of as a switch, operated from the square wave at f_0, which inverts all negative parts of the signal. In the frequency domain, its effect is to move the sidebands both up and down the spectrum by an amount f_0. The $f_0 + f$ component now becomes f, while the $f_0 - f$ component becomes $-f$. This is folded about the zero-frequency axis and is added to the existing component at frequency f. The f_0 carrier signal is transposed into a DC component and a new signal at $2f_0$. Finally, the low-pass filter removes all high-frequency signals and high-frequency noise.

It is not therefore possible to dynamically calibrate an AC bridge in the same manner as a DC bridge because the circuit is only tuned to signals

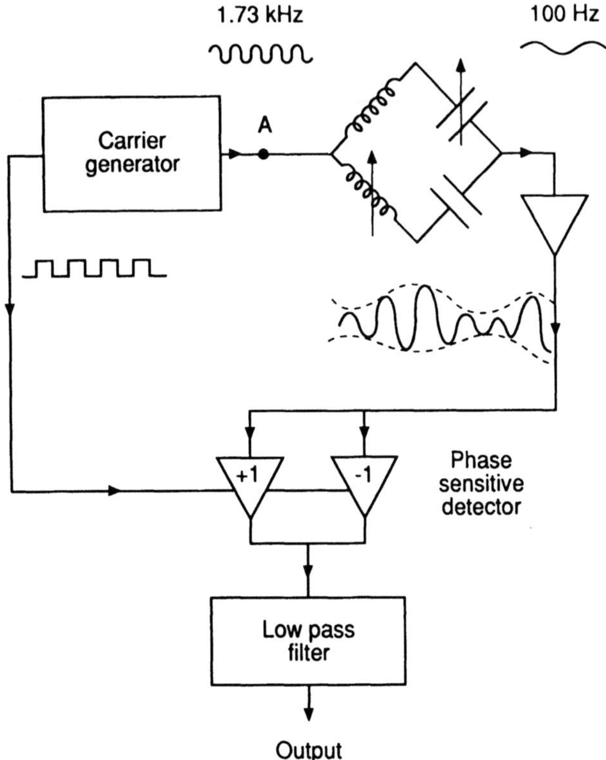

Fig. 4.C.2 The capacitance gauge instrumentation.

at or near the carrier frequency. (It is this characteristic that makes AC bridges relatively insensitive to pick-up from the mains supply at 50 or 60 Hz.) Calibration may, however, be achieved by determining a calibration factor for each sideband frequency individually and then by averaging the calibration factors for each pair of sidebands. This average of any pair of sidebands represents the calibration factor C for the frequency of operation of the transducer that produced those two sidebands.

The procedure is therefore to break the circuit at point A (Fig. 4.C.2) and apply a sinusoid from a calibrated source to the bridge at the point. For every frequency applied, the output frequency is the difference between the input and the carrier. If the frequency applied were exactly that of the carrier the output from the complete circuit would be DC. In practice, the frequency resolution and stability of the source will usually limit the lowest frequency that can be generated: if the frequency from

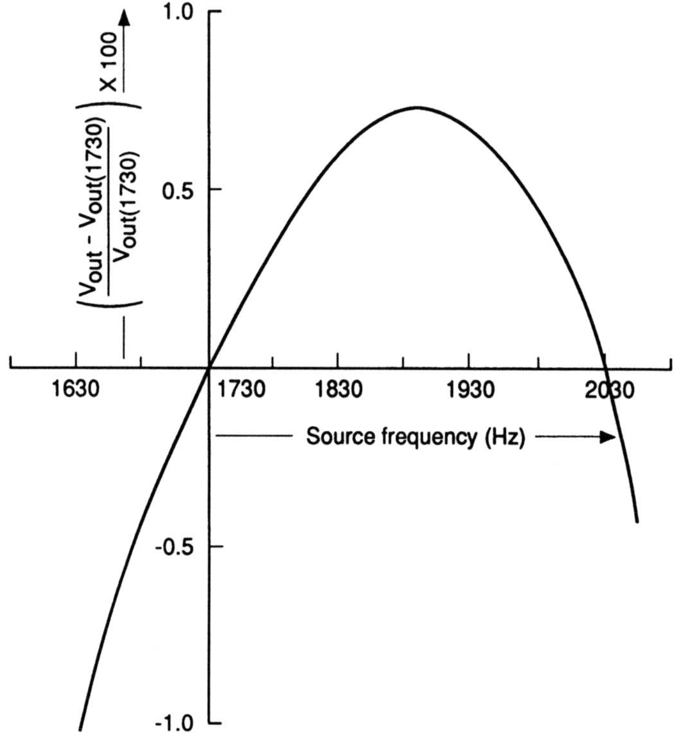

Fig. 4.C.3 The sideband calibration factors.

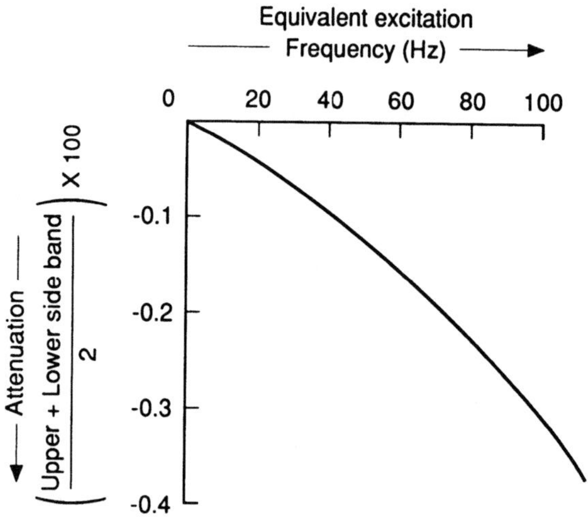

Fig. 4.C.4 The calibration curve derived from the sideband factors.

the source can only be set to within 1 Hz of the carrier, then the lowest frequency that can be generated will be 1 Hz.

In Fig. 4.C.3, the sideband calibration factors for the capacitance gauge instrumentation are plotted for a calibration range of 1–100 Hz, normalized against the lowest frequency generated (less than 1 Hz). In Fig. 4.C.4 the average for the pairs of sidebands at each frequency is plotted: these are the values of C as defined above.

Note that for this type of calibration a voltage source of the magnitude and frequency of that produced by the carrier generator is required. This is not a difficult requirement as the magnitude is typically in the range 1–10 V, but, as in the static case, the voltage must be stable to at least a factor of 10 better than the accuracy required from the calibration. The frequency resolution will also determine the number of calibration points that can be taken as this determines the increment for each sideband.

APPENDIX 4.D   EQUIPMENT SUPPLIERS

ASL, 28 Blundells Road, Bradville, Milton Keynes, MK13 7HS. 0908 320666

Bruel & Kjaer, 92 Uxbridge Road, Harrow, HA3 6BZ. 081 954 2366

Micro-Measurements, Stroudley Road, Basingstoke, Hants, RG24 0FW. 0256 462131

Nicolet Instruments Ltd, Budbrooke Road, Warwick, CV34 5XH. 0926 494111

REFERENCES

ASTM : E467 : 1976 (1982)   Verification of Constant Amplitude Dynamic Loads in an Axial Load Fatigue Testing Machine.

British Standards Institution (1971) *Dynamic Force Calibration of Axial Load Fatigue Testing Machines by Means of a Strain Gauge Technique*, draft for development DD2.

British Standards Institution (1986) *Dynamic Force Calibration of Axial Load Fatigue Testing Machines by Means of a Strain Gauge Technique*, document 85/44083.

BS 1610 : Parts 1 and 2 : 1985   Materials Testing Machines and Force Verification Equipment.

Collier, P., Care, C. M. and Hilyard, N. C. (1986) An automated dynamic mechanical spectrometer. *Journal of Physics E, Scientific Instruments*, **19**, 342–7.

Darrouj, M. N. and Faulkner, R. G. (1989) A new method for measuring dynamic response in polymers over a wide range of frequency and temperature. *Journal of Physics E, Scientific Instruments*, **22**(4), 218–22.

Dixon, M. J. (1988) Errors in dynamic force measurement. *Strain*, **24**(4), 139–42.

Dixon, M. J. (1989) A dynamic force measurement system, in *Proceedings of the SEM Spring Conference on Experimental Mechanics*, Cambridge, MA.

Dixon, M. J. (1990a) A traceable dynamic force transducer. *Experimental Mechanics*, **30**(2).

Dixon, M. J. (1990b) Dynamic calibration methods for transducer instrumentation. *Experimental Techniques*, December, 51–4.

Dixon, M. J. (1991a) Development of a load cell compensation system. *Experimental Mechanics*, **31**(1).

Dixon, M. J. (1991b) Errors in dynamic strain measurement. *Strain*, **27**(3).

Dixon, M. J. (1991c) *Dynamic Force Calibration for Materials Testing Machines*, draft British Standard, available from the author.

Geers, W. J. (1982) *An Analysis of the Dynamic Response of Servo-Hydraulic Testing Machines*, Instron Ltd technical report ISD 013.

ISO 4965 : 1979   Axial Load Fatigue Testing Machines. Dynamic Force Calibration. Strain Gauge Technique.

McConnell, K. G. and Park, Y. S. (1981) Electronic compensation of a force transducer for measuring fluid forces acting on an accelerating cylinder. *Experimental Mechanics*, **21**(4), 169–72.

Milz, U. (1978) Measurement of force and mass under influence of unwanted acceleration, in *Proceedings of the International Conference on the Measurement of Force and Mass*, VDI, Dusseldorf.

Weibull, W. (1961) *Fatigue Testing and the Analysis of Results*, Pergamon Press, pp. 3–5.

CHAPTER 5

Strain measurement by contact methods and extensometry

D. J. Walters, with Appendix by M. S. Loveday

5.1 INTRODUCTION

Modern demands for strain measurement vary enormously in range, test environment, test material and testing regime. From plastics at room temperature to ceramics at 1500 °C, materials are pulled, compressed, twisted and fatigued while the strain response is simultaneously recorded. It is essential that such recordings are repeatable, accurate and obtained in a manner that allows direct comparison with other sources of data. Material testing is therefore subjected to an increasing demand for traceability and compliance with national standards. For those unfamiliar with the relevant standards, a list is presented as an Appendix to this chapter, together with a recommended calibration procedure.

To begin a review of strain measurement, it is important to establish the definition of strain used throughout the chapter. In tension or compression, **strain** is defined as the change in length per unit length, and the conventional measure of engineering strain is described by

$$\varepsilon = \frac{l - l_0}{l_0} \tag{5.1}$$

where l is the gauge length at any time and l_0 is the original gauge length. This expression is satisfactory for elastic strains where $l - l_0$ is small. However, for plastic deformation the gauge length will change considerably, and 'true' or 'natural' strain ε_n gives better account of the instantaneous changes in gauge length:

$$\varepsilon_n = \ln l / l_0 \tag{5.2}$$

$$\varepsilon_n = \ln (\varepsilon + 1) \tag{5.3}$$

The two measurements of strain give similar results for strains less than 0.1. Below this value it is therefore reasonable to measure a simple surface deflection between two points to quantify strain. For larger strains, and depending upon the application, the measuring device may be required to take account of natural strain when interpreting output signals.

The measurement of strain between two surface points is a convenient procedure and forms the basis of most modern extensometers. However, such methods assume the straining to be uniform within the gauge length and, for metals, it has been the experience of the author that such an assumption is not always valid. At high temperatures in particular, localized straining within the gauge length may occur and the averaging effect of the gauge length selected must be carefully considered.

In contrast, the electrical wire resistance gauge uses a relatively small gauge length and is completely bonded to the test surface. Resistance strain gauges have an excellent record for ambient studies and their technology is now well proven, making these gauges the most widely used for direct strain measurement. However, when they are used in isolation, great care is required to ensure that the axis of the gauge is coincident with the principal strain axis. Deviation from this alignment will cause errors in the measured strain. When strain gauges are incorporated in the design of extensometers, multiple gauging with a fully active bridge circuit is generally used to ensure that maximum sensitivity and full thermal compensation are achieved.

Although widely used, resistance strain gauges are not individually calibrated. The procedure described in British Standard 6888 (1988) is only applied to a 'sample' gauge selected at random from each pack of gauges used.

Where gauges are used in conjunction with a flexible beam, the beam itself may be deflected relative to a standard source to obtain a grading for the instrument. Load cells and pressure transducers are common examples of such applications. It should be noted that when the flexible beam forms part of an extensometer, the calibration procedure specified by BS 3846, ISO 9513 and EN 10002/4 requires the complete extensometer to be attached to a calibration apparatus in the same manner as it is normally used. Calibration of the beam section in isolation is therefore not acceptable.

Whilst numerous extensometers have achieved a high standard of accuracy at room temperature, the development of high-temperature instrumentation presents many problems. Early designers were forced to transmit the required displacement to a cool region before measurement could be made. Furnace construction in which heating elements were wound on to solid tubular formers also imposed restrictions and forced designers to transmit movement vertically over substantial lengths. Although it is still common practice to transmit the displacement to a cooler region, modern furnace designs now permit horizontal access to the testpiece providing a new and exciting approach to extensometer design. These developments will be discussed in the course of this chapter.

5.2 CONTACT METHODS OF DETERMINING STRAIN

Strain measuring techniques using direct contact with the testpiece surface require a high degree of skill by the user. Surface preparation of the testpiece is particularly important since heavy grinding will introduce significant compressive stressing to the surface material. Alignment and bonding of the strain gauge also require a high degree of skill and the process of installing a gauge is time consuming. In materials testing there is a general trend towards reusable instruments which are installed quickly. Whilst this approach is suited to laboratory applications, the demand for strain measurement in remote locations still creates a need for contact methods to be developed.

From the well-established technology of wire resistance gauges at ambient temperatures, research has continued to develop a high-temperature strain gauge. Bonding has only been successful to 300 °C and therefore gauge construction has been considerably modified to proceed to higher temperatures (Burgess, 1986; Rolik, 1986). In summary, the resistance wires are brazed or spot welded to a tension member within the sensor. The complete assembly is then attached to the structure by multiple spot welds. Various gauge configurations and temperature compensation circuits are used to provide stability at the testing temperature.

The use of multiple spot welds along the edge of the sensor when attaching the gauge permits prior calibration of the device since, by increasing the flange width, the gauge may be mounted, calibrated and then 'cut off' the test section to permit a second installation. This method complies with the requirement for traceable standards and may encourage the use of such instrumentation.

The introduction of capacitance to strain gauge design also opened up new possibilities for surface contact instrumentation. Both Planer and Interatom gauges (Hammond, 1985; Hofstötter, 1986) are attached by only two spot welds, the distance between them comprising the gauge length. The stringent requirements of codes such as BS 3846 and EN 10002/4 for measurement of the gauge length may present problems to users of this instrumentation. For a gauge length of 20 mm, a grade A device requires an accuracy of ± 0.25% or 0.05 mm. Grade C and all other grades require ± 1% or 0.2 mm. The width of spot welds may vary between 0.5 mm and 0.8 mm, exceeding all grades. Nevertheless, prior calibration is simple and measurements at test temperatures up to 600 °C can be achieved. If the centres of the spot welds are accurately determined,

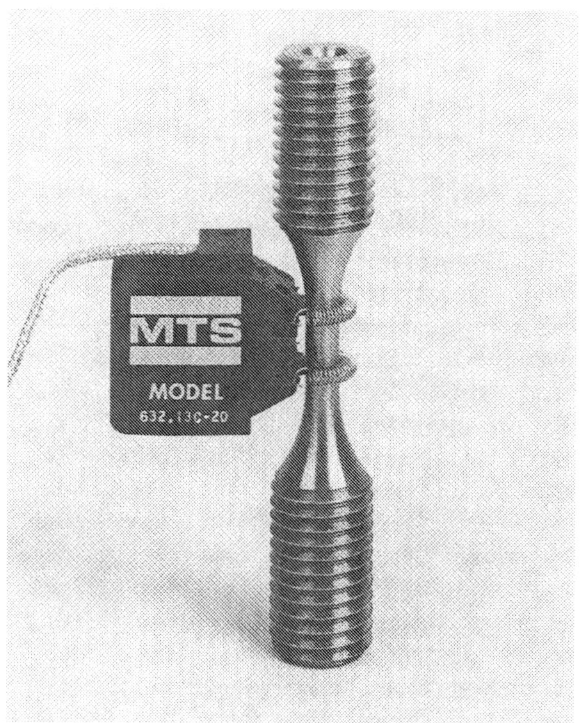

Fig. 5.1 Example of an extensometer using bonded resistance-type strain gauges.

grade C may optimistically be claimed for these strain measuring devices.

As with wire resistance gauges, evidence on the performance of capacitance gauges is contradictory, but work by Walters (1981) has led to the conclusion that installation and cables are key issues in gauge performance. Under laboratory conditions, with short controlled cable routes, most devices produce acceptable results. Where multiple installations using long cable lengths are undertaken, cross-talk, resistance, external capacitance and internal cable capacitance are all factors in determining the performance of gauges.

To conclude this section on surface contact devices, the author acknowledges the use of grids, interferometry and optical systems based upon surface targets and surface replica techniques. These non-contact methods will not be considered here, but some have been reviewed by McEnteggart (1992).

5.3 AMBIENT TEMPERATURE EXTENSOMETERS

The essential difference between ambient and high-temperature instrumentation is the requirement in high-temperature studies to transmit the displacement to a cooler region. At ambient temperature, a wide selection of spring and clip attachments may be used to hold the extensometer to the test section. Contact between the extensometer and the testpiece is made by a knife edge pulled against the testpiece by a spring or clip reacting from the opposite side. In the past, highly polished testpieces and heavy extensometry, particularly unsupported cables, produced slip and ratcheting effects. However, modern lightweight construction of extensometers has in general eliminated these problems.

The proven technology of foil strain gauging now plays a leading role in the development of extensometry for both ambient and high-temperature applications. Incorporating a strain gauged beam deflected by horizontal arms produces a device that is simple to install, accurate and linear and has very repeatable outputs (Fig. 5.1). The construction of these devices permits uniaxial or simultaneous biaxial measurements to be made. The Instron 2620 series biaxial extensometer (Fig. 5.2) and the

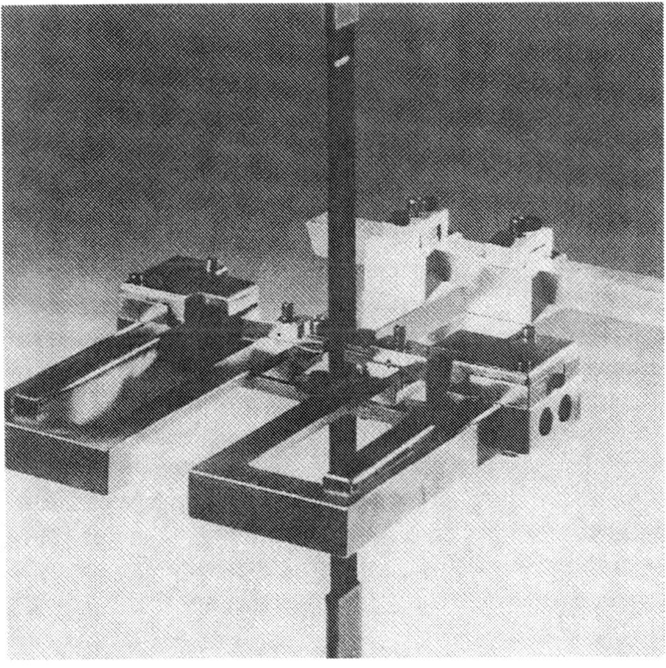

Fig. 5.2 The Instron 2620 biaxial extensometer.

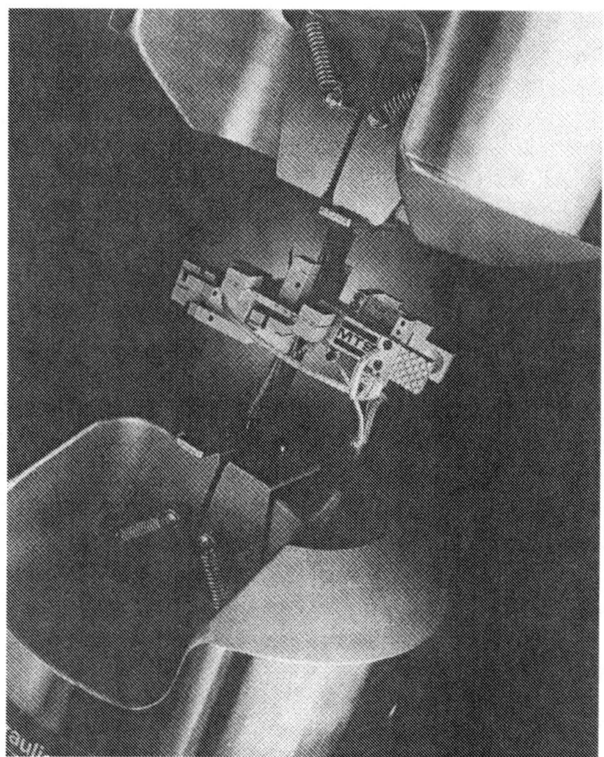

Fig. 5.3 The MTS 632 biaxial extensometer.

MTS 632.85 (Fig. 5.3) are examples of this modern generation of extensometer. However, the measurement of diametral strain warrants special
consideration and this is discussed fully in section 5.4.1.

Whilst new innovative designs extend instrumentation to biaxial and
triaxial conditions, the need for regular and precise calibration of the
instruments must be re-emphasized. As instruments become more complex, the greater are the demands of the calibration procedure. For many
experimentalists, precise calibration of an extensometer moving in a
single direction presents sufficient difficulties. Simultaneous calibration
in another plane or with rotation requires very specialized testing equipment and expertise.

Similarly, the Instron 2665 series automatic extensometer (Fig. 5.4),
whilst excellent in concept, leaves the user with uncertainties in calibration. The instrument is a dual-mode servo-controlled automatic extensometer which is able to measure short travel extension at high resolution

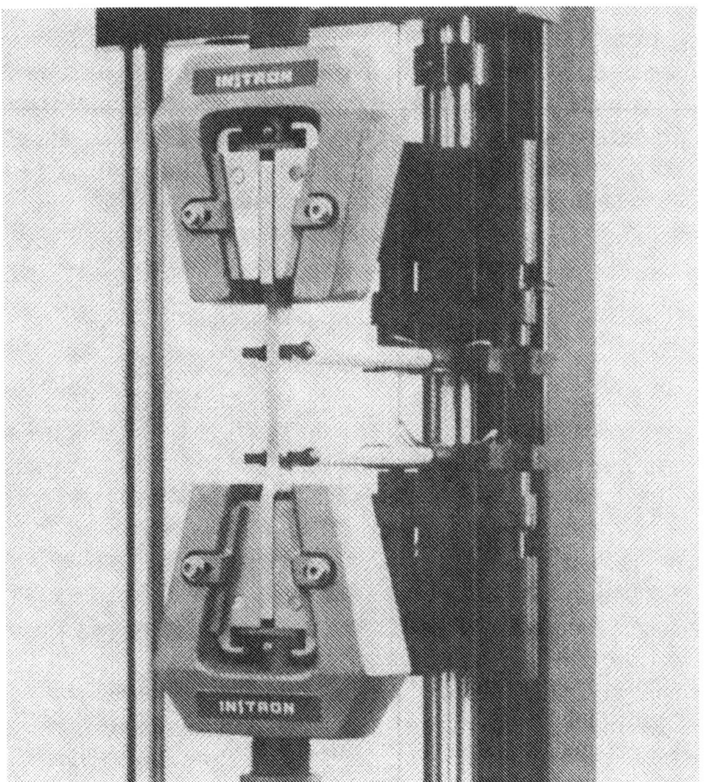

Fig. 5.4 The Instron 2665 extensometer.

before auto-ranging to continue strain measurement throughout the plastic region to testpiece fracture. The axial automatic extensometer measures the average axial strain on either side of the testpiece. For calibration, the manufacturer removes the strain sensor from the machine to compare with a standard source. To comply fully with EN 10002/4, the complete extensometer should be calibrated in several positions of the long-term travel. Reference by the manufacturer to automatic internal calibration does not, unfortunately, exclude the user from the task of verification. This feature of the machine refers only to an internal resistor used as a comparator for the strain bridge. The classification and verification of extensometers must follow an approved procedure, namely BS 3846, ISO 9513 or EN 10002/4, with the accuracy of displacements traceable to the national measurement system (see Chapter 9).

Whilst automatic loading, ranging and averaging are excellent concepts and demonstrate the tremendous advances in modern instrumentation, they also demonstrate the increasing demands and difficulties placed upon users to comply with verification procedures.

Finally, from a review of literature, the term, 'high-resolution digital electronics' is now appearing in association with precision measuring circuits. It is sometimes implied that such circuits eliminate signal noise and so provide greater resolution and stability. However, the elimination of noise relates only to transmission effects, and noise generated within the measuring circuit will still be transmitted – albeit with greater precision!

5.4 HIGH-TEMPERATURE EXTENSOMETRY FOR METALS AND CERAMICS

Few strain measuring techniques are readily adapted to high-temperature conditions. The use of springs, coils and elastic beams is restricted as temperature increases and it has become an accepted practice to first transmit the displacement mechanically to a cooler region. Before the development of modern split furnaces, these movements were transmitted vertically using a tube and rod or flat parallel beams (Fig. 5.5). Transducers were then attached in the cool region to record the relative deflection transmitted.

For tensile and creep testing the inclusion of ridges on the testpiece provided both a convenient means of attaching the extensometer and a predetermined gauge length. Although numerous ridge shapes and attachments have been proposed, there is little evidence to suggest that angle contact (Fig. 5.6(a)) is superior to point contact (Fig. 5.6(b)). Providing the ridge is substantial and unlikely to undergo significant dimensional change, the errors resulting from collet to ridge contact are relatively small. Errors are more likely to occur from inadequate clearance at either the top or the root of the notch (Fig. 5.6).

One significant source of error from this type of instrumentation is the differential expansion between the transmission rods. Tubes may heat faster than the centre rod or draughts may cause only one leg to vary in temperature. Whilst the relative displacement is small, of the order of 0.005 mm, as a percentage of total displacement in the testpiece, errors may be 2–4%. The design of such instruments may be improved by the inclusion of a dummy leg to balance these variations (Fig. 5.7).

Two additional legs are introduced at 90 degrees to the strain recording legs terminating at the lower end of the testpiece. Strain within the testpiece will not be transmitted by these legs, but all expansion effects from the rod and tube assembly are transmitted to the displacement sensor. By carefully matching the displacement sensors and giving the compensation leg a negative output, expansion effects may be minimized by summing an active and a dummy gauge. Further information on this type of compensation may be obtained from Walters and Grady (1970).

Whilst numerous ridge and collet configurations give excellent results for tension and creep testing, the same is not true for fatigue applications. As a generalization, materials which fatigue harden cannot have ridges, whilst materials which fatigue soften can. In Fig. 5.8(a), a 316 stainless

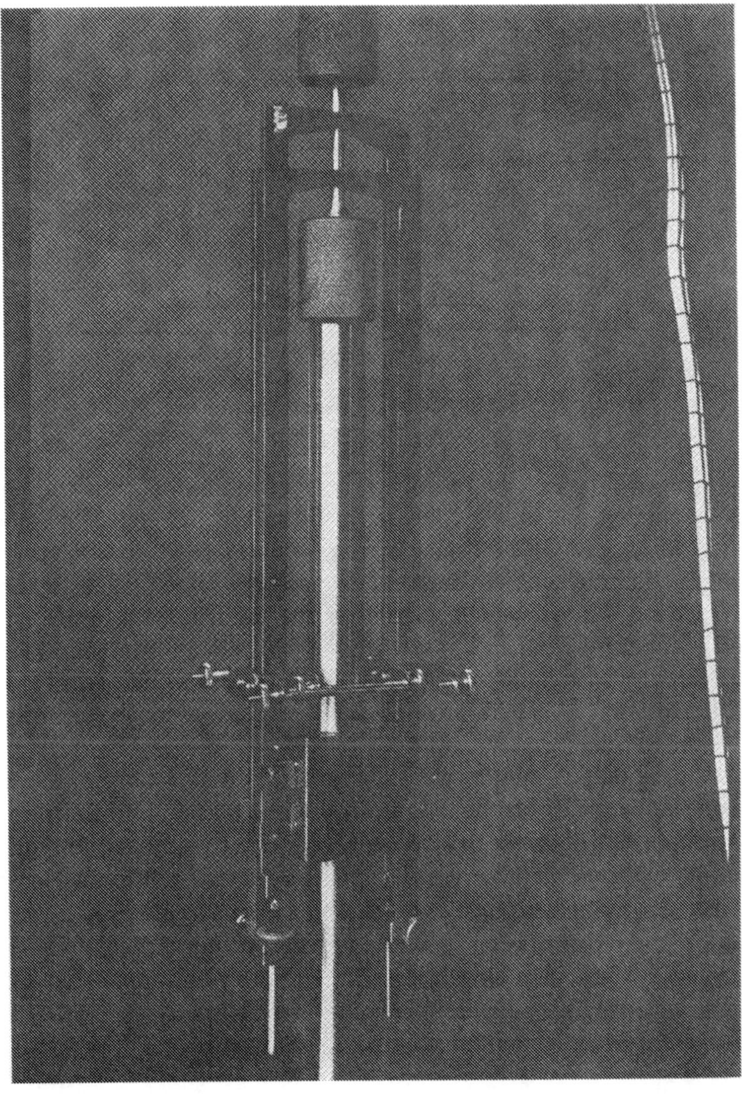

Fig. 5.5 The W. H. Mayes longitudinal extensometer.

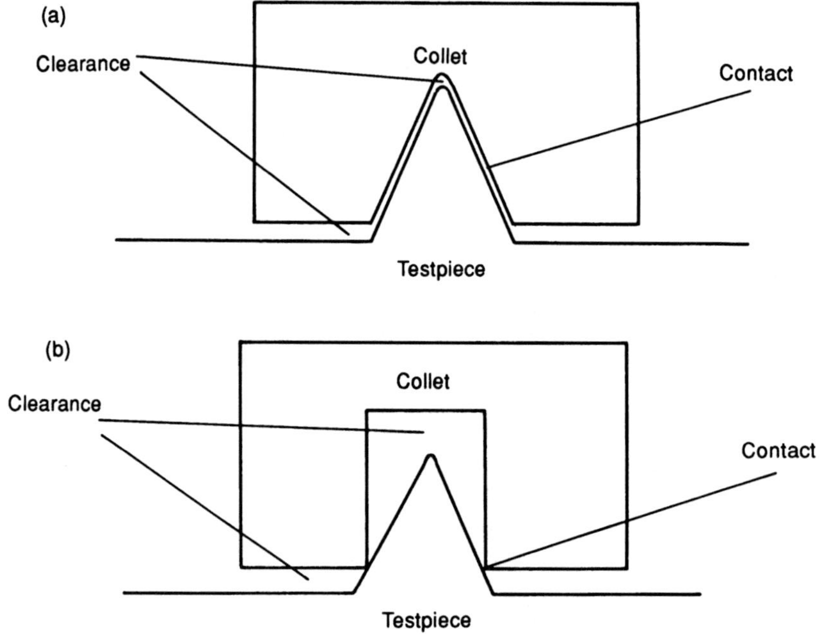

Fig. 5.6 Variations in contact between collet and testpiece.

steel testpiece fatigued at room temperature failed within the gauge length; the material was shown to cyclically soften. In Fig. 5.8(b) a similar piece tested at 600 °C failed at the ridge; this material exhibited cyclic hardening characteristics. A third piece, tested without ridges and with a side-loading extensometer, increased the recorded endurance from 800 to 10 000 cycles (Fig. 5.8(c)). It was for this reason that side-loading extensometry was developed by the author (Walters, 1981) and similar developments have occurred in parallel from several machine manufacturers world-wide.

Side-loading extensometers may be divided into two categories: instruments measuring displacement within the hot region, and instruments transmitting displacement to a cooler location before measurement. It is the author's preference that displacement be converted to an electrical signal as close to the testpiece as is possible, an example being the Mayes side-loading extensometer (Fig. 5.9). In this way, mechanical errors from the support system, arm length, rotation etc. are not transmitted. Of the high-temperature transducers available, only capacitance provides a practical transducer for such applications.

Where displacement is first transmitted to a cool region, arm length, thermal effects and possibly angular effects must be considered. When extensometers are not presented at right angles or strain displacement is excessive, errors of 1–2% may be experienced.

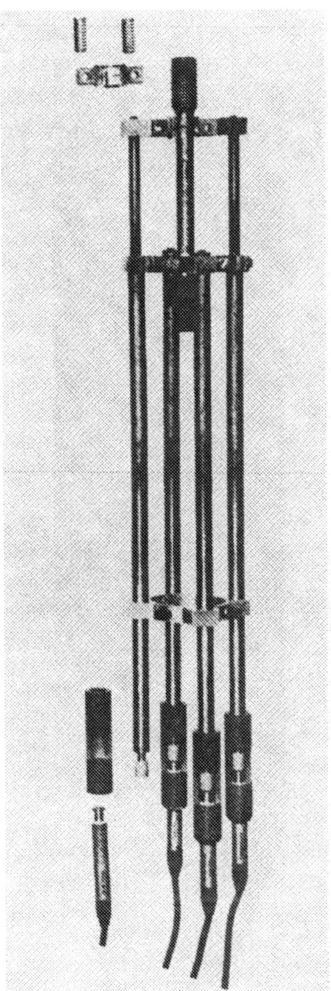

Fig. 5.7 A method of compensating thermal effects in a longitudinal extensometer.

The solution to these and other user problems lies in the type and quality of calibration used with the extensometer. The importance of repeated and reliable calibration of the complete extensometer assembly cannot be overemphasized. Manufacturers for their part have responded

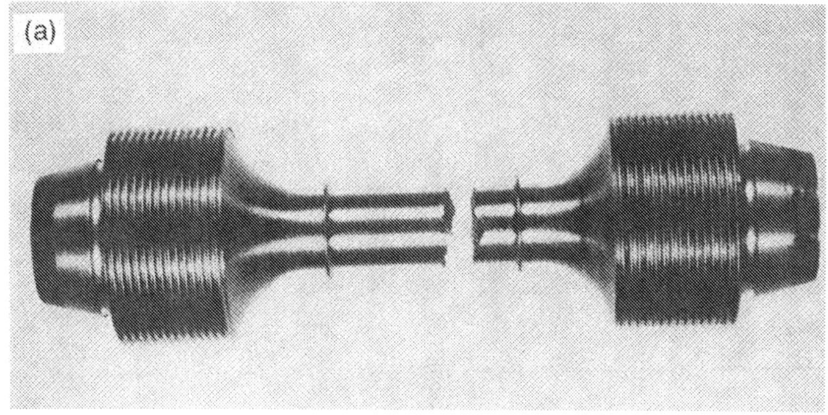

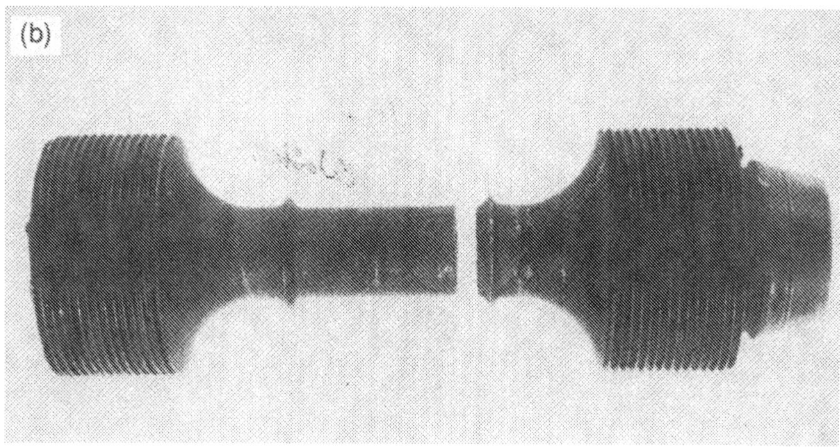

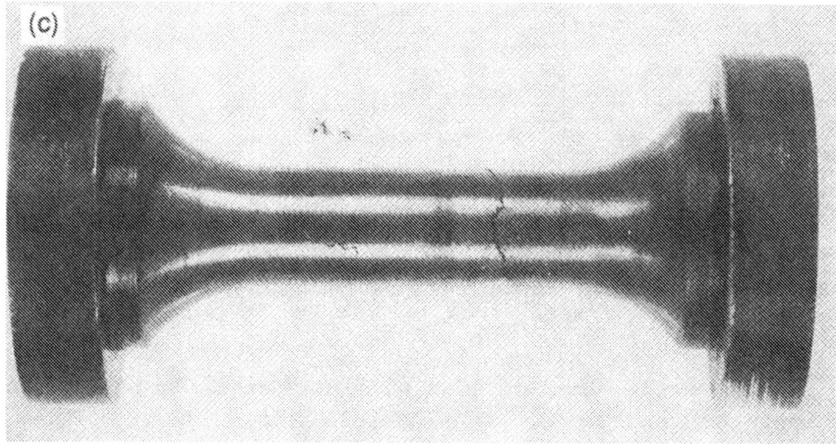

Fig. 5.8 Location of cracks in fatigue testpieces.

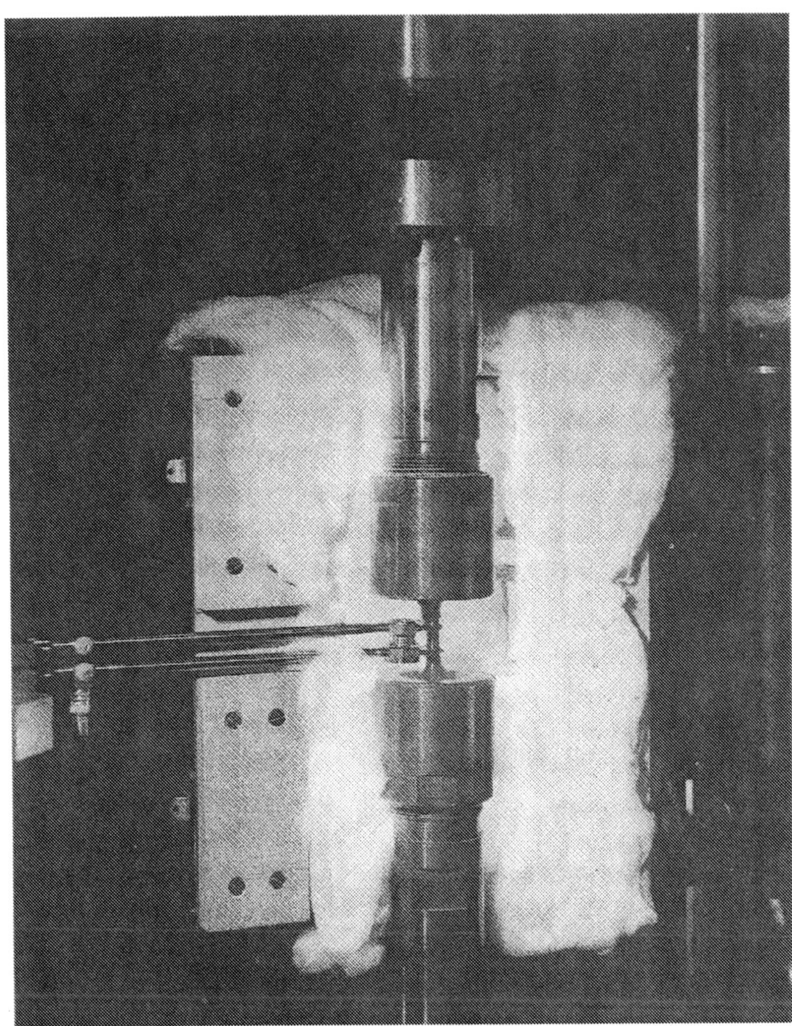

Fig. 5.9 The W. H. Mayes side-loading extensometer.

by including calibration devices in their sales (Fig. 5.10). This response is welcome, although users must be careful to select the correct instrument for their needs. The device shown achieves a grade C; hence it would be unsuitable for fatigue testing of testpieces with short gauge lengths. The acquisition of such a calibration facility is important to all regular users of high-temperature extensometry and, indeed, provision of this facility at test temperature may become essential to comply with European Standards.

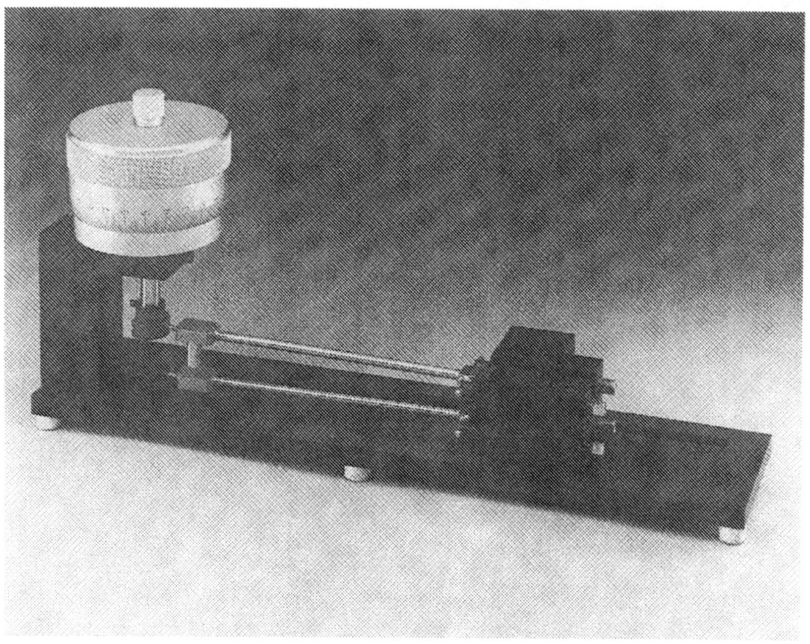

Fig. 5.10 Capacitec calibration device.

5.4.1 Diametral extensometry

Whilst numerous commercial devices are available to record diametral changes, the interpretation of strain from such instruments must be carefully reviewed.

(a) Parallel-sided testpieces

A diametral extensometer measures the displacement occurring on a testpiece in a plane perpendicular to the axis of loading. For a parallel-sided testpiece of diameter D, the extensometer records a total displacement comprising an elastic component of strain ε_e and a plastic component of strain ε_p. To relate the measured displacement to strain in the axial direction, the elastic and plastic components of strain in the axial direction must be multiplied by an appropriate value of Poisson's ratio μ_e or μ_p. The displacement d recorded by the extensometer is therefore

$$d = (\varepsilon_{eL}\,\mu_e + \varepsilon_{pL}\,\mu_p)\,D \tag{5.4}$$

where ε_{eL} and ε_{pL} are the elastic and plastic strains in the direction of the axis of loading.

The interpretation of strain becomes extremely important when a strain-controlled test is undertaken in an axially loaded machine. In the case of a fatigue test, if the material cyclically hardens or softens, the ratio of elastic to plastic strain can vary for any given load. Since total displacement is controlled by changing the applied load, both elastic and plastic strains cannot be held constant by this method of displacement control. To overcome this problem, the instantaneous load signal may be fed to a microprocessor that determines the elastic displacement from knowledge of load and Young's modulus. The elastic strain ε_{eL} may then be used to separate the elastic and plastic components of strain and determine the equivalent longitudinal strain.

Several unique problems may be encountered when diametral measurements are taken:

- Since the loading plane is perpendicular to the axis of measurement, only the Poisson effect is measured; hence sensitivity is reduced by a factor of two to three.
- The gauge length will also be significantly less; hence sensitivity is further reduced.
- Since the movements we require to measure are extremely small, of the order of 0.01 mm, indentations resulting from point contact of the extensometer will be greater than the displacements to be measured. Deformation of the extensometer point itself may produce similar errors in the measurement.
- Oxide growth on the diameter may also be greater than the displacement to be measured.
- As a material approaches failure, necking will occur at some arbitrary point in the gauge length. It would be extremely fortuitous if the extensometer were positioned at this precise location unless an hourglass specimen is used.
- Where contact is made on two sides of the testpiece, extreme care is required to ensure that a true diameter and not a chord is being measured.

(b) Notched testpieces

The problems of diametral strain measurement across the throat of Bridgman blunt circumferential notches in uniaxial testpieces have been considered by Loveday (1986a). Such notches are employed to generate a triaxial stress state in a simple and economic manner, although the interpretation of the resulting data is rather complex. Guidelines for the use and interpretation of such data have recently been given in a code of practice issued under the auspices of the High Temperature Mechanical Testing Committee (Webster *et al.*, 1991). A patented diametral extensometer (Furse and Loveday, 1981) has been successfully used at temperatures up to 900 °C for the measurement of strain across the notches of a variety of nickel-base superalloys, details being cited in Loveday (1986a).

5.5 DISPLACEMENT TRANSDUCERS FOR HIGH-TEMPERATURE EXTENSOMETERS

In the current study we shall consider devices that convert displacement into an electrical output. Whilst several modern optical systems can perform this operation with considerable precision, direct strain measurement appears limited to 250 °C and these systems have not been considered in detail. It is acknowledged, however, that this technology could operate outside a furnace in a similar manner to many of the devices described. However, the transducers in most common use are:

- strain gauged beams;
- linear variable differential transformers (LVDTs);
- capacitance transducers.

5.5.1 Strain gauged beams

Precision resistive-type foil gauges are bonded to a flexural beam to form a four-arm fully active Wheatstone bridge. DC excitation is normally used to provide greater stability for the output signal. Displacements in the testpiece are transmitted through horizontal arms to cause bending of the beam, thus giving an output in the region of $2\,\mathrm{mV}\,\mathrm{V}^{-1}$ from the instrumentation. Providing the beam is strained elastically, repeatability and linearity are excellent from such devices and the complete assembly is relatively easy to calibrate or install on a testpiece. These devices are extremely accurate and now occupy a significant part of the commercial instrumentation available.

5.5.2 LVDTs

Linear Variable Differential Transformer (LVDT) devices were used extensively with longitudinal extensometers in the 1970s and early 1980s. With good linearity and low cost, they became commonplace in creep laboratories. As a result of their long exposure to testing, considerably more information is available on their performance. The author has experienced three recurring faults with these devices:

- Rubber bellows protecting the instrument from dust became hard during long-term testing and often ceased, with disastrous consequences, in fatigue tests.
- Internal spring loading of the device produced an error when the direction of loading was changed. In the time/displacement graph

presented in Fig. 5.11, the error indicated when a calibration rig was reversed in direction was for a long time considered to be backlash in the calibration instrument. However, a non-contact capacitance transducer and an LVDT were calibrated simultaneously and no error occurred from the non-contact gauge. When superglue was used to attach the LVDT probe to the non-rotating spindle of the calibration rig, the error again disappeared: thus it was concluded that the internal spring of the LVDT was the source of this error.

- LVDT devices are also sensitive to ambient changes and the presence of magnetic fields.

5.5.3 Capacitance transducers

Since capacitance is a function of the dielectric, area and gap, it can be used in three separate ways to measure displacement:

- dielectric: placing an object between two parallel plates;
- area: retaining the distance between separate plates and displacing the relative area;
- gap: changing the gap.

These methods have been examined in detail in Walters (1986), and for the purposes of this chapter only direct changes in the gap are considered.

Since capacitance is inversely proportional to the gap, a non-linear output would result. However, capacitive reactance is also inversely proportional to capacitance; hence a linear relationship exists between capacitive reactance and displacement.

The advantages of such capacitive devices are that they are constructed with no moving parts and are easily adapted to high-temperature environments. The conversion of displacement to an electrical signal may therefore take place in close proximity to the testpiece, which avoids the necessity to first transmit the movement to a cooler region. A 5 mm linear displacement can easily be achieved from a sensor less than 10 mm in diameter.

The disadvantages centre around the choice, length and routing of cables associated with the instrument. These technical problems have been addressed by several commercial suppliers, and capacitance now offers a reliable and accurate method of displacement measurement.

5.6 DISCUSSION

In order to design structures and components using modern engineering materials, it is essential to have a good understanding of the materials' stress and strain characteristics. Since strain is directly related to a

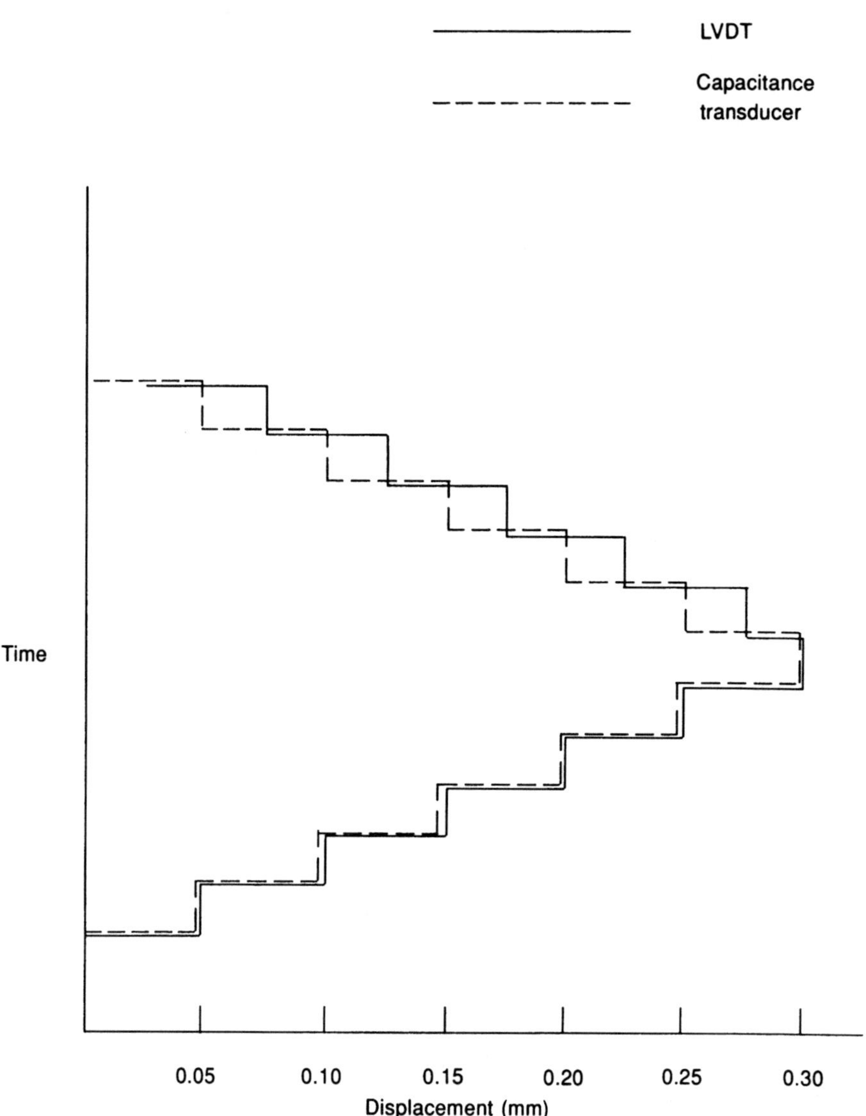

Fig. 5.11 Time–displacement calibration graph of LVDT and capacitance transducers.

dimensional change, it can be measured on small testpieces in the laboratory, or on large structures under full operational conditions. Regardless of how the measurement is made, it is essential that the accuracy of the measuring device used is recorded and is also traceable to an approved calibration procedure. It is not sufficient to state that a device complies, for example, with BS 3846. The accuracy of the device is defined within this standard by a grade; hence the grade obtained and the primary standard used are the relevant information required.

In reviewing strain measurement and extensometry, it was interesting to note that foil strain gauges, probably the most widely used strain measuring devices, are also the most difficult to calibrate or refer to a traceable standard. The user relies entirely upon information supplied by the manufacturer, i.e. the gauge factor. Foil gauges used in isolation are also very dependent upon the skill of the user, since any misalignment of the gauge relative to the principal strain axis will result in errors in the measured strain. It is to avoid such errors that rosettes are generally applied.

When foil gauges are used in conjunction with an elastic beam, traceable calibrations can be obtained. It is important, however, that such devices are not removed from the extensometer for calibration. Accuracy should be determined for the complete assembly to encompass any mechanical movements or magnification of errors by the arms.

As discussed earlier, strain gauged beams are now used in both side loading and multiaxial extensometers. However, as measuring devices become more complex, the problems of calibration and traceability also increase, discouraging rather than encouraging in-house verification of the calibration.

When using multiaxial devices, extreme care is also required in the interpretation of strain. The displacement occurring in the plane perpendicular to the axis of loading is of opposite sign to the loading axis and will be a combination of elastic and plastic displacements. Since Poisson's ratio changes from 0.33 to 0.5 for these cases, the distribution of elastic and plastic displacement is significant in determining strain. This issue is discussed in full in section 5.4.1, with special emphasis being placed upon the use of diametral strain control.

For high-temperature applications, most extensometers still transfer displacement from hot to cold regions before measurement is made. Whilst this is done to accommodate foil gauge technology, errors may be introduced from arm length, movements and expansion. The conversion of the mechanical displacement to an electrical signal should ideally occur in close proximity to the testpiece, thus avoiding expansion and lever effects.

Of the sensors available for extensometry, only capacitance transducers are suited to operation within the high-temperature region. Since capacitance is a function of area, thermal expansion will introduce errors and such devices should be calibrated at the testing temperature. Indeed, it is the recommendation of many codes that high-temperature calibration is undertaken for all instrumentation.

The use of strain gauged elastic beams in extensometry is increasing, particularly for side loading and multiaxial devices. The technology for the resistance gauges used is well established and, providing the complete extensometer is used in the calibration, grade C or sometimes B is achieved. However, few manufacturers identify grades in their advertising literature.

The long-standing LVDT still provides good services to instrumentation, although attempts to introduce this device inside the furnace using primary and secondary couplings as compensation were not impressive. It may be concluded that the LVDT is best used in its more traditional role in the cool region. The spring effects reported in section 5.5.2 when load was reversed are unusual and may be attributed to only one manufacturer. Further investigation may be warranted, however, and extended to any device using springs.

Finally, whether a foil gauge or a capacitance strain gauge is used, the observation is that laboratory testing produces far greater consistency in results than instruments on site locations. Multi-gauge installations utilizing long cable lengths require special calibration and installation procedures to achieve optimum results. Calibration of such devices should ideally occur after installation to include any environmental effects upon the cables.

ACKNOWLEDGEMENTS

The author wishes to acknowledge information supplied by MTS Systems, Instron, W. H. Mayes and the Central Electricity Generating Board.

APPENDIX 5.A NOTATION

d	displacement
D	diameter
ε	linear strain
ε_n	true strain
ε_{pL}	plastic strain in longitudinal plane
ε_{eL}	elastic strain in longitudinal plane
l	gauge length at any time
l_0	original gauge length
ln	natural logarithm
μ_p	Poisson's ratio, plastic
μ_e	Poisson's ratio, elastic

APPENDIX 5.B CALIBRATION OF EXTENSOMETERS

The completion of the Single European Market by the end of 1992 in principle required that common standards exist throughout the

Community. BS 3846 will therefore be superseded by EN 10002 : Part 4. In some respects, the new proposed European Standard is less stringent than BS 3846 outlined in this Appendix. A detailed comparison of standards is given by Loveday in Appendix 5.C (see also Loveday, 1991).

5.B.1 Requirements of BS 3846

An extensometer must be calibrated over the range and at the gauge length at which it is required to work. A series of 10 equally spaced increments is recommended over the working range selected and extensometers are graded according to the following properties:

- gauge length;
- repeatability;
- discrimination;
- accuracy.

Extensometers designed to operate at several magnifications and gauge lengths must be graded at each magnification.

(a) Gauge length

In three trials the gauge length must be reproducible to within $\pm 0.25\%$ for A, $\pm 0.5\%$ for B, $\pm 1.0\%$ for all other grades.

(b) Repeatability

The difference between the highest and lowest values of indicated extension at the same displacement value must be recorded during a series of repeated tests.

(c) Discrimination

The smallest change of extension which may be indicated or estimated from the extensometer scale must be recorded.

(d) Accuracy

The arithmetic mean of errors obtained at any extension during a repeated series of observations must be recorded.

Grading of extensometers is determined by the formula

$$Y = Mx + C$$

where Y is the maximum permissible error, M is the proportional error, x is the strain being measured and C is the lowest practical limit of measurement.

It is also recommended that a calibration of the extensometer be undertaken at the temperature at which testing will be undertaken.

APPENDIX 5.C STANDARDS FOR THE CALIBRATION OF EXTENSOMETERS

M. S. Loveday

5.C.1 Introduction

The completion of the Single European Market at the end of 1992 in principle required that common standards exist throughout the Community. Thus publication of the new series of European Standards (EN) requires the withdrawal of the superseded equivalent British Standards. Hence British Standard 3846 'Methods for Calibration and Grading of Extensometers for Testing of Metals' is replaced by BS EN 10002 : Part 4 'Metallic Materials – Verification of Extensometers Used in Uniaxial Testing', which has the same title as ISO 9513 and has been extensively based thereon. In these standards the term 'extensometer' is understood to mean a displacement measuring device together with the system used for indicating or recording the displacement.

It should be appreciated that international standards are achieved normally by harmonious compromise on technical details which may lead to slightly less stringent standards than any individual nation's own standards. At the ISO level, it was often necessary to accept broader tolerances on testing parameters to achieve agreement, bearing in mind that the tolerances in the national standards would be encompassed by the ISO Standard. This was not generally perceived as a difficulty since the ISO Standards were not mandatory, and testing in compliance with a national standard would invariably comply with the testing conditions specified in the ISO Standard. With the advent of the Single European Market, the ISO Standards, where they exist, have been adopted as the working draft for the new European Standards which will become mandatory throughout the member nations of the European Community.

When viewed against this background, it may be understood that in certain regimes the proposed European Standard is less stringent than BS 3846. However, in many respects the proposed new standard offers many advantages, not only in a reduction in the effort required and a simplification of the analysis for the verification of extensometers, but also in a more logical approach to the calibration of the extensometer over the range for which it will be used.

The classification of extensometers will be based on the concepts of **global errors**, i.e. all the readings must comply with specified tolerances, unlike BS 3846 which employs separate requirements for repeatability and accuracy (bias), with the requirement that it is only the mean value of the four accuracy readings at any given strain that needs to comply with the specified tolerance.

5.C.2 Comparison of standards

(a) Gauge length

All the standards require that the extensometer gauge length can be defined within specified limits, and that the gauge length can be independently verified within the specified limits shown in Table 5.C.1, with traceability to the national measurement system.

Table 5.C.1 Specified limits for grades/classes in standards

BS 3846	grade	A	B	C	D	E	F
	$\pm\,\mu$m	62	125	250	250	250	250
EN 10002/4	class	0.2	0.5	1	2	–	–
	$\pm$%	0.2	0.5	1.0	2.0	–	–
ISO 9513	class	–	0.5	1	2	–	–
	$\pm$%	–	0.5	1.0	2.0	–	–

It should be recognized that in some cases the extensometer has no fixed gauge length, and that the gauge length is determined by the attachment points which are an integral part of the testpiece; under such circumstances it would be necessary to demonstrate that the testpiece gauge length complies with the requirements in the table by an independent verification. In the case of an extensometer having several fixed gauge lengths, the extensometer should be verified at each gauge length required by the user.

(b) Discrimination and resolution

The various standards specify the discriminations and resolutions in Table 5.C.2.

(c) Repeatability

BS 3846
The repeatability at each extension is the maximum permissible difference between the highest and lowest indicated extensions, expressed in strain units.

Table 5.C.2 Discrimination and resolution in standards

BS 3846

Discrimination	*Grade*	*Discrimination*
The maximum permissible	A	3×10^{-6}
value of the smallest change	B	6×10^{-6}
of extension which can be	C	12×10^{-6}
indicated or estimated from	D	30×10^{-6}
the scale, expressed in strain	E	60×10^{-6}
units	F	120×10^{-6}

EN 10002/4

Resolution	*Class*	*Resolution*	
The resolution r is the		Maximum	Maximum
smallest quantity which can		percentage	absolute
be read on the instrument		of reading* r / l_i (%)	value r
			(μm)
	0.2	0.1	0.2
	0.5	0.25	0.5
	1	0.50	1.0
	2	1.0	2.0

ISO 9513

Resolution

Definition as in EN 10002/4	The values for the resolution are the same as in EN 10002/4 except as yet ISO 9513 does not include a class 0.2 device

* Where l_i is the displacement indicated by the extensometer.

The permitted values are 0.6Y, where Y is the specified accuracy for a given grade of extensometer (see section 5.C.2(d)).

ISO 9513 and EN 10002/4
There are no specified criteria for repeatability in the ISO and European Standards since they use the concept of global error and all the readings must comply with the specified grading values for any particular class of instrument (Loveday, 1986b).

(d) Accuracy/bias

The British Standard differs from the other standards in that it is only the average of the four calibration readings at any given displacement which must comply with the specified value to achieve the grading criterion. The ISO and European Standards followed the ASTM approach and all the readings must comply with the specified value to achieve the classification criterion.

In the British Standard the accuracy grading is governed by the maximum permissible error Y, expressed in strain units, where Y is given by the formula

$$Y = Mx + C$$

where M is the proportional error, x is the strain being measured and C is the lower practical limit of measurement.

The following relationships apply for the various grades:

$$Y_A = 0.002x + 10 \times 10^{-6}$$
$$Y_B = 0.004x + 20 \times 10^{-6}$$
$$Y_C = 0.008x + 40 \times 10^{-6}$$
$$Y_D = 0.020x + 100 \times 10^{-6}$$
$$Y_E = 0.040x + 200 \times 10^{-6}$$
$$Y_F = 0.080x + 400 \times 10^{-6}$$

In the ISO and European Standards the accuracy criterion is determined by the relative bias error q for a given displacement l_t, and is calculated from the formula

$$q = \frac{l_i - l_t}{l_t}$$

where l_t is the true displacement and l_i is the indicated displacement. Values for q are given in Table 5.C.3.

Table 5.C.3 Values of relative bias error in the ISO and EN standards

Standard	*Class*	*Bias*	
		Max. relative error q (%)	Max. absolute error $l_i - l_t$ (μm)
EN 10002/4	0.2	± 0.2	± 0.6
ISO 9513	0.5	± 0.5	± 1.5
and	1	± 1.0	± 3.0
EN 10002/4	2	± 2.0	± 6.0

(e) *Range of verification*

The British Standard requires that the extensometer is calibrated over its full operational range, and that the grading criteria apply to each calibration step, including the zero point.

The ISO and European Standards introduce the concept of the range of verification being dependent upon that over which the extensometer will actually be used. For example, if it is to be used for measuring Young's modulus it should be calibrated over a small strain range, say 0.05–0.5%. This approach ensures that sufficient calibration readings are recorded over the appropriate range.

In the new European Standard, the maximum and minimum limits E_{max} and E_{min} of the verification range must lie within

$$5 \leqslant \frac{E_{max}}{E_{min}} < 10$$

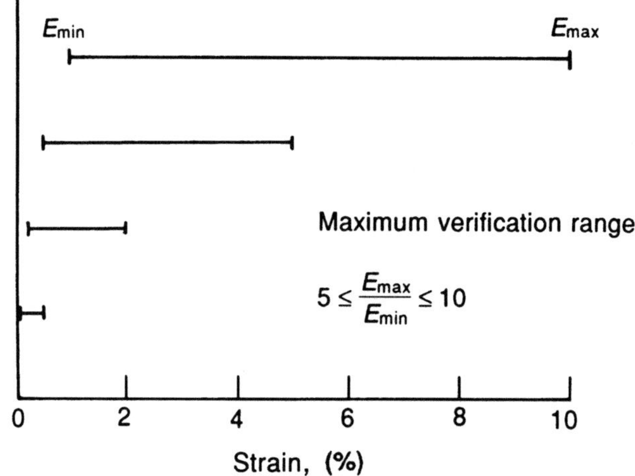

Fig. 5.C.1 Example of verification ranges of an extensometer (ISO 9513 and EN 10002/4).

It should be noted that this criterion is slightly different from that specified in ISO 9513. However, it is anticipated that the ISO Standard will be aligned with the European Standard when it is next revised.

Depending upon the applications for which the extensometer will be used, it may be necessary to calibrate over several ranges, as shown in Fig. 5.C.1.

(*f*) *Frequency of reverification*

BS 3846 states that extensometers used for short-term tests, e.g. tensile and stress rupture tests, must be recalibrated annually. For those used on long-term tests, e.g. creep tests, the calibration certificate is valid for three years, unless the test has a duration exceeding three years in which case it must be reverified at the earliest opportunity.

The ISO and European Standards recommend that the extensometer is reverified at intervals of approximately 12 months, and for short-term tests the reverification period must not exceed 18 months. For tests lasting longer than 18 months, e.g. creep tests, the extensometer must be reverified before commencing the next test.

All the standards contain a statement to the effect that an extensometer must be reverified if it undergoes any major repairs or adjustments to any of its component parts.

5.C.3 Calibration procedure

(*a*) *General*

The calibration procedure normally involves the careful attachment of the extensometer to a special calibration apparatus in the same manner in which it is normally used, and sufficient time must be allowed for the calibrator and the extensometer to equilibrate to a similar stable temperature. The calibrator usually incorporates a divided testpiece, one part of which can move relative to the fixed part such that known displacements can be introduced with specified accuracy and traceability to the national measurement system. As a pre-calibration procedure the extensometer must be exercised over its working range at least three or four times (BS 3846) or at least twice (ISO 9513 and EN 10002/4).

Full details of a practical procedure used for the calibration of creep and tensile extensometers calibrated in accordance with BS 3846 are given in the NPL *Creep Laboratory Manual* (Osgerby and Loveday, 1992).

(b) Number of calibration readings

The British Standard requires that four series of calibration readings are undertaken, with the extensometer being dismounted from the calibrator between the second and third series of readings. The ISO and European Standards require only two series of readings, with dismounting of the extensometer between the two series. All the standards require that each series of readings comprises at least 10 readings which must be approximately equally spaced over the verification range.

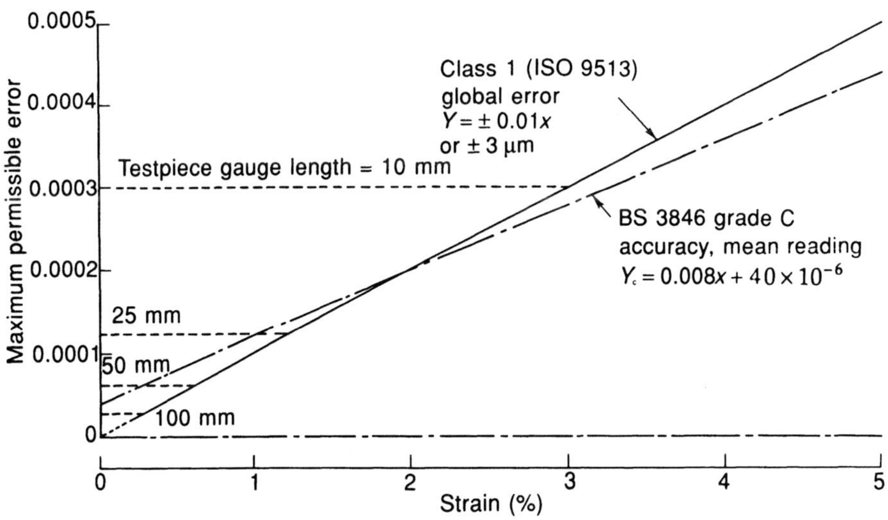

Fig. 5.C.2 Comparison of the bias (accuracy) classification envelope for an extensometer verified in accordance with BS 3846 or ISO 9513 (EN 10002/4).

5.C.4 Comparison of grading

A comparison of the bias (accuracy) for a class 1 (grade C) extensometer for various gauge lengths classified in accordance with ISO 9513 (EN 10002/4) and BS 3846 is shown in Fig. 5.C.2. To a first approximation the class 1 and grade C extensometers are equivalent; however, at low strain readings the differences may become appreciable depending upon the extensometer gauge length. Since the ISO Standard has a 3 μm threshold reading, at low strains large errors in measurement may be encountered (Fig. 5.C.3), and for high-precision measurements such as modulus determination it may be appropriate to specify a more accurate class of extensometer.

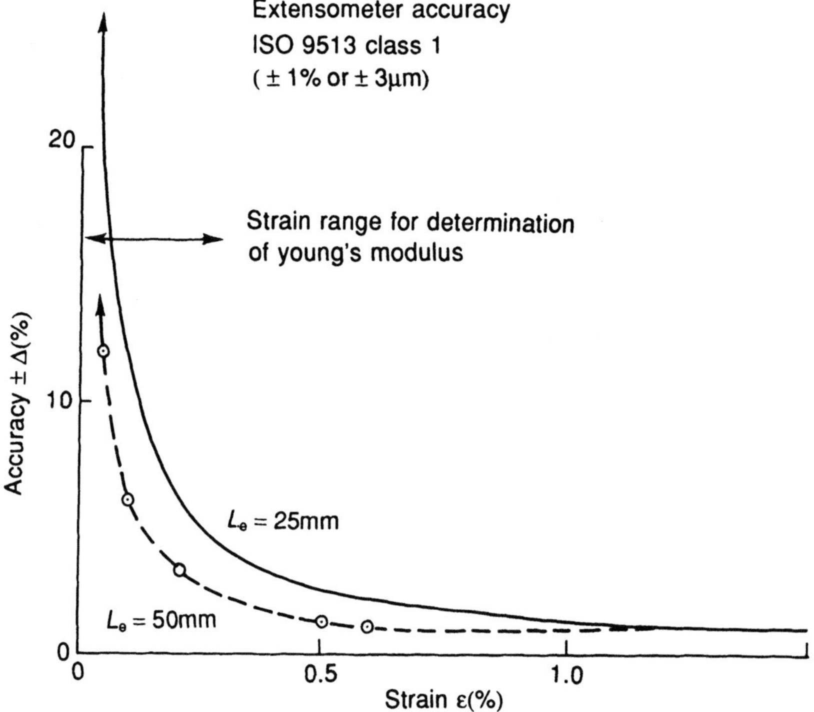

Fig. 5.C.3 Errors likely to be encountered in determining Young's modulus as a function of strain using a class 1 extensometer classified in accordance with ISO 9513: error due to 3 μm lower threshold limit.

The difference between the global approach and that adopted in BS 3486 may be more easily appreciated by consideration of the data shown in Fig. 5.C.4. In the case of the data graded in accordance with BS 3846 (upper diagram), for the four data points recorded at the lower strain (position 1) the mean value $\bar{x}$ falls inside the grading envelope represented by the chain-dotted line, and similarly the repeatability

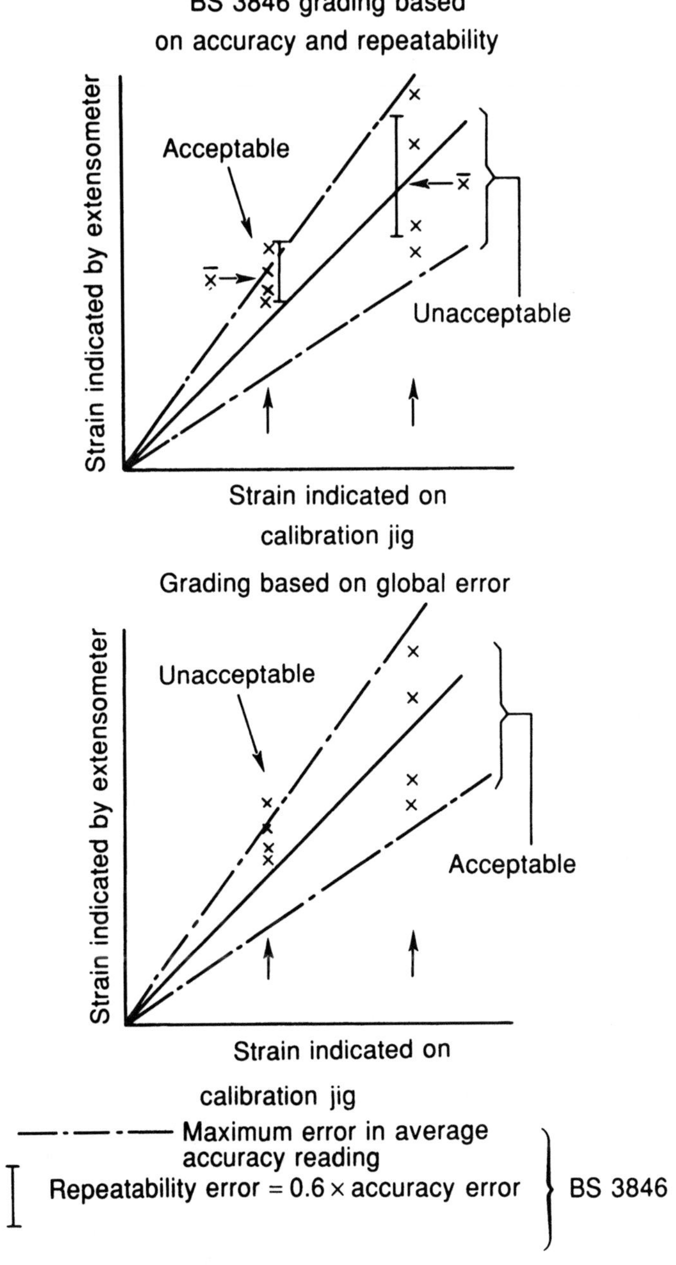

Fig. 5.C.4 Comparison of verification readings based on separate accuracy and repeatability criteria (BS 3846) with global error classification (ISO 9513 or EN 10002/4).

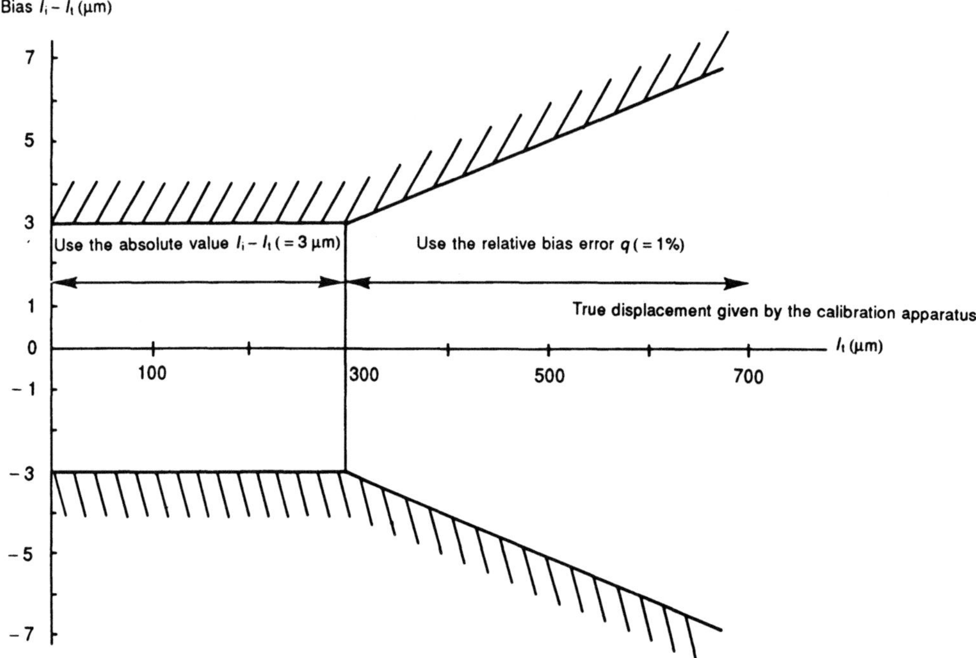

Fig. 5.C.5 Limits of bias error of a class 1 extensometer (ISO 9513).

criterion is also satisfied. However at the higher strain (position 2), although the mean value of the four readings clearly lies within the accuracy envelope, the scatter in the readings means that the repeatability criterion is not achieved. Now consider the same data plotted in the lower diagram of Fig. 5.C.4, where the classification of the extensometer is based on the global error concept, which implies that *all* the readings must satisfy the classification criterion as specified in ISO 9513 or EN 10002/4. Clearly at the lower strain (position 1) one of the readings falls outside the required accuracy envelope and hence the extensometer would only meet the criterion of a lower classification. However at the higher strain (position 2), all the readings achieve the classification criterion (Loveday, 1986b).

Thus these two examples show that the same group of readings may either satisfy the grading of BS 3846 and fail that of ISO 9513, or vice versa, depending upon their scatter and accuracy. Hence it is not possible to categorically state that the British Standard is superior to the ISO Standard; one can merely observe that they are broadly similar. However it may be argued that the ISO Standard offers greater confidence to the materials scientist since all the readings need to satisfy a given classification criterion, which is a closer analogy to the situation encountered when using the extensometer for materials testing. An example of the limit of bias error envelope for a class 1 extensometer is shown in Fig. 5.C.5, taken from ISO 9513. In addition, since only two series of verification readings are required, the analysis is more straightforward and less effort is involved in the verification procedure.

5.C.5 Traceability to the national measurement system

Traceability can be provided in the UK to the national measurement system (NMS) via the NAMAS (calibration) accredited organizations. These organizations use an extensometer calibration device which must be verified periodically (two-year intervals) by the National Physical Laboratory, which can provide traceability to the wavelength of light using a laser interferometer. Figure 5.C.6 shows the NPL laser extensometer calibrator being verified by the primary standard interferometer to provide traceability to the NMS.

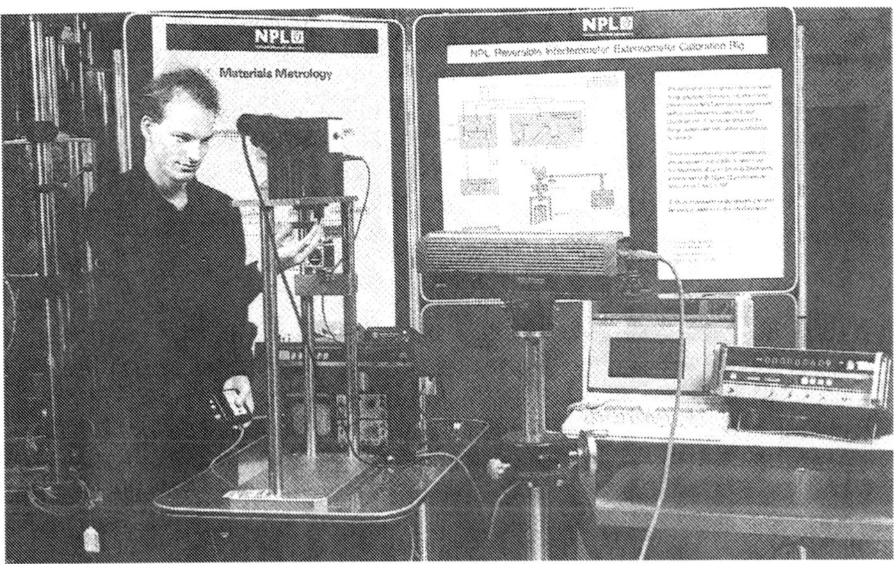

Fig. 5.C.6 The NPL laser extensometer calibrator being verified by the primary standard interferometer to provide traceability to the national measurement system.

In Section 4 of BS 3846 the requirements for calibration devices, including interferometer-based systems, are outlined in considerable detail. There is a general requirement that the error of the calibrating rig must not exceed one-fifth of the permissible error of the extensometer being calibrated. This implies that a grade A extensometer or transducer with its associated readout system could be used as a transfer standard for the verification of grade C extensometers.

In the ISO and European Standards the requirements are less stringent for the calibration device in that the resolution is only required to be one-half of the particular class of extensometer being verified, whilst the bias error (accuracy) must be one-third that of the extensometer.

5.C.6 The future

Apart from the introduction of BS EN 10002 : Part 4, there are a number of other matters concerning strain measurement and processing of the data that will need to be considered in the not too distant future:

- Consideration will need to be given to the optimum means of using these standards for the classification of non-contacting extensometers. Although in principle the requirements will be similar to those for contacting extensometers, special care will perhaps be needed to demonstrate that the gauge length classification criterion is met.
- Computers are now widely used to remove the tedium from the analysis and grading/classification process, and various computer programs have been written to undertake this task.
- The standards discussed in this appendix have of course been concerned with the static calibration of extensometers, and at some stage in the future the matter of dynamic calibration of extensometers used for fatigue testing should be considered.

5.C.7 Conclusion

As a result of the need to harmonize standards for the implementation of the Single Market, a new standard EN 10002/4 for the classification of extensometers supersedes BS 3846. Although there are some significant technical differences, the new standard should be easier to use for the verification of extensometers, and it would therefore be wise to make preparation for the introduction of the new standard.

REFERENCES

BS 3846 : 1970 (1985) Methods for Calibration and Grading of Extensometers for Testing Metals.
BS 6888 : 1988 Methods for Calibration of Bonded Electrical Resistance Strain Gauges.
Burgess, M. C. (1986) Performance testing of bonded and weldable strain gauges at elevated temperatures, in *Strain Measurement at High Temperature* (ed. R. C. Hurst), Elsevier Applied Science, pp. 105–11.
EN 10002 : Part 4 : 1994 Metallic Materials – Verification of Extensometers Used in Uniaxial Testing.
Furse, J. E. and Loveday, M. S. (1981) Improvements in or relating to extensometers. Patent 8131641 J X 5782.

Hammond, V. C. (1985) High temperature capacitance strain transducers, in *Strain Measurement at High Temperature* (ed. R. C. Hurst), Elsevier Applied Science, pp. 129–39.

Hofstötter, P. (1986) The use of electrical high temperature strain gauges in West Germany, in *Strain Measurement at High Temperature* (ed. R. C. Hurst), Elsevier Applied Science, pp. 157–71.

ISO 9513 : 1989 Metallic Materials – Verification of Extensometers Used in Uni-axial Testing.

Loveday, M. S. (1986a) Practical aspects of testing circumferential notch speci-mens at high temperature, in *Multiaxial Creep Testing* (eds D. J. Gooch and I. M. How), Elsevier Applied Science, Chapter 10, pp. 177–97.

Loveday, M. S. (1986b) High temperature axial extensometers: standards, calibra-tion and usage, in *Strain Measurement at High Temperature* (ed. R. C. Hurst), Elsevier Applied Science, pp. 31–47.

Loveday, M. S. (1991) *Standards for the Calibration of Extensometers*, NPL report DMM(A)41.

McEnteggart, I. (1992) Contacting and non-contacting extensometry for ultra high temperature testing, in *Proceedings of the International Symposium on Ultra High Temperature Mechanical Testing* (eds R. D. Lohr *et al.*). To be published, Woodhead Publications, 1994.

Osgerby, S. and Loveday, M. S. (1992) *Creep Laboratory Manual*, NPL report DMM(A)37.

Rolik, G. P. (1986) A strain gauge transducer sensor for use at elevated tempera-tures, in *Strain Measurement at High Temperature* (ed. R. C. Hurst), Elsevier Applied Science, pp. 121–8.

Walters, D. J. (1981) *A High Temperature Extensometer*, CEGB technical disclosure bulletin 352.

Walters, D. J. and Grady, A. (1970) *A High Temperature Extensometer for Stress Relaxation Testing*, CEGB report RD/B/N1483.

Webster, G. A., Cane, B. J., Dyson, B. F. and Loveday, M. S. (1991) *A Code of Practice for Notched Bar Creep Rupture Testing: Procedures and Interpretation of Data for Design*, NPL report. Also published in *Harmonisation of Testing Practice for High Temperature Materials*, 1992 (eds M. S. Loveday and T. B. Gibbons), Elsevier Applied Science.

CHAPTER 6

Dynamic strain measurements

J. Albright

6.1 INTRODUCTION

Dynamic measurements are a required part of any fatigue test. Two frequently measured parameters are applied load and testpiece strain. Load cells and extensometers are frequently used to make those measurements, respectively. Detailed standards for static calibrations of these devices have been available for many years. These standards provide a method for calibrating a transducer, with traceability to national and international standards.

Consideration of dynamic effects has only recently been addressed. Standards are now available for dynamic calibration of load measurement devices (Chapter 4). No widely accepted standard exists, however, for dynamic calibration of extensometers. In fact basic knowledge of the actual dynamic characteristics of extensometers is quite limited.

This chapter presents a basic description of how strain gauged extensometers work. Potential sources of dynamic errors are discussed. The chapter also summarizes the initial findings of an ongoing study of extensometer dynamics. A key finding is that the system is quite sensitive to waveform quality, and is test machine dependent. It briefly reviews some applicable standards. In the final section, a general framework for a dynamic extensometer standard is presented.

6.2 EXTENSOMETER FUNDAMENTALS

6.2.1 Static considerations

The majority of extensometers in use today are believed to be based on strain gauges. For these extensometers, the strain gauges are placed on thin flexures, bonded to two attachment arms. These arms also mount directly on to the testpiece being tested. The end of the arm that contacts the testpiece is usually very sharp. A separate part called a knife edge is often used to provide a sharp, durable edge. This reduces frictional hysteresis at the contact point, and improves measurement accuracy by measuring the testpiece at a specific point. The distance between the two contact points of the extensometer on the testpiece is called the gauge length. A simplified drawing of an extensometer is shown in Fig. 6.1.

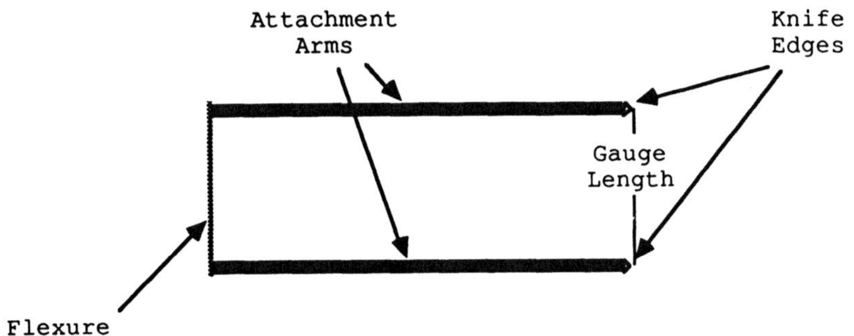

Fig. 6.1 Simplified extensometer.

Transducer output occurs owing to bending of the flexure. When there is strain in the testpiece, within the gauge length of the extensometer, bending of the flexure occurs. This is the desired situation. Figure 6.2 shows this case.

Two 'static' forces are important factors in the use of an extensometer. They are contact force and activation force. Contact force is the force applied by the knife edges to the testpiece. Activation force is the vertical force required at a knife edge to operate the extensometer through the desired range of motion. This force is due to the bending of the flexure. Units with stiffer flexures require higher activation forces to operate. This implies a higher contact force requirement. For low activation forces a soft flexure is desired.

Contact force can be created in two general ways which can be called testpiece reacted and frame reacted. Testpiece reacted is the simplest method. A spring is wrapped around the testpiece, clipping the extensometer to the testpiece. This provides a high contact force with a minimum of complexity. However, sometimes it is not possible to wrap a spring around the testpiece, in which case the contact force must be provided by pushing on the flexure end of the attachment arms. In that case, the extensometer is mounted from the frame; thus the term 'frame reacted'.

If there were no limitation to the amount of contact force that could be applied, extensometers would be very robust transducers. Unfortunately, pressing a sharp knife into the testpiece with a great deal of force is often unacceptable. Also, with frame reacted extensometers, the contact force adds a bending moment to the testpiece, which may be unacceptable. Thus

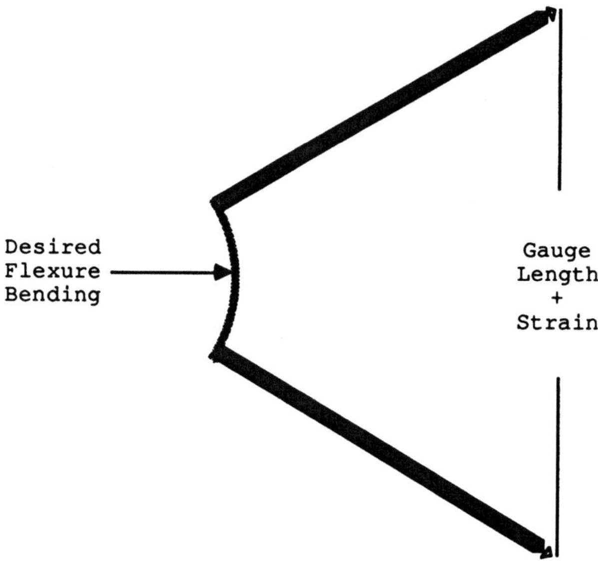

Fig. 6.2 Flexure bending example, pure strain input.

it is frequently desirable to keep contact forces very low. This desire to limit the contact force creates a struggle to find enough force to keep the extensometer from slipping without providing excessive contact force.

6.2.2 Inertial considerations

In addition to the fundamental operating forces described above, there are also inertial 'body' forces acting on the extensometer. They create vertical forces which attempt to slide the extensometer up or down the testpiece, or rock the extensometer about the knife edge contacts. Even in static tests, there is the force of gravity. The extensometer resists vertical forces through the friction of the knife edges on the testpiece. Higher contact forces provide higher frictional forces. For quasi-static operation, the friction only needs to overcome the force of gravity and the activation force of the extensometer. During dynamic operation, the extensometer must also resist inertial forces due to the testpiece moving vertically and due to the motion of the attachment arms. Failure to resist these forces means a loss of zero at a minimum. In the worst case, the extensometer is shaken completely off the testpiece.

In addition to a loss of zero, inertial forces have the potential to introduce unwanted 'strain' readings in the extensometer. Inertial forces attempt to rotate the extensometer around the knife edges, without any testpiece strain. This creates the need for a bending moment in the extensometer. Some or all of this is provided by the same strain gauged flexure used to measure strain. This provides the possibility for extensometer output without testpiece strain. Figure 6.3 shows an example of this type of input.

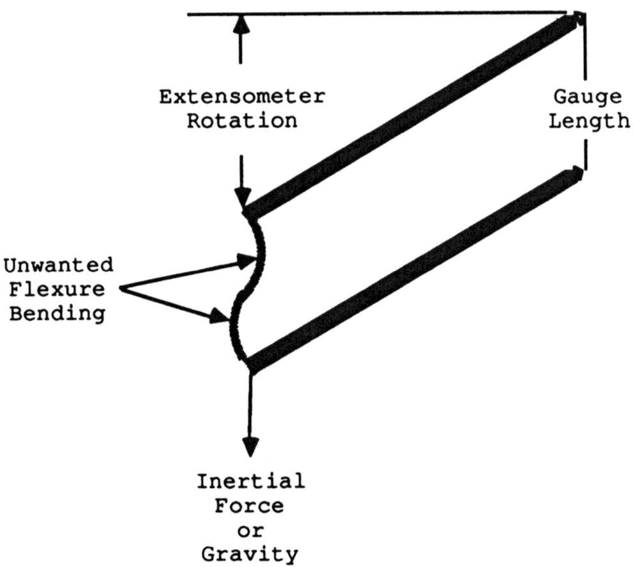

Fig. 6.3 Flexure bending example, pure inertial input.

The amount of extensometer output that occurs for this type of input depends on a variety of design features. Exact placement of the gauges and Wheatstone bridge cancellation effects influence the amount of sensitivity to this type of input. Also, symmetry of the flexure will affect sensitivity. Lastly, relative stiffness of the extensometer flexure affects the amount of bending strain for a given inertial or gravity force input.

The forces created by inertial effects are dependent on the amount of mass in the extensometer. The bending moment required of the flexure depends on that mass, the length of the attachment arms, and the gauge length. Shorter arms and longer gauge lengths reduce the required moment. Though shorter arms are preferred, longer arms are frequently required. They keep the extensometer out of the way of other devices, such as heaters and furnaces. They also may improve linearity. Thus high-temperature test applications with externally mounted extensometers are generally very low-frequency devices. These devices can also get more frame input if any stiff links to the frame are present. Also, strain control will be difficult owing to the extensometer's high sensitivity to lightly damped resonances in the test machine.

6.2.3 Resonant considerations

The final area of transducer consideration relates to the resonant frequencies of the extensometer. Resonant situations are frequencies where an inordinate amount of part motion is created for a small stimulus. The attachment arms and the flexure will have finite resonant frequencies. At these frequencies significant bending of the flexure could occur for very small amounts of strain. Also, the rotational mode described in the inertial considerations will have a fundamental resonant mode. At that frequency, even if there is low sensitivity of the extensometer to the rotational mode, there may be appreciable signal due to large rotations. The rotational mode is generally the lowest-frequency mode. Either of these conditions would be undesirable if it occurred within the desired operating range.

This concludes the discussion of extensometer fundamentals. The technical information above is based on product observation, discussion with MTS transducer engineers, and speculation based on the author's experience with other mechanical structures. In the following section, dynamic test results are reviewed for a specific extensometer.

6.3 TEST RESULTS

Testing of some extensometers was conducted to determine if the sources of error described in section 6.2 do indeed exist. Six different extensometer designs taken from several different manufacturers were tested for dynamic characteristics. There were three specific objectives of the design evaluations:

1. Determine if extensometers have dynamic characteristics.
2. Determine if different designs have different characteristics.
3. Determine if the dynamics matter in general test applications.

Test methods were initially developed on a single extensometer. An MTS 632.25 extensometer was evaluated for its dynamic characteristics. It was found to have several significant dynamic characteristics in the frequency range below 1000 Hz. It is important to note that the extensometer was tested well beyond its rated operating frequency of 20 Hz. The choice of extensometer was intentional, however. Of the extensometers available for test, this had the longest attachment arms. It was chosen since long attachment arms were believed to be the worst case for dynamic effects.

The extensometer was mounted between two metal extension rods on a 1000 Hz material test machine. One rod was attached to the actuator piston rod. The other was attached to the frame mounted load cell. The two rods were not attached to each other, thus effectively forming a divided testpiece. This permitted creating 'strain' inputs without creating loads in the pull rods. This method is identical to the method described in current calibration standards for static calibration. For the purpose of this exercise, the actuator linear variable differential transformer (LVDT) was used as the reference standard for 'strain'; however, it is recognized that the LVDT would not be a suitable transfer standard for extensometer calibration. Since these were to be comparative tests, the accuracy of the LVDT was not as important as its repeatability. Definite dynamic differences were seen between the various designs during testing. This implied that the dominant dynamic characteristics were extensometer related, rather than LVDT related. Thus, use of the LVDT was viewed as an appropriate reference.

The actuator was operated in stroke control. For the initial frequency response test, a low-level pseudo-random excitation signal was input to the stroke controller. The program level was kept sufficiently low to prevent extensometer slippage for this test. The actuator LVDT and extensometer outputs were monitored. They were fed into a Scientific Atlanta SD380 spectrum analyser, and the transfer function was computed.

Figure 6.4 shows the test data. The LVDT output is channel A; the extensometer output is channel B; and the amplitude ratio is B/A. The LVDT sensitivity was nominally $0.4 \, \text{V mm}^{-1}$. The extensometer sensitivity was nominally $4 \, \text{V mm}^{-1}$. Significant dynamics are clearly seen. The lower plot is the amplitude ratio. The upper plot is the phase lag of the extensometer relative to the actuator LVDT. The frequency is plotted linearly, with 100 Hz per division. The amplitude ratio is a log scale with a decade per major division.

For accurate strain measurements this plot should be a flat line. Quick inspection shows that the transfer function is only flat to about 50 Hz. The amplitude ratio is only slightly more than 10 in that region. This is the expected value, considering the sensitivity of the LVDT and extensometer. Beyond that frequency the plot is very irregular, and almost always larger in amplitude than the ideal value. This confirms that extensometers do indeed have dynamic effects.

The sharp peaks are due to resonances. Four resonances are labelled (a)–(d). The first three are noticeably different from resonance (d) for two reasons. First, there is not a lasting amplitude ratio change on either side of the first three resonances. If an imaginary line is drawn through the resonance point, connecting the curve on either side of the resonance, it is seen that the underlying curve continues in an upward, increasing direction. Resonance (d) is distinctly different in this regard. After the resonant peak, the amplitude ratio trend is definitely downward.

In order to better understand what is happening in the extensometer, a model based on four spring–mass pairs was created. This model was run on a Macintosh using Extend™ (registered trade mark of Imagine That, Inc.). Similar transfer functions could be obtained with the simulation.

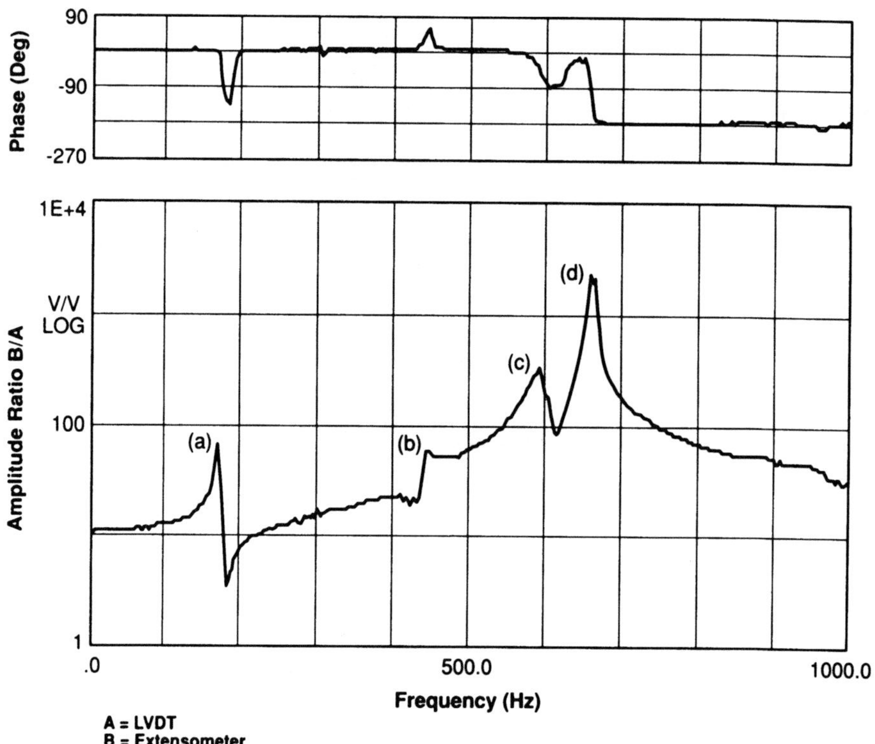

Fig. 6.4 MTS 632.25 frequency response, actual strain input.

The different resonance types depended on whether a spring–mass pair was directly in the measurement path, or external to it. A representative transfer function from the simulation is shown as Fig. 6.5. The line with the upward peaks is the amplitude ratio, plotted against the left-hand axis. The other curve is the phase lag, plotted against the right-hand axis. Note that in addition to the amplitude ratio difference, the resonances have different types of phase change.

For the purposes of distinguishing between the two types of resonances, extraneous resonances will refer to resonances without a sustained phase change, and fundamental resonances will have a sustained phase change. In Fig. 6.4, resonances (a)–(c) are extraneous; resonance (d) is a fundamental resonance. For the data from the simulation (Fig. 6.5), the two resonances below 30 are extraneous. The other two are fundamental. What is important to know now is that fundamental resonances are believed to alter a transducer's sensitivity over a wide range of frequencies. Extraneous resonances are believed to alter the response over a much narrower ranger of frequencies.

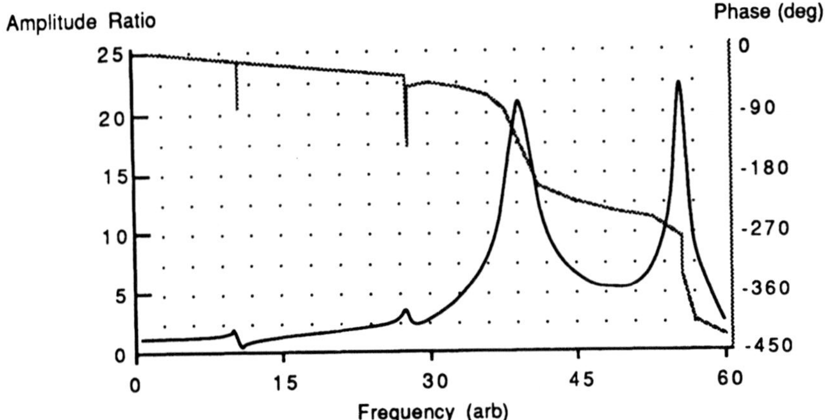

Fig. 6.5 Extend™ simulation result, sample transfer function.

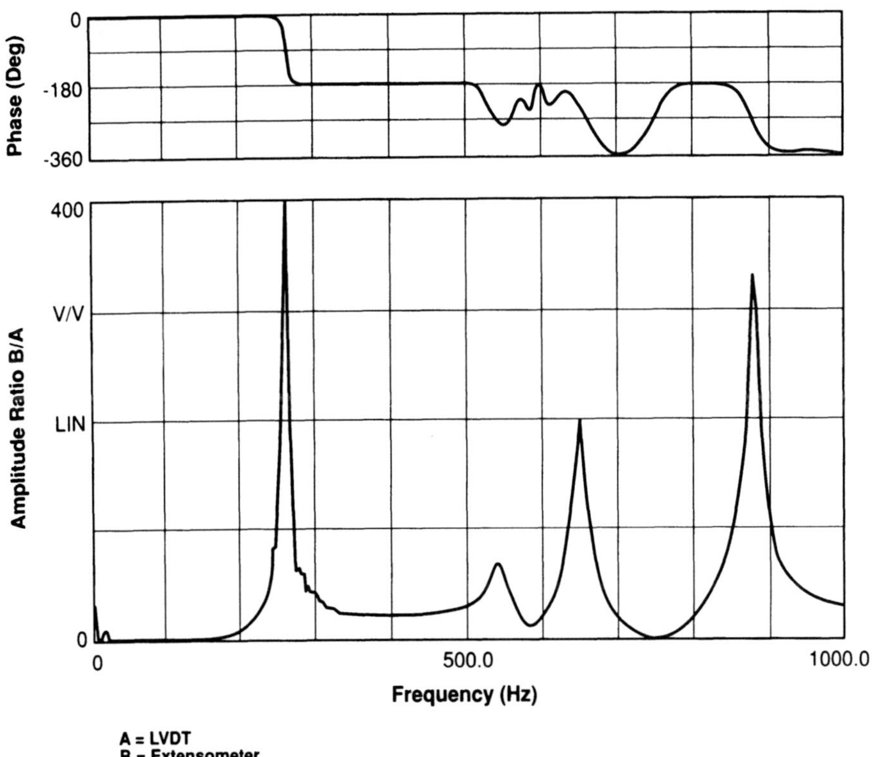

Fig. 6.6 MTS 632.25 frequency response, common mode input only.

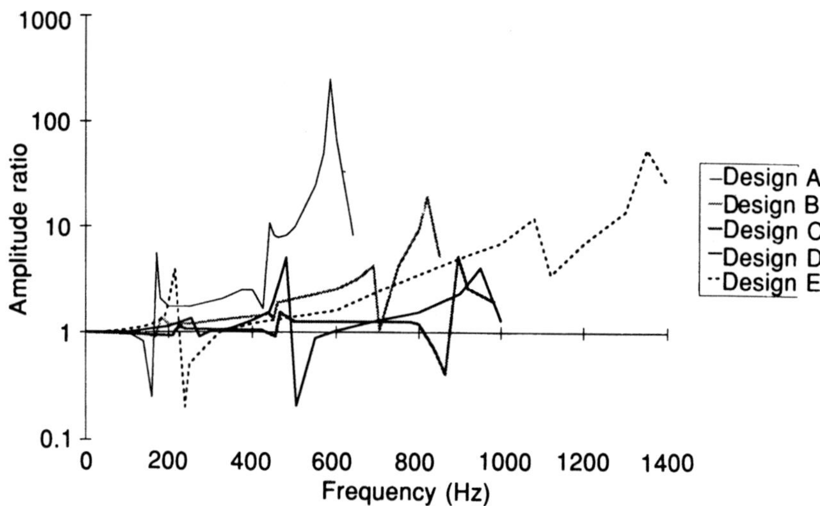

Fig. 6.7 Extensometer frequency response, design comparison, all designs.

One possible source of extraneous resonance is common-mode-induced signals. Common mode signals are a result of motions of the knife edge where no strain occurs but both edges move together. To test that possibility, the extensometer was attached to only the moving extension rod. This is the type of test proposed as an addition to ASTM Standard E83 (Amaral, 1990). Now, for actuator motion there will be acceleration of the extensometer, but basically no strain input. Figure 6.6 shows the frequency response to this 'common mode only' input. A perfect transducer would have absolutely no output. In this case, for frequencies above 250 Hz, the extensometer has more output than it had when it was measuring 'strain'! Obviously, there is a large amount of common mode sensitivity at these frequencies. There does appear to be a change in frequency between pure common mode input and combined strain and common mode input. Further work may reveal an explanation. However, for now it is sufficient to note that common mode output is a real possibility.

At this point, the first of the testing objectives was concluded. These tests confirmed there are extensometer dynamics which can be measured.

The next area of investigation regarded differences between designs. Six designs were evaluated, from four vendors. The designs were generally intended for maximum frequency use between 20 and 100 Hz. One did not provide sufficient signal amplitude for a good comparison. However, with the other five designs it was very easy to compare frequency response. Only the extensometer's frequency response was evaluated. Figure 6.7 compares the results of the five designs. As can be seen, generally all have non-resonant response below nominally 150 Hz. Also, they all have resonant responses between 150 Hz and 1400 Hz. Lastly, the resonant frequencies and amount of damping are quite different between the designs. Thus it is readily apparent that typical extensometers will have dynamics in the frequency range below 1000 Hz, and that they will be distinctly different between designs.

6.4 ADDITIONAL CONSIDERATIONS

6.4.1 Instrumentation

Thus far the discussion has centred on the 'pure' effects of the transducer. These factors are related only to the transducer. Two other factors are very important to consider. They are the conditioning/readout instrumentation and the influence of the test machine waveform quality. Less study has been done in these areas to date. However, many basic issues have been noticed and can be mentioned.

The first is the frequency response of the conditioning electronics. These electronics convert the physical strain into an analogue voltage. Strain-gauge-based extensometers require low-noise, high-gain preamplifiers to condition the signal. Because of the large amplification, it is generally necessary to limit the bandwidth of the amplifier. Otherwise, the local radio station can be as large a signal as the test result! For static testing, the lower the bandwidth the better as this reduces the electrical noise and improves the extensometer readout accuracy. For dynamic testing a balance must be struck between bandwidth and noise. Excessive bandwidth results in excessive noise. This reduces the measurement accuracy owing to noise-induced uncertainties. Inadequate bandwidth reduces the measurement accuracy owing to changes in sensitivity at the higher test frequencies.

Once the analogue voltage is generated it must be output. The output device will have the same bandwidth versus noise tradeoffs. When using mechanical readout devices such as $X-Y$ recorders this will essentially always be less than the conditioning electronics. Additionally, the bandwidths of the X channel and the Y channel are probably not the same.

Digital output devices will also have significant input filtering. This is required because digital equipment samples the analogue data at a finite rate, often called the sampling rate. These anti-aliasing filters are essential to reduce higher-frequency noises to prevent aliasing. Aliasing is the phenomenon of creating a 'fictitious' low-frequency signal by sampling a high-frequency signal at a lower frequency. The sampling rate must be significantly greater than the maximum frequency content of the maximum test frequency desired. To put this in perspective, to accurately measure the peak of a sine wave to within 0.2% requires sampling at least 50 points per cycle. For testing at 100 Hz this requires a 5 kHz sample rate at a minimum! If the waveform is more triangulated, even higher sample rates are needed.

Use of digital computers also introduces quantization errors. Quantization errors result because analogue to digital converters (ADCs) can only measure voltages in finite steps. Common ADC resolutions are 1 part in 4096 (12 bit units) and 1 part in 65 536 (16 bit units). Not only are they finite steps, but the steps are not the same size! They are at least generally monotonic, meaning that an increasing analogue voltage will result in a stable or increasing digital output.

Practically speaking, the above does not mean that all extensometer users must acquire intimate knowledge of the dynamics of instrumentation. Rather, it is essential that users have an accurate method of electrically exciting the intended conditioning/readout electronics. By comparing the instrumentation output with the standard it is possible to verify the accuracy of the instrumentation. According to Dahlberg (1991), however, 'Not just any signal source can be used, due to the high accuracy requirements.' Though more information and study are needed in this area, it is beyond the scope of this discussion.

6.4.2 Test machine influence

As shown above, the sensitivity of an extensometer is both frequency dependent and acceleration dependent. Each of these factors is influenced by the test machine being used. The somewhat more obvious test machine influence is its vibrations. A lightweight machine is likely to have greater vibrations than a heavy machine. A test machine using heavy grips is likely to have greater vibrations than the same machine with light grips. Lastly, a machine with lower frequency or lightly damped structural resonances will disturb the extensometer more than one with higher frequency and/or well-damped resonances.

The more subtle test machine factor is the waveform type and quality. The extensometer undergoes vertical motion dependent on the strain profile, regardless of the type of test machine. If the strain profile is a perfect sine wave, then the extensometer will be excited only at the test frequency. However, if the waveform is distorted in the slightest, then it will be composed of sine waves at the fundamental frequency and at harmonics of the fundamental. That is, the waveform can be created by summing sine waves of different amplitudes and frequencies. The frequencies will be the test frequency and integral multiples of the test frequency. Odd multiples such as three, five and seven often have the largest amplitudes after the fundamental. For example, a distorted 36 Hz sine wave will usually have considerable 'amplitudes' at 108, 180 and 252 Hz. For our previous example of the 632.25, the 180 Hz signal coincides with the first resonance of the extensometer. That resonance had an amplification factor of at least five. Thus, if the strain distortion amplitude was 0.2% of the fundamental then the extensometer would have distortion of 1%. If the resonance was much more lightly damped, such as the one at 660 Hz, then the amplification would be more severe.

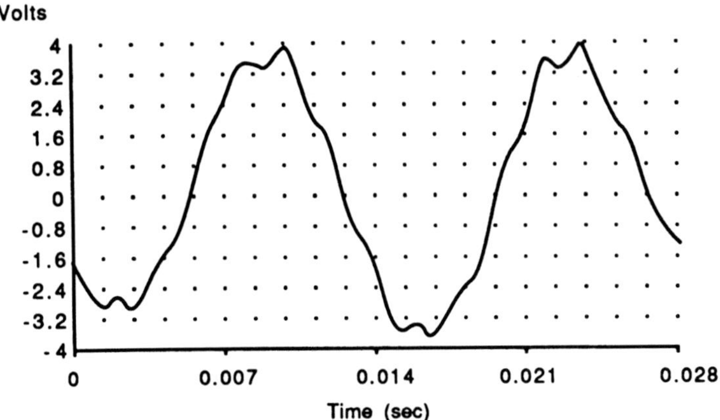

Fig. 6.8 Extensometer response, 70 Hz.

This distortion is very evident in Figs 6.8 and 6.9. They show extensometer response for a 100 Hz extensometer when run at 70 and 100 Hz
respectively. The extensometer had a resonance at roughly 500 Hz. At
70 Hz, the seventh harmonic is roughly 500 Hz. Notice that there are
seven peaks in the distorted sine wave per single cycle of the fundamental.
For the 100 Hz test, the fifth harmonic is 500 Hz. The distortion is quite
severe at this point. Five peaks in the distorted sine wave are clearly
evident per single test cycle. The amplitude of the distortion is easily
greater than 10% of the fundamental amplitude.

How repeatable is the distortion? A frequency response test was repeated three times for a different extensometer. Instead of the pseudo-
random/FFT method used previously, the system was operated at
discrete frequencies. The peak to peak amplitude of the extensometer
was recorded, as well as the peak to peak load measurement. The value
plotted is the ratio of the extensometer reading to the load reading. It has
been non-dimensionalized by referencing the measurements to the 4 Hz
result. The system was operated at the same load command level every
2 Hz between 2 and 100 Hz. By taking the peak to peak value, the
influence of the distorted waveform is measured. The previous method
of using a pseudo-random signal and taking the FFT of the result only
measures the response to a 'pure' sine wave at each frequency.

The result is shown in Fig. 6.10. Four curves are plotted. The three
lightweight lines show results for a single test. The dark line is the
average of the three tests. Notice the large peaks in the frequency range
between 50 and 100 Hz. Although there is considerable scatter of
the data, it is quite apparent that there are definite frequencies where

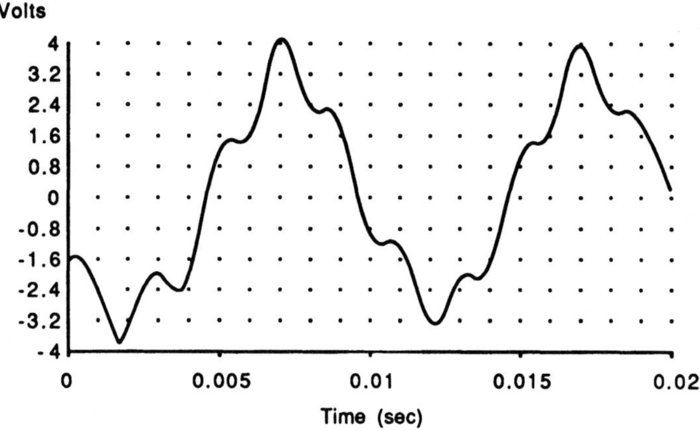

Fig. 6.9 Extensometer response, 100 Hz.

peaking of the error will occur. These peaks are believed to correspond to subharmonics of extensometer or frame resonances as described above. Frame resonances are now included because frame resonances that excite the extensometer can produce significant output. If the desired accuracy was 5%, then use of this extensometer would need to be limited to something less than 65 Hz.

These errors are considerably higher than those that would be obtained using the pseudo-random/FFT method. Using that method with this same extensometer, error increased linearly with frequency, not reaching 5% until approximately 180 Hz. This is significantly higher in frequency than the 65 Hz result obtained by the peak to peak method! Based on these tests, the pseudo-random/FFT method would appear to give an unrealistic estimate of usable bandwidth.

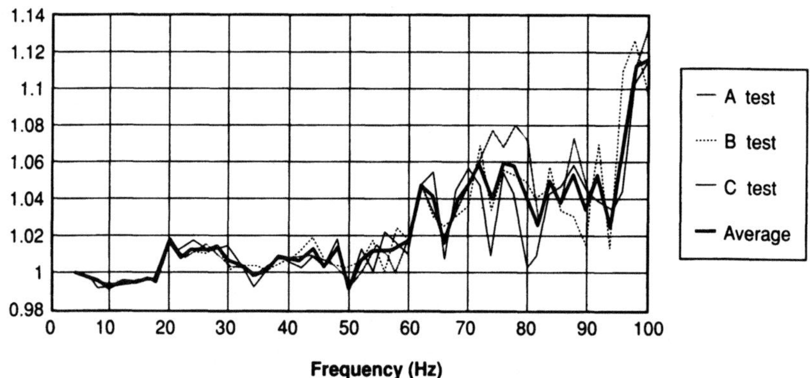

Fig. 6.10 Extensometer dynamic response, comparison of three tests.

Obviously then the test machine does influence the usable frequency range of an extensometer. High-temperature extensometer applications are particularly difficult. It is common practice in those situations to mount the extensometer external to the heated region, and attach the extensometer to the testpiece via long arms. The mounting style is usually frame reacted. Typically it is attached to the load frame columns or baseplate. This results in an extensometer with significantly more mass, measurably lower resonant frequencies, and substantially greater coupling of the test machine's vibrations to the extensometer: basically, the worst of all worlds! To compound matters, these conditions tend to lead to a lower servoloop gain. This results in greater distortion of the test waveform, again creating more distortion. This reduces some high-temperature extensometers to little more than 'seismic microphones'. (This phrase was coined by Harry Winslow, a seasoned test machine operator, while evaluating a prototype design high-temperature extensometer. Control with it was nearly impossible, and the distortion was generally larger in magnitude than the signal. The problem has since been remedied, but the phrase lives on.)

6.5 EXTENSOMETER STANDARDS

6.5.1 Applicable calibration standards

A variety of standards are available for calibration of force cells and extensometers. The list at the end of the chapter covers applicable BS, ISO and ASTM standards I have read. There are certainly other standards in the world that may also apply: I would be grateful if people would send them to me. In Table 6.1 the standards are grouped by transducer type and operating frequency (i.e. static or dynamic). This section attempts to summarize the key aspects of these standards, and to show what issues must be considered in a standard for dynamic calibration of extensometers.

All of the standards are nominally composed of the seven basic components listed below. Six of the seven components are directed at providing accurate, repeatable evaluations of transducers. The seventh item, classification, attempts to group products based on their measured errors.

The seven parts of a standard

Scope	What devices are covered?
Terminology	What words are used?
Verification apparatus	What will be the standard?
Procedure	How is the test conducted?
Classification	How to group products?
Verification frequency	How and when to recheck?
Report format	How to report the results?

Table 6.1 Groupings of calibration standards

	Static	*Dynamic*
Transducer/verification apparatus		
Force		
BS	1610 : Part 2 : 1985	Draft 30 May 1991 : Part 2
ISO	376 : 1987 (E)	
ASTM	E74 : 1983	
Strain		
BS	3846	
ISO	9513 : 1989 (E)	
ASTM	E83 : 1990	E467 (proposed)
Test machine		
Force		
BS	1610 : Part 1 : 1985	Draft 30 May 1991 : Part 1
ISO	7500 : Part 1 : 1986 (E)	4965 : 1979 (E)
ASTM	E4 : 1989	E467 : 1991

6.5.2 Simplified standard

The different standards apply to different types of products, and in different operating environments. This section illustrates the common structure of these standards.

Calibration must be done using a reference measurement system (transducer and instrumentation), traceable to a national or international standard. This reference system must have a measurement error significantly less than the measured error of the measurement system (transducer and instrumentation) to be calibrated. Perform the following four-step procedure:

1. Assemble the two transducer systems in a machine capable of creating force/displacement.
2. Adjust the zero-output points of the transducers to some reference. Force transducers are set to zero with no testpiece, removing the influence of attached grip weight. Extensometers are set to zero at a specified gauge length, which itself has a calibration requirement.
3. Move the machine to a force/displacement at a specific rate or frequency. The length of time to maintain the level before doing step 4 is variable.
4. Record the output of the two transducers.

Repeat steps 3 and 4 at different levels to obtain sufficient data points to permit accurate estimation/interpolation of data between test points. Repeat some or all of the steps to check for repeatability of measurement at a particular point. This includes checking a particular output during an ascending series versus the output during a descending series. Repeat step 1 at least once to show errors due to mounting and orientation of the transducer.

Following the procedure are instructions on evaluating the data. Different classification categories are specified, based on the worst-case error. Guidelines are provided as to when the complete verification process must be performed, as well as partial verifications that can be done between times. Lastly, instructions are provided as to how to report the data from the testing.

Included with each standard are cautions related to possible sources of error. Minimizing readout error by proper reading of dials and charts is a common caution. Care to avoid axial misalignment when performing load calibrations is another common caution. Allowing for temperatures to stabilize is included in all of the standards. The standards generally are intended for use at room temperature. Amaral (1990) proposed expanding ASTM Standard E83 to address the use of extensometers at elevated temperatures.

Dynamic operation introduces two new categories of potential errors. As described previously, they are inertially induced errors and instrumentation frequency response errors.

6.5.3 Basis for dynamic calibration standard

The previous simplification was done in order to highlight what issues must be resolved when developing a standard for dynamic calibration. The specific product (i.e. load cell or extensometer) has intentionally not been specified, because it is felt that one standard can be applied to both products. This is due to the fact that for an elastic testpiece, force and strain are linearly related. This fact has been used for many years with calibration standards. That is the basis of using elastic devices such as proving rings for calibration of load cells (BS 1610). The relatively large deflection and the linear relationship with load permit use of a more commercially available displacement transducer to make accurate load measurements.

Two challenges are associated with a dynamic calibration standard. The first is defining a suitable reference system that can show traceability back to national and international standards. The current approach for force calibration is to create a new working standard by strain gauging a testpiece and calibrating it to another standard (either primary or working). The rationale for this approach is that since the major mechanical errors are inertially induced, it is necessary to duplicate the motions of the test machine during calibration. This implies that the test machine must include a testpiece similar to that used in normal testing. Also, for greatest accuracy, the transducer should be of minimal mass. This leads to the approach of measuring the strain directly at the testpiece. It is proposed that extensometer dynamic calibration can use the exact same reference. The strain measurement will be as accurate for the extensometer as it is for the load cell. Since the strain gauged part must have the same deflections as the proposed testpiece, it should have sufficient strain for calibrating the extensometer on this machine. This may not be the device's full range, but owing to the heavy influence of the test machine and testpiece on the extensometer's accuracy it is only reasonable to calibrate the extensometer on a testpiece by testpiece basis.

Defining a reference standard for the measurement system will be a more difficult task. However, its needs apply equally to load cells and extensometers. Fundamentally, what is required is to apply an accurate, cyclic voltage to the electrical inputs of the conditioning/measurement system. The output is then calibrated against the reference input. Definition of that accurate source is beyond the scope of a dynamic calibration standard for extensometers and load cells. The proposed direction is to find or create a standard that covers accuracy of cyclic voltage sources. This standard would address the noise/frequency-response tradeoffs mentioned previously. Then the dynamic calibration standard for extensometers and load cells would simply require that measurement systems be calibrated against that standard.

The other major challenge is the seemingly large amount of effort involved to do a single dynamic calibration. The calibration is more time consuming and expensive owing to the need to make a strain gauged testpiece, and to the significantly greater number of test points. To do a full calibration, multiple amplitude levels must be recorded at multiple frequencies. Depending on the number of intended test frequencies, the calibration task effort can grow astronomically. Also, the need to perform a dynamic calibration or verification for each type of testpiece implies much greater calibration frequency than for static checks.

It has been determined that the greatest mechanically induced load error is due to acceleration of the grip attached to the load cell. Happily, acceleration effects are very dependent on frequency, increasing generally with the square of the operating frequency. This suggests that for a specific acceleration limit, there will be a minimum frequency below which inertial load errors are not an issue.

It is proposed that the dynamic calibration standard uses acceleration of the test machine load cell grip as a measure of dynamic accuracy of the test machine load cell. At frequencies and amplitudes where the acceleration error is sufficiently below the desired error of the load cell, the requirement for detailed dynamic calibration of the load cell should be waived. Since the acceleration is a direct measure of error, the accuracy of acceleration measurement can be considerably lower than the desired error of the load cell. This should permit use of economical commercially available accelerometers or other means of acceleration measurement. This does not eliminate the need for dynamic calibration of the measurement system, however.

It is further proposed to permit use of the test machine load cell as the reference for the extensometer dynamic calibration in certain cases. This would be allowed where the acceleration errors are low enough that the implied dynamic error of the load cell is significantly lower than the desired error of the extensometer. It is not apparent that there is an equivalent acceleration error check that could be done for extensometers. This is due to the large influence of the test machine vibrations on the extensometer output, and the variety of ways that the vibration can couple into the extensometer. Detailed calibration of the extensometer would still need to be done at the intended test frequencies and amplitudes. This is because of the strong influence of the resonances in the extensometer. However, at least the need for a calibrated, strain gauged testpiece is eliminated.

Also, this invites automation of the dynamic calibration process directly within the test machine controls, since both parameters will be present within the test machine controller. This would help reduce the time needed for dynamic calibration and improve the quality of calibration documentation. This view is supported by Perlov and Martin (1991), who made the following comment about the value of computer-aided calibration in static calibration of load transducers: 'A major benefit of automated verification is the speed with which documentation is available at the end of the procedure and the accuracy of reporting and record keeping.'

The proposed method obviously has introduced an additional standard layer between the extensometer and the national or international standard. This is not a strong point of the approach. Its strength is that if the process becomes simple enough, then users may actually do it. It has been my impression, while working with ASTM Subcommittee E904.02 to revise ASTM E467, that users were not performing dynamic calibrations owing to the amount of effort required, and the doubt that it is even necessary. This proposed method would hopefully make the process straightforward enough to be done. Then only in the isolated cases where severe dynamic errors are present will a more detailed calibration be required.

6.5.4 What about the extensometer vendors?

What would this standard imply for extensometer vendors? Owing to the strong influence of the test machine on the extensometer output, it is unrealistic for an extensometer vendor to accurately specify the dynamic range of the device. However, initial studies (Annala, 1991) have shown the behaviour of a variety of extensometers to be very similar in shape, but not frequency. Table 6.2 shows a comparison of the performance of four extensometers when using the pseudo-random/FFT test up to 1 kHz (Annala, 1991). Resonant frequencies and amplitudes are tabulated. Also listed is the lowest frequency at which there was a 5% error between extensometer output and a reference load cell. This type of testing looks like a potentially useful way to make comparisons of extensometer dynamic performance. Of course, final dynamic accuracy cannot be determined until the unit is installed in the actual machine. More work by vendors could refine the technique to create a uniform method of comparison of dynamic characteristics.

Table 6.2 Performance comparison of four extensometers

	Design A	Design B	Design C	Design D
Differential mode				
Change in output of 5%	310 Hz	95 Hz	240 Hz	136 Hz
1st resonance frequency	410 Hz	232 Hz	326 Hz	293 Hz
1st resonance magnitude	1.80:1	3.71:1	2.80:1	1.82:1
2nd resonance frequency	720 Hz	735 Hz	710 Hz	743 Hz
2nd resonance magnitude	2.80:1	9.31:1	3.12:1	2.76:1
3rd resonance frequency	915 Hz	–	955 Hz	–
3rd resonance magnitude	23.1:1	–	43.0:1	–
Common mode				
1st resonance frequency	458 Hz	345 Hz	400 Hz	425 Hz
1st resonance magnitude	5.61:1	18.0:1	4.03:1	5.97:1
2nd resonance frequency	915 Hz	–	980 Hz	–
2nd resonance magnitude	11.3:1	–	41.7:1	–

6.6 CONCLUSION

In conclusion, three simple questions have been answered.

- Do extensometers have dynamics?   Yes
- Are there differences between designs? Yes
- Does it matter in actual testing?   Yes

Additionally, the influence of the test machine on the extensometer accuracy was shown to be dramatic, owing to reasons of waveform distortion. This basically necessitates some form of dynamic calibration or verification of an extensometer in each configuration used. Since to perform a detailed extensometer dynamic calibration each time it is used would be prohibitively time consuming and expensive a simplified approach is proposed. The next step is for users and vendors to test these ideas for practicality. If they can be found to be reasonable, a standard could be prepared, and test machine suppliers could begin to integrate them into their machines/controllers. Although not a simple task, it is realizable now. Let us make the 1990s the decade of dynamic accuracy.

ACKNOWLEDGEMENTS

Many people helped to make this chapter possible. Rick Meyer provided valuable insight into the actual workings of today's extensometers, as well as providing 'the book' on standards. Jay Annala, Jeff LaBore and Mike Langer were invaluable in providing test data. Dave Burruss and Jane Torpey-Tuma kept us supplied with testpieces, which we seemed to break at an alarming rate. And of course, a very big thank you to Sharon Mantifel, who never fails to turn my last-minute ramblings into a finished product, and always with a smile.

REFERENCES

Amaral, J. E. (1990) *Effects of Dynamics and Temperature on the Classification of Extensometers*, Instron Corporation memorandum.
Annala, J. (1991) *632.11B-20 Data that is Comparable to 632.31F-24, 632.31D-24*, MTS Systems Corporation memorandum.
Dahlberg, G. L. (1991) *Dynamic Calibration of MTS Electronics in the Field*, MTS Systems Corporation memorandum.
Perlov, A. and Martin, J. (1991). Calibration and verification of materials testing machines and systems. *ASTM Standardization News*, 38–43.

STANDARDS

ASTM Standards

ASTM E4	Standard Practices for Load Verification of Testing Machines.
ASTM E74	Practice for Calibration of Force Measuring Instruments for Verifying the Load Indication of Testing Machines.
ASTM E83	Standard Practice for Verification and Classification of Extensometers.
ASTM E467	Practice for Verification of Constant Amplitude Dynamic Loads in an Axial Load Fatigue Testing Machine.

British Standards

BS 1610	Materials Testing Machines and Force Verification Equipment. Part 1: Specification for the Grading of the Forces Applied by Materials Testing Machines. Part 2: Specification for the Grading of Equipment Used for the Verification of the Forces Applied by Materials Testing Machines.
BS 3846	Calibration and Grading of Extensometers for Testing of Metals.
Proposed	Draft 30 May 1990 Dynamic Force Calibration of Materials Testing Machines. Part 1: Calibration of the Testing Machine. Part 2: Calibration of the Force Proving Equipment.

ISO Standards

ISO 376	Metallic Materials. Calibration of Force-Proving Instruments Used for the Verification of Uniaxial Testing Machines.
ISO 4965	Axial Load Fatigue Testing Machines. Dynamic Force Calibration. Strain Gauge Technique.
ISO 7500	Metallic Materials. Verification of Static Uniaxial Testing Machines. Part 1: Tensile Testing Machines.
ISO 9513	Metallic Materials. Verification of Extensometers Used in Uniaxial Testing.

CHAPTER 7

Uncertainties in uniaxial low-cycle fatigue measurements due to load misalignments

F. A. Kandil and B. F. Dyson

7.1 INTRODUCTION

Scatter in low-cycle fatigue (LCF) data has for the most part been attributed either to intrinsic inhomogeneities in the microstructure within a single batch of material or to microstructural variations between different batches caused by the usual complex fabrication processes. Evidence for this view is invariably apocryphal, since little if any account has been taken of measurement errors associated with testing practice and no attempt seems to have been made to quantify their extent. The importance of errors associated with testing practice has recently been brought to the fore during an analysis of the results of a large international LCF round robin exercise sponsored by the Community Bureau of Reference (BCR) of the European Community. Twenty-six laboratories from Europe and Japan participated by testing four engineering alloys at 550 or 850 °C. The most significant feature of this extensive database was the relatively small scatter within each laboratory (i.e. good repeatability) compared with the very large interlaboratory scatter (i.e. poor reproducibility). The data were compiled and analysed comprehensively by Thomas and Varma (1992), who concluded that the cause of the poor reproducibility was systematic errors in testing practice between the different laboratories, and not material scatter.

Scatter in LCF data due to testing practice is mainly a consequence of:

- unquantified misalignment of the testpiece with respect to the loading train;
- uncertainties in the measurement of strain range;

- uncertainties in the measurement and control of temperature (in the case of LCF at elevated temperatures).

These quantities could all be well controlled within a single laboratory and thereby lead to good repeatability of data, but at the same time could give rise to systematic differences between laboratories and thus to the poor reproducibility observed in the BCR exercise. The objectives of this chapter are to take one of these possible sources of error – load misalignment – and to model its influence on data scatter by identifying and analysing several mechanisms of bending in rigid grip systems, and to show that the magnitude and bias of the data scatter depend upon the type of extensometer used, in conjunction with the fatigue crack initiation mechanism.

7.2 DEFINITION OF PARAMETERS USED TO ASSESS BENDING

Consider the cylindrical bar in Fig. 7.1 deformed elastically in tension to a nominal axial strain ε_0. Assuming that any arbitrary cross-section remains plane during deformation, the originally horizontal cross-section U will rotate in the maximum bending plane MM′ by an angle Φ about its neutral axis NN′. The strains measured at the two opposite locations M and M′ on the testpiece perimeter are, respectively, equal to ε_1 and ε_2.

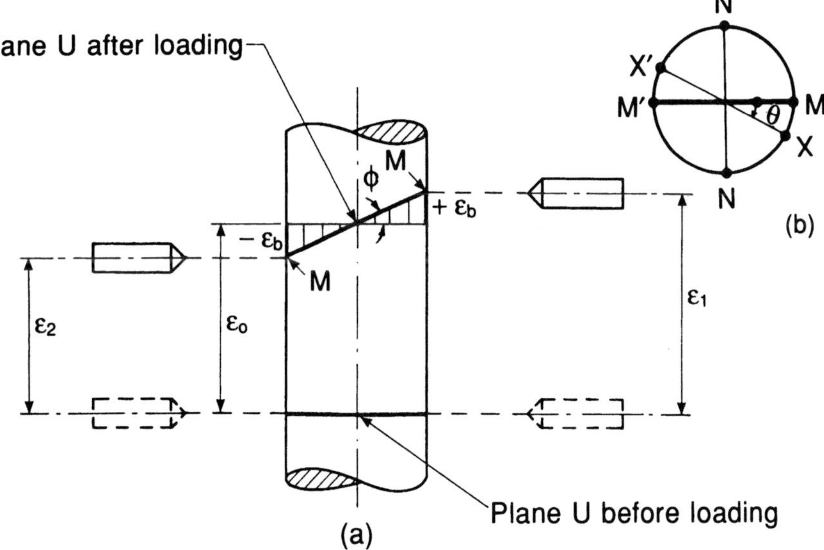

Fig. 7.1 Definition of bending parameters for a maximum bending strain ε_b, superimposed on the average elastic strain ε_0: (a) longitudinal section along the diameter MM′ (b) circular cross-section showing the angular position θ of a single extensometer located at point X with respect to point M in the maximum bending plane.

The bending strain ε_b is maximum in amplitude at points M and M′ and is given by

$$\varepsilon_b = \frac{\varepsilon_1 - \varepsilon_2}{2} \qquad (7.1)$$

At a given load, the bending due to misalignment is usually represented by the percentage bending B, given by

$$B = \frac{\varepsilon_b}{\varepsilon_o} \times 100 \qquad (7.2)$$

where ε_o is the average axial strain defined by $\varepsilon_o = (\varepsilon_1 + \varepsilon_2)/2$. Since, in general, the bending strain on the testpiece surface will vary along the gauge length in both magnitude and direction, ε_b must be the largest value anywhere on the cylindrical surface of the testpiece.

Equation 7.2 may also be expressed as

$$B = \frac{\varepsilon_1 - \varepsilon_2}{\varepsilon_1 + \varepsilon_2} \times 100 \qquad (7.3)$$

The use of the angle of rotation Φ was suggested by Christ (1973) as an alternative measure of bending in the so-called 'precision of alignment' parameter P, defined as

$$P = \tan \Phi = \frac{\varepsilon_b \ell}{r} = \frac{\sigma_b \ell}{Er} \qquad (7.4)$$

where ℓ and r are the gauge length and radius of the testpiece, σ_b is the maximum bending stress and E is Young's modulus.

In the circular cross-section shown in Fig. 7.1(b), θ is the angle between the two surface points M and X, where the maximum bending strain and the strain measuring device, respectively, are located. The bending strain ε_{bx} at any point X is related to ε_b by

$$\varepsilon_{bx} = \varepsilon_b \cos \theta \qquad (7.5)$$

The strain ε_x, measured using a single extensometer (for example, a conventional side-entry axial extensometer or clip gauge) placed arbitrarily at point X, is given by

$$\varepsilon_x = \varepsilon_o + \varepsilon_{bx} \qquad (7.6)$$
$$= \varepsilon_o + \varepsilon_b \cos \theta$$

Thus the strain measured by this type of extensometer will only equal ε_o when θ is either 90° or 270°.

In contrast, the strain measured by a dual extensometer (double-sided extensometer) located at arbitrary points X and X′ will always equal ε_o, since

$$\varepsilon_x = \varepsilon_o + \varepsilon_{bx}$$
$$\varepsilon_{x'} = \varepsilon_o - \varepsilon_{bx}$$
$$\varepsilon_{av} = \frac{\varepsilon_x + \varepsilon_{x'}}{2} = \varepsilon_o \tag{7.7}$$

There is therefore a considerable advantage in using a dual extensometer.

7.3 BASIC MECHANISMS OF MISALIGNMENT IN RIGID GRIP SYSTEMS

Bending results from eccentric loading of the testpiece. Sources of departure from the ideal situation in any test are caused by a combination of:

- inaccurate machining of the testpiece;
- poor conformance of the testpiece centre-line to top and bottom grip centre-lines;
- poor alignment in the loading train;
- misalignment in the machine frame.

A further complication is that bending due to non-axiality, measured in terms of either the percentage bending B or the precision of alignment P, is usually load-dependent as a result of straightening of the load-train as load increases and also the deformation of its components including the testpiece. Consequently, it is impossible to predict exactly the magnitude of the bending component for a given test system and testpiece configuration at the test load without carrying out a set of measurements.

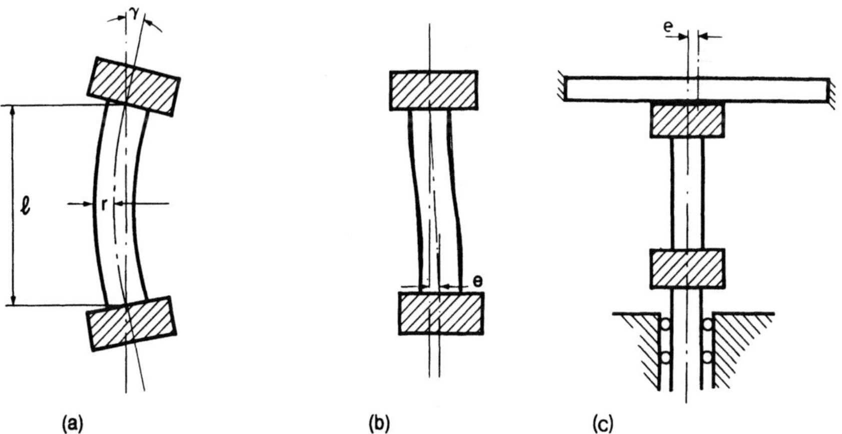

Fig. 7.2 Basic bending mechanisms due to load misalignment in rigid grip systems: (a) angular offset (b) lateral offset (c) load train offset in a non-rigid system.

In the following, some simple calculations of the bending strain will be presented. These are based on only a few assumptions which relate to simple misalignment situations due to: an angular offset of the loading base; a lateral offset of the loading bars; and an offset in the load train assembly with respect to the machine frame. These are depicted in Fig. 7.2 and represent the possible primary causes of misalignment for the rigid grip systems usually encountered in fatigue testing.

7.3.1 Angular offset

In Fig. 7.2(a) the testpiece is fastened to ideally rigid grips showing an angular offset γ, but without lateral displacement of either centre-line. When the grips are only allowed to move vertically, then the testpiece will deform during the gripping operation into a circular arc, and the bending formula is

$$\varepsilon_b = \sigma_b / E = 2\gamma r / \ell \tag{7.8}$$

where γ is the angle of deflection and r and l are the radius and length of the testpiece.

Given a typical testpiece with dimensions $r = 3.5\ \text{mm}$ and $\ell = 15\ \text{mm}$, the angular offset for a maximum bending strain $\varepsilon_b = 2.5 \times 10^{-4}$ (corresponding to the maximum bending allowed by testing standard ASTM E606 : 1980 assuming a typical lowest total strain range in a test programme of 0.5%) is $\gamma = 5.4 \times 10^{-4}$ radians.

7.3.2 Lateral offset

In Fig. 7.2(b) the centre-lines of the ideally rigid grips are offset with respect to each other by an amount e. The bending formula (Bressers, 1982) now becomes

$$\varepsilon_b = \sigma_b / E = 4er / \ell^2 \tag{7.9}$$

Using the same testpiece dimensions as in section 7.3.1, a lateral offset of 4×10^{-3} mm will again give $\varepsilon_b = 2.5 \times 10^{-4}$.

7.3.3 Load train offset

Lateral offsets of the centre-lines of the load train with respect to the machine's frame can cause bending of the testpiece. Figure 7.2(c) considers the top grip, testpiece and bottom grip to be well aligned but with the top grip off-centre with respect to the machine's cross-beam. Provided that the machine frame is not ideally rigid, deformation in the load frame can cause the testpiece to deform under load into a banana-like shape as depicted in Fig. 7.2(a). However, reversing the load will change the sense of curvature in the case of Fig. 7.2(c) only: that is, during the fatigue cycle, the sign of the bending strain at any point in the testpiece will alternate between the tension and compression modes. Load train offset in a non-rigid system is expected therefore to cause the largest effect on fatigue life owing to testpiece bending.

In this chapter, the bending due to angular and lateral offsets will be termed non-reversed bending, while that due to a load train offset will be termed reversed bending (Kandil and Dyson, 1991). An ideally rigid system with all centre-lines parallel would not of course cause any bending at all. It should be stressed that the cases shown in Fig. 7.2 are all idealized and that bending geometries in practice are likely to be combinations of these simplified ones.

7.4 MODELLING DATA SCATTER CAUSED BY BENDING

Consider a fatigue test performed with a controlled strain limit ε_o and no bending: both single and dual types of extensometer will give the same number of cycles to failure N_{fo}. If bending is now introduced, in this case by load train offset, there are two cases to be considered depending upon the mechanism of fatigue crack initiation. In both cases, scatter due to material will be assumed to be negligible.

7.4.1 Case 1

It is assumed that fatigue failure is due to a crack emanating always from the point of maximum bending strain. This may, for example, be the case either when surface finish effects are non-existent or when they and other crack initiation sites are totally uniform in size and distribution over the gauge length of the testpiece.

Figure 7.3(a) illustrates the prediction that a dual extensometer will produce a fatigue life $N_f(\varepsilon_o + \varepsilon_b)$ uniformly lower than N_{fo} and that there will be no repeatability or reproducibility scatter attributable to bending.

In sharp contrast, a single extensometer will produce a fatigue life dependent on θ as shown in Fig. 7.3(b). The maximum is N_{fo} which occurs when $\theta = 0$, while the minimum is $N_f(\varepsilon_o + 2\varepsilon_b)$ at $\theta = 180°$. For a given θ, it is again predicted that there will be no data scatter due to bending. Since, to a first approximation, θ will vary only slightly within a single laboratory, repeatability scatter from this source will be negligible.

7.4.2 Case 2

It is assumed that the major crack which causes failure emanates from the single deepest machining groove on the surface of the testpiece or the

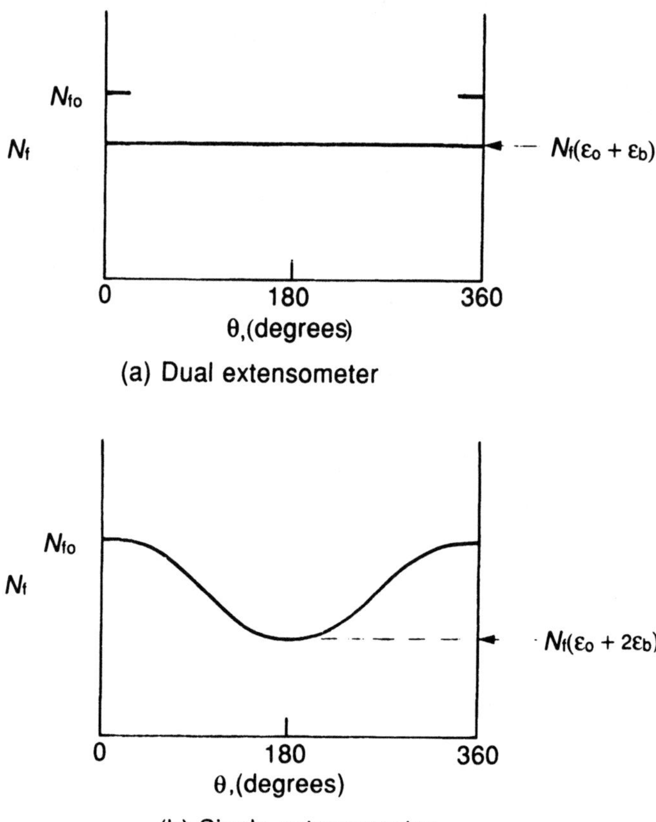

Fig. 7.3 Schematic prediction for variation in the number of cycles to failure with the angle θ according to case 1 using (a) a dual extensometer (b) a single extensometer.

largest subsurface inclusion. Its location can be anywhere and not necessarily in the maximum bending plane.

In this case, a dual-extensometer arrangement will produce a fatigue life within a scatter band (Fig. 7.4(a)) whose lower and upper limits occur, respectively, when the crack starts either at the point of maximum bending or at the opposite point to it (M or M' in Fig. 7.1). The range of the scatter band may be expressed as

$$\Delta N_f = N_f (\varepsilon_o - \varepsilon_b) - N_f (\varepsilon_o + \varepsilon_b) \qquad (7.10)$$

with a mean value approximately equal to N_{fo}.

The use of a single extensometer will produce a data scatter band in the manner shown in Fig. 7.4(b). Both the range and the mean of the scatter band are dependent on θ. Thus, at a given θ, the scatter band is given by

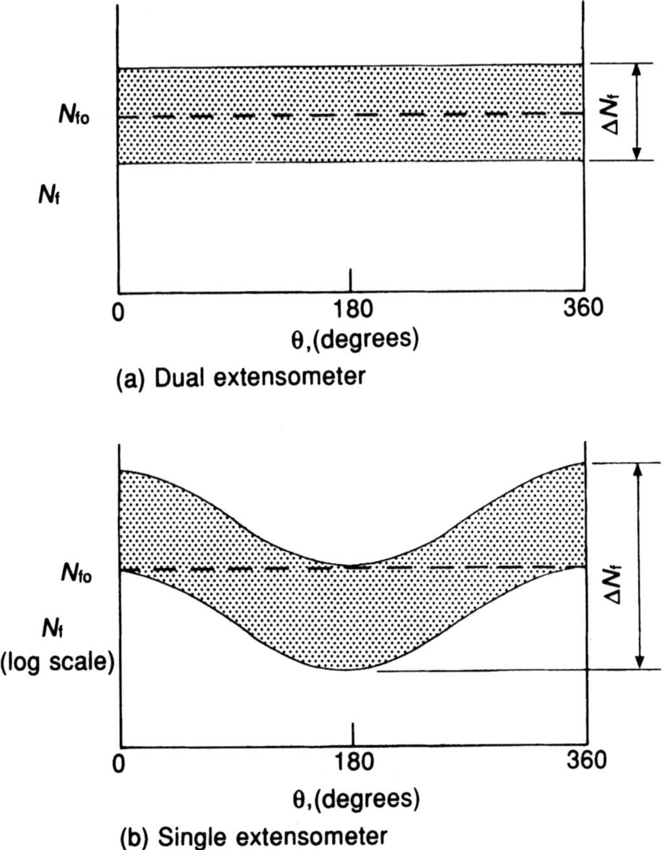

(a) Dual extensometer

(b) Single extensometer

Fig. 7.4 Schematic predictions as in Fig. 7.3 but according to case 2.

$$\Delta N_f = N_f(\varepsilon_2) - N_f(\varepsilon_1) \tag{7.11}$$

where ε_1 and ε_2 are the maximum and minimum strains anywhere on the testpiece surface (Fig. 7.1) and are given by

$$\varepsilon_1 = \varepsilon_o + \varepsilon_b(1 - \cos\theta) \tag{7.12}$$

$$\varepsilon_2 = \varepsilon_o - \varepsilon_b(1 + \cos\theta) \tag{7.13}$$

Over the whole range of θ (the case of interlaboratory data) the overall scatter band in fatigue life using a single extensometer would be

$$\Delta N_f = N_f(\varepsilon_o - 2\varepsilon_b) - N_f(\varepsilon_o + 2\varepsilon_b) \tag{7.14}$$

The overall mean is again equal to N_{fo}.

7.5 PREDICTION OF LIFETIME SCATTER BANDS DUE TO LOAD MISALIGNMENTS

Two methods for quantitative prediction of scatter have been developed for this chapter. For simplicity we will refer to them as the differential method and the incremental method.

7.5.1 The differential method

LCF is usually described by an equation relating the number of cycles to failure N_f to the plastic strain range $\Delta\varepsilon_p$ in the form

$$\Delta\varepsilon_p N_f^\alpha = C \tag{7.15}$$

where α and C are constants. Since the total strain range $\Delta\varepsilon = \Delta\varepsilon_p + \Delta\varepsilon_e$, where $\Delta\varepsilon_e = \Delta\sigma / E$ is the elastic strain range and $\Delta\sigma$ is the stress range, equation 7.15 may be rewritten as

$$N_f = C^v \left(\Delta\varepsilon - \frac{\Delta\sigma}{E} \right)^{-v} \tag{7.16}$$

where

$$v = 1/\alpha$$

or as

$$N_f = C^v \left(1 - \frac{E_c}{E} \right)^{-v} (\Delta\varepsilon)^{-v} \tag{7.17}$$

where

$$E_c = \Delta\sigma / \Delta\varepsilon \tag{7.18}$$

is a cyclic modulus that can be measured from the cyclic stress–strain curve. Differentiation of equation 7.17 gives the error in N_f due to an error in $\Delta\varepsilon$ as

$$\frac{dN_f}{N_f} = v \left[\frac{\Delta\varepsilon}{(E - E_c)} \frac{dE_c}{d\Delta\varepsilon} - 1 \right] \frac{d\Delta\varepsilon}{\Delta\varepsilon} \tag{7.19}$$

where $d\Delta\varepsilon / \Delta\varepsilon$ is the component of the error in the total strain range due to bending. The term in the square brackets in equation 7.19 has a minimum value of -1 at high value of $\Delta\varepsilon$ and goes to ∞ as $\Delta\varepsilon \to \Delta\varepsilon_e$.

Thus, errors in $\Delta\varepsilon$ due to bending will always lead to greater errors in N_f as $\Delta\varepsilon$ decreases. It must be pointed out that both E and E_c should ideally be obtained from tests with no bending.

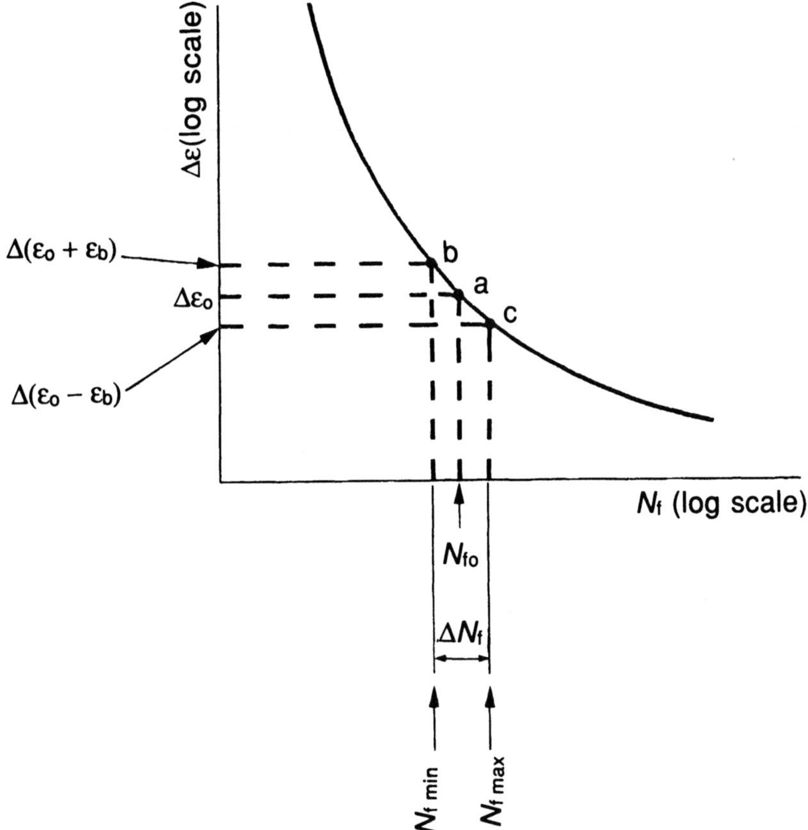

Fig. 7.5 Relationship between the total strain range $\Delta\varepsilon$ and number of cycles to failure N_f, depicting some of the parameters used with the incremental method of analysis.

7.5.2 The incremental method

It is desirable from the point of view of engineering design to use fatigue curves in terms of the controlled total strain range $\Delta\varepsilon$. The relationship between $\log \Delta\varepsilon$ and N_f is, however, generally non-linear.

For the purpose of the present analysis, we assume the following relationship:

$$\Delta\varepsilon\, N_f^{\beta} = K \tag{7.20}$$

β is the slope of the tangent line to the fatigue curve at point a in Fig. 7.5. Although β and K are both functions of $\Delta\varepsilon$, it is reasonable to approximate the short curve between the two points b and c in Fig. 7.5 by a straight line. Thus for the total strain ranges between the two limits $\Delta(\varepsilon_o + \varepsilon_b)$ and $\Delta(\varepsilon_o - \varepsilon_b)$, the values of β and K are considered constants.

In order to obtain expressions for the limits of interlaboratory data scatter, the following parameters will be defined. $\Delta\varepsilon_o$ and N_{fo} are the total strain range and number of cycles to failure in the case of no bending. f is an extensometer factor: $f = 1$ for a dual extensometer and $f = 2$ for a single extensometer. The bending parameter ψ_B is the ratio between the highest total strain range anywhere on the surface of the testpiece and the controlled total strain range:

$$\psi_B = [\Delta(\varepsilon + \varepsilon_b)/\Delta\varepsilon] - 1 \tag{7.21}$$

For reversed bending ψ_B is simply equal to $(\varepsilon_b/\varepsilon)$, but for non-reversed bending ψ_B is zero. N_{fmax} and N_{fmin} are the extreme values of the scatter band, that is $\Delta N_f = N_{fmax} - N_{fmin}$. Finally, F is a scatter factor defined as the ratio N_{fmax}/N_{fmin}.

Using the case 2 model (section 7.4.2) and equations 7.20 and 7.21, the following relationships for interlaboratory data can be derived:

$$N_{fmax} = N_{fo}\,\omega_{max} \tag{7.22}$$

$$N_{fmin} = N_{fo}\,\omega_{min} \tag{7.23}$$

$$\Delta N_f = N_{fo}\,(\omega_{max} - \omega_{min}) \tag{7.24}$$

$$F = \omega_{max}/\omega_{min} \tag{7.25}$$

where

$$\omega_{max} = \left[\frac{1}{1 - f\psi_B}\right]^{1/\beta} \tag{7.26}$$

$$\omega_{min} = \left[\frac{1}{1 + f\psi_B}\right]^{1/\beta} \tag{7.27}$$

It is clear from equations 7.26 and 7.27 that the error in fatigue life is dependent on the slope β. The smaller the slope, the greater the error (or scatter) in fatigue life. Substituting β with α and α' (for the plastic and elastic relationships, respectively) will give the minimum and maximum scatter limits for any material and test system combination.

Figure 7.6 shows schematically the predictions for the scatter band during a test programme. Two cases are considered in this figure: (a) when $\Delta\varepsilon_b$ is constant; (b) when $\Delta\varepsilon_b/\Delta\varepsilon$ is constant. The magnitude of scatter at lower strain ranges is predicted to be greater in (a) than in (b).

The parameter ψ_B should be evaluated by bending measurements at loads corresponding to the peak values during the fatigue cycle. The worst case is predicted to be that for non-reversed bending. In such a case, $\psi_B = \varepsilon_b/\varepsilon$.

The slope β should, ideally, be obtained from tests with no bending. However, in practice, any set of data with a reasonable level of bending

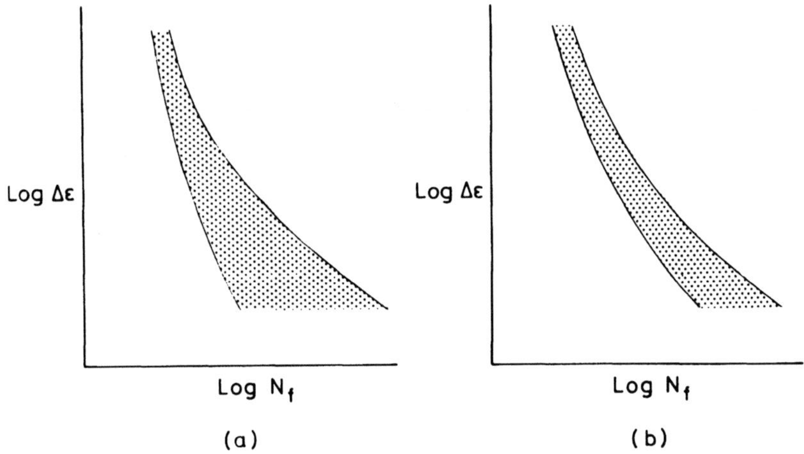

Fig. 7.6 Schematic representation of predicted errors in the number of cycles to failure in tests when (a) $\Delta\varepsilon_b$ is constant (b) $\Delta\varepsilon_b / \Delta\varepsilon$ is constant.

is believed to be adequate for its determination. Generally, small variations in the value of β should have negligible effects on the accuracy of predictions considering the many interactive factors involved.

7.6 DISCUSSION

There are many ways to conduct a low-cycle fatigue test, but a general consensus is emerging that the state-of-the-art technique for elevated temperature LCF uses a parallel-sided testpiece with a single extensometer equipped with pointed or chisel-edged probes to delineate the gauge length accurately. In view of the ideas expressed here, this appears to be the best procedure proposed to date, but a better (but also more expensive) method would be to use a pair of these extensometers to minimize interlaboratory scatter due to bending. The only comprehensive data set available to test the results of the present model-based analysis of bending is the BCR exercise analysed by Thomas and Varma (1992). The most tested material in that programme was 316L stainless steel at 550 °C. Although 21 laboratories in Europe and Japan were involved in testing this particular material, only seven could be considered to have used state-of-the-art techniques, and the latter group reported values of the percentage bending B (equation 7.2) less than or equal to 5%. However, as discussed in Kandil and Dyson (1991), comparative values of B (and also the more recent precision of alignment parameter P) are of little value unless load conditions relevant to the

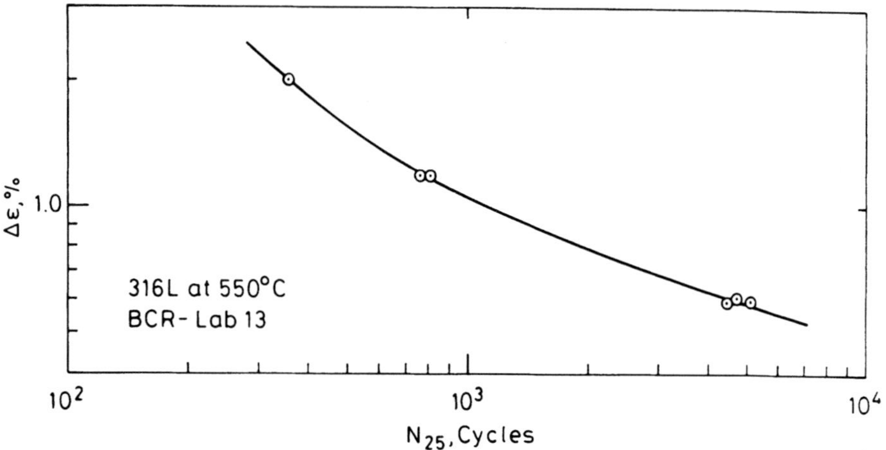

Fig. 7.7 Life as a function of the total strain range for 316L stainless steel at 550 °C. Data are for laboratory 13 (Thomas and Varma, 1992).

fatigue test are used for its determination. All seven laboratories tested three testpieces at each of the three levels of strain range, $\Delta\varepsilon = 2.0\%$, 1.2% and 0.6%. The repeatability within a single laboratory was good and is illustrated in Fig. 7.7, where the laboratory chosen is that reporting the lowest bending, $B \sim 3\%$. The number of cycles to failure is represented by N_{25}, defined as the number of cycles required to achieve a 25% decrease in force from the 'stabilized' level of the tensile force. Since the angle θ (Fig. 7.1) was not measured in the BCR exercise, the curve which represents the mean of the data in Fig. 7.7 does not necessarily represent the behaviour under no bending.

Data from all seven laboratories are plotted in Fig. 7.8 as dotted bars, where it can be seen that maximum lifetime at a given strain range can be about a factor of four larger than the minimum. Predictions of data scatter due to the bending component in terms of the scatter factor F were made using equation 7.25, assuming material model case 2 (section 7.4.2) and the following input parameters

- $\psi_B = \varepsilon_b / \varepsilon = 0.1$. This criterion is in agreement with the standard ASTM E606 : 1980 only at the lowest strain range, $\Delta\varepsilon = 0.6\%$. At higher strain ranges, using this criterion will give more conservative predictions than the ASTM one (Kandil and Dyson, 1993a).
- $f = 2$. This represents a single extensometer;
- β is derived using equation 7.20 from the slope of the curve in Fig. 7.7.

The predicted values of the scatter factor F ranged from 1.66 at $\Delta\varepsilon = 2.0\%$ to 3.86 at $\Delta\varepsilon = 0.6\%$. These are shown for comparison purposes as square brackets in Fig. 7.8, and indicate the large proportion of interlaboratory

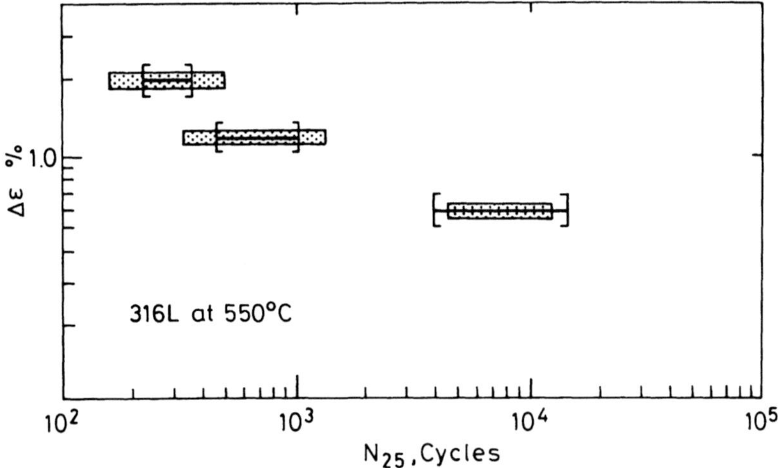

Fig. 7.8 The experimental and predicted interlaboratory scatter bands in fatigue life for 316L stainless steel at 550 °C. Experimental data are shown as dotted bars; the predictions are the square brackets.

scatter which can be attributed to bending caused by load misalignment. At the lowest strain range, the predictions show that the experimental scatter can be accounted for exclusively by bending.

A recent analysis (Kandil and Dyson, 1993b) of effects due to uncertainties in the measurement of strain range, assuming grade C extensometers conforming to BS 3846 : 1985, shows that lifetime scatter due to this effect was within ± 4%. Likewise the analysis for temperature effects assuming an error of ± 2 °C, and utilizing the experimental data due to Conway, Stentz and Berling (1975), has shown that the predicted error in life due to uncertainty in temperature was only about ± 1%. These additional calculations strongly suggest that for this material bending due to load misalignment is the main cause of data scatter.

7.7 CONCLUSIONS

Sources of misalignment in LCF test systems have been identified and their influence analysed. The three most critical factors influencing the quantitative analysis are (a) the mechanism of bending, (b) the mechanism of fatigue crack initiation and early growth and (c) the type of extensometer.

Modelling studies demonstrate that there are sharp distinctions between the predicted data scatter produced with single and dual types of extensometer operating under the same test conditions. Dual extensometers show much less scatter due to bending than do the single ones and are therefore preferable.

Finally, mathematical expressions have been developed for predicting error scatter bands due to bending. Comparison with an internationally acquired database on 316L stainless steel has shown that bending was the primary cause of the observed interlaboratory scatter.

ACKNOWLEDGEMENT

The research reported in this chapter was carried out as part of the Improvement and Development of Measurement Methods for High Temperature Materials Properties, a programme of underpinning research financed by the United Kingdom Department of Trade and Industry.

APPENDIX 7.A   NOTATION

α, α', C	constants in the Coffin-Manson equations
β	tangent to $\log \Delta\varepsilon$ and $\log N_f$ curve
ν	$1/\alpha$
ε_b	maximum bending strain
ε_x	axial strain at point X
ε_{bx}	bending strain component at point X
ε_o	axial strain at centre of testpiece
$\varepsilon_1, \varepsilon_2$	absolute maximum and minimum strains
$\Delta\varepsilon$	total strain range (controlled parameter)
$\Delta\varepsilon_o$	total strain range under no bending
$\Delta\varepsilon_p$	plastic strain range
$\Delta\varepsilon_e$	elastic strain range
$\Delta\sigma$	stress range
σ_b	maximum bending stress
θ	angle in the circular cross-section between the maximum bending plane and the extensometer plane
γ	angular offset
e	lateral offset
ℓ, r	testpiece length and radius
Φ	angle of rotation of cross-section due to bending
Ψ_B	bending parameter
B	percentage bending
E	Young's modulus
E_c	cyclic modulus $= \Delta\sigma / \Delta\varepsilon$
N_f	number of cycles to failure
N_{fo}	number of cycles to failure under no bending
N_{fmax}, N_{fmin}	limits of lifetime scatter band
ΔN_f	range of scatter band $= N_{fmax} - N_{fmin}$

f	extensometer factor: 1 for dual, 2 for single extensometer
F	scatter factor $= N_{fmax} / N_{fmin}$
K	parameter in equation 7.20

REFERENCES

ASTM E606 : 1980 Standard Recommended Practice for Constant-Amplitude Low-Cycle Fatigue Testing, in *1991 Annual Book of ASTM Standards*, Part 3.01, 609–21, American Society for Testing and Materials, Philadelphia.

Bressers, J. (1982) Axiality of Loading, in *Measurement of High Temperature Mechanical Properties of Materials* (eds M. S. Loveday, M. F. Day and B. F. Dyson), HMSO, London, Chapter 17, pp. 278–95.

BS 3846 : 1970 (1985) Methods for Calibration and Grading of Extensometers for Testing of Metals.

Christ, B. W. (1973) Effects of misalignment on the premacroyield region of the uniaxial stress strain curve. *Metallurgical Transactions*, **4**, 1961–5.

Conway, J. B., Stentz, H. and Berling, J. T. (1975) *Fatigue, Tensile, and Relaxation Behaviour of Stainless Steels*, Mar-Test, Inc., Cincinnati, Ohio, report TID-26135.

Kandil, F. A. and Dyson, B. F. (1991) *Systematic Study of Effects of Load Misalignment in Uniaxial Low Cycle Fatigue Testing at High Temperatures*, NPL report DMM(A)27, National Physical Laboratory, Teddington, Middlesex.

Kandil, F. A. and Dyson, B. F. (1993a) The influence of load misalignment during low-cycle fatigue testing. I: Modelling. II: Applications. *Fatigue and Fracture of Engineering Materials and Structures*, **16**(5), 509–37.

Kandil, F. A. and Dyson, B. F. (1993b) Prediction of uncertainties in low cycle fatigue data, in *Proceedings of the 5th International Conference on Creep and Fracture of Engineering Materials and Structures* (eds B. Wilshire and R. W. Evans), 28 March to 2 April 1993, University College Swansea, Institute of Materials, London, pp. 517–25.

Thomas, G. B. and Varma, R. K. (1992) Review of the BCR/VAMAS low cycle fatigue intercomparison programme, in *Harmonisation of Testing Practice for High Temperature Materials* (eds M. S. Loveday and T. B. Gibbons), Elsevier Applied Science, London, pp. 155–85.

CHAPTER 8

Aspects of modulus measurement

G. D. Dean, M. S. Loveday, P. M. Cooper,
B. E. Read and B. Roebuck, with appendix by R. Morrell

8.1 INTRODUCTION

A value for a modulus of a material may be defined by the ratio of any component of a stress field applied to the material to any component of the resulting strain field. For an isotropic elastic material, two modulus values are sufficient to characterize its linear elastic behaviour. A knowledge of any two moduli therefore means that all components of the strain field may be calculated for any applied stress field. For anisotropic materials, such as crystalline solids containing some alignment of the crystal regions or oriented fibre composites, the elastic behaviour is not isotropic and more than two moduli are needed to relate arbitrary stress and strain components (Read and Dean, 1978, Chapter 1).

The properties most commonly used to characterize isotropic materials are the Young's modulus and the shear modulus or Poisson's ratio (Adams, 1989). Young's modulus is conventionally determined using tensile or flexural modes of deformation. In the flexure test, the stress varies through the thickness of the testpiece. The test may therefore be used for the measurement of modulus at small strains only where the elastic behaviour is linear. This mode of deformation is commonly used with stiff, brittle materials such as ceramics, hard metals and unidirectional fibre reinforced composites because accurate stress and strain values are determined more conveniently than by loading in tension.

With materials such as tough polycrystalline metals and plastics, that are significantly non-linear up to failure, it is necessary to use a uniaxial tensile test to generate data for design. Such data are usually represented by a tensile stress–strain curve from which various strain-dependent moduli can be defined as shown in Fig. 8.1.

The shear modulus of stiff materials is commonly determined by the torsion of rectangular bars, rods or tubes. Of these, the thin walled tube

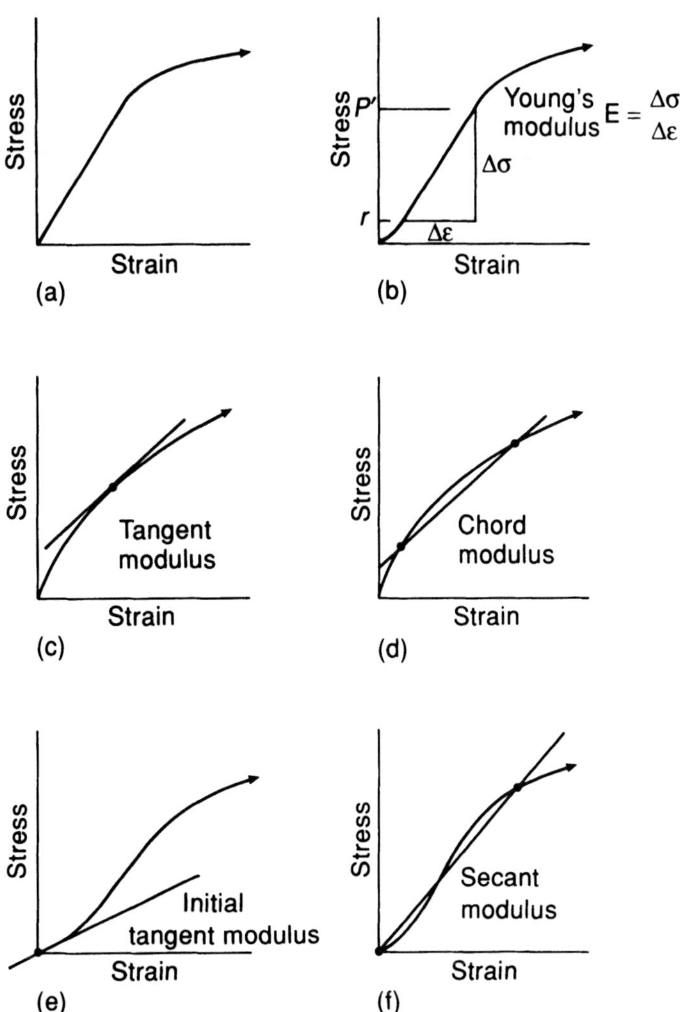

Fig. 8.1 Idealized stress–strain curve and illustration of various terms used to specify a modulus value.

is the only testpiece geometry for which the shear stress is essentially uniform through the testpiece cross-section and is therefore most conveniently used for the determination of stress–strain curves describing non-linear behaviour to failure. Alternative testpiece geometries defined by the Iosipescu (Iosipescu, 1967; Broughton, Kumosa and Hull, 1990) and the Arcan (Weissberg and Arcan, 1988) test methods have also been proposed for determining shear modulus. For more flexible materials such as soft plastics and rubbers, suitable testpieces can be loaded under simple shear and the shear properties are derived following the application of a small correction for bending.

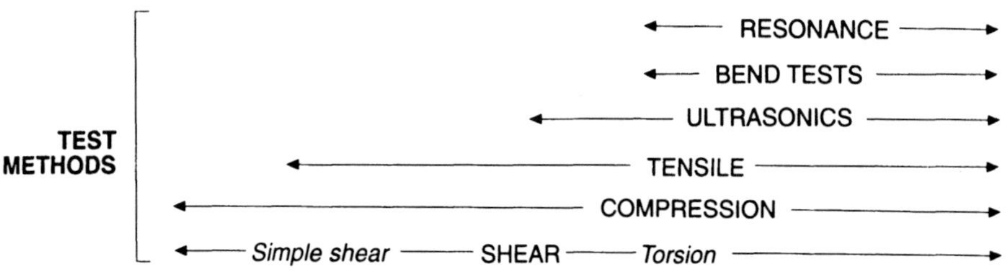

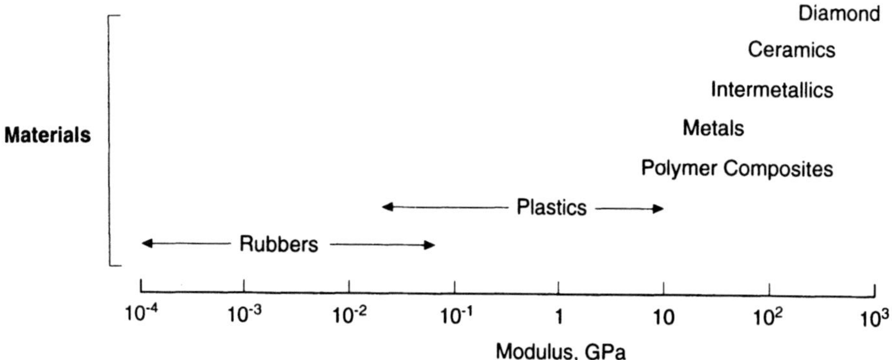

Fig. 8.2 Schematic diagram indicating the range of properties of engineering materials and the various techniques for measuring modulus.

A simplified diagram showing modulus ranges for different classes of material and the regions over which various measurement methods operate is shown in Fig. 8.2.

Certain materials, such as plastics and rubbers at ambient temperatures and engineering metals at elevated temperatures, have moduli that are time and rate dependent. This means that the measured value of a modulus depends on the rate at which the stress is applied or the time after the application of a load or deformation at which measurements of stress and strain are made. This time-dependent, or viscoelastic, behaviour is characterized by creep or stress relaxation experiments. In a creep test, a constant load is applied to a testpiece and the time-varying deformation is measured. In a stress relaxation test, it is the deformation that is kept constant and the time-dependent stress is measured.

The viscoelastic behaviour of polymeric materials is commonly determined by dynamic mechanical tests. Here a sinusoidally varying load or deformation is applied to the testpiece. There is then a phase difference between stress and strain cycles, implying that the modulus is a complex quantity (in the mathematical sense). The real and imaginary parts of the complex modulus are derived from measurements of the amplitudes of the stress and strain cycles and the phase angle between them. The real part of the complex modulus of a viscoelastic material is analogous to the modulus of an elastic material, whereas the imaginary part is related to the material's damping capacity which is a measure of its ability to dissipate vibrational energy as heat. The damping capacity is usually expressed as a dimensionless loss factor which is the ratio of the imaginary to the real parts of the complex modulus (Gittus, 1975).

In this chapter, selected methods are described for measuring modulus values of a variety of materials to a high and quantifiable accuracy. Reference is made to the status of existing standards for these methods and to current developments in the preparation of new ones. A brief summary of some of the existing standards is shown in Table 8.1.

Table 8.1 Summary of standards for modulus measurement

Area	National standard	Technique	International standard	Technique
General	ASTM Com. E28	(D) FR, U		
	ASTM E494	(D) U		
Metals	ASTM E111	(S) T(C)		
Ceramics				
Cermets,	ASTM C623	(D) R	ENV 843	(S) B
graphite	ASTM C747	(D) R		(D) U
	ASTM C848	(D) R		FR TR LR NR
Polymers	ASTM D3039	(S) T	Pr EN 2561	(S) T
and			Pr EN 2562	(S) B
reinforced	BS 2782 : Part 3	(S) T	ISO 178	(S) B
plastics			ISO 527 : Part 1	(S) T
			Part 2	
			ISO 6721 : Part 1	(D) General
			Part 2	(D) NR
			Part 3	(D) FR

B	static flexure (three- and four-point bend test)
C	compression
D	dynamic
FR	flexural resonance
LR	longitudinal resonance
NR	natural resonance
S	static
T	tensile
TP	torsion pendulum
TR	torsional resonance
U	ultrasonics

In conventional tensile testing, a tensile force is applied to a testpiece and the resulting strain or deformation is measured. The precision that can be achieved in the derived value for Young's modulus is critically dependent upon the accuracy with which the strain can be measured. In section 8.2, the factors influencing this accuracy in the testing of metallic materials are discussed in detail.

Certain materials display only a very limited range where stress is linearly related to strain and this can lead to significant uncertainties in derived modulus values, which are additional to those discussed in section 8.2. Metal matrix composites are a particularly important example of this. In section 8.3, a procedure is described which can reduce these uncertainties provided there are digitized records of values of stress and strain.

Sections 8.2 and 8.3 demonstrate that it is difficult to obtain an accuracy as high as ± 1% in modulus by measurements of stress and strain. In sharp contrast, such levels of accuracy are readily achievable in many materials using acoustic resonance and ultrasonic wave techniques (Read and Dean, 1978; Wolfenden, 1990). The acoustic resonance method is particularly attractive for materials having a very high modulus and a low failure strain and which therefore pose problems with load introduction in testing machines. Such materials include aligned carbon fibre reinforced plastics and ceramics. Apparatus for making resonance measurements is described in section 8.4 and emphasis is given to sources of error and the analysis of results for the flexural mode. A summary of the equations and corrections required to derive properties using other modes of deformation is given in Appendix 8.A.

Ultrasonic wave propagation methods are also highly suited to the measurement of properties of stiff or brittle materials that are difficult to test by conventional methods. They also offer an accurate and convenient method for measuring all the elastic constants of materials whose properties are anisotropic. A method based on the measurement of the velocity of ultrasonic pulses travelling along known directions in a test-piece is described in section 8.5.

Probably the greatest limitation of the ultrasonic and resonance methods is that they operate at very small strains (in the region of 10^{-6} or less). Nevertheless they make valuable contributions to projects aimed at evaluating the precision of conventional techniques for determining modulus and at establishing the properties of reference materials. There are also severe experimental problems in making measurements at temperatures much greater than 100 °C, which limits their usefulness since many of today's high-performance materials are operative at > 1000 °C and tomorrow's will require up to 2000 °C.

A limited comparison of data obtained using the various techniques is given in section 8.6.

8.2 STATIC DETERMINATION OF MODULUS

8.2.1 Standards for tensile testing of metals

Young's modulus has traditionally been determined from the linear elastic region of a uniaxial tensile stress–strain curve usually measured at a constant strain rate, or constant crosshead speed, in a universal tensile testing machine. An idealized curve is shown in Fig. 8.1(a), and the modulus is determined from the slope of the curve using the relationship

$$E = \frac{\text{stress}}{\text{strain}} = \frac{Pl}{Ae} \tag{8.1}$$

where E is Young's modulus, P is the load, l is the original gauge length, A is the cross-sectional area and e is the extension.

Owing to experimental difficulties, e.g. elastic follow-up of the loading train in load control mode, on testpiece bending it is often found that the initial portion of the curve is non-linear and thus the Young's modulus is determined between a pre-set stress value r and a stress p' which is below the proportional limit (Fig. 8.1(b)). Two other methods for determining tensile modulus from stress–strain curves are also defined in ASTM E111 : 1992, namely tangent modulus (Fig. 8.1(c)) and chord modulus (Fig. 8.1(d)), both of which are suitable for materials exhibiting non-linear elastic behaviour. Because of the difficulty in defining the origin of the stress–strain curve, the use of either the initial tangent modulus (i.e. the slope of the stress–strain curve at the origin, Fig. 8.1(e)) or the secant modulus (Fig. 8.1(f)) is not recommended by ASTM E111 : 1992. Although it should be noted that some standards relating to polymeric materials refer to measurement of a secant modulus, in reality what is actually specified is a chord modulus between designated load values (EN 2561) or strain values (BS 2782).

Young's modulus was traditionally determined as a best fit 'by eye' from a straight line drawn on the stress–strain curve, which may easily give rise to uncertainties in the region of ± 5% (see section 8.3). However, today's testing machines invariably determine the slope of the line by computer, usually using a least-mean-squares fit of the digitized data points. If there is no scatter in the data points and the material truly exhibits linear behaviour, then clearly the chord modulus and Young's modulus are identical.

Although design engineers traditionally use modulus values in the concept design of most components or structures, there has not been a separate standard for the determination of Young's modulus for metallic materials, other than ASTM E111. Similarly there are no standards which rigorously define the proportional limit. In BS 18 : 1987 'Methods for Tensile Testing of Metals' (at room temperature), Young's modulus of elasticity is defined as the ratio of stress to strain during the elastic behaviour of the testpiece. However, this standard has now been superseded by EN 10002 : Part 1 (based on ISO 6892) which, although referring to modulus of elasticity in the text, gives no formal definition of the parameters, nor specifies a method for determining its magnitude. A separate standard for the determination of modulus was recently considered by ISO TC 164; for the process to proceed under the new ISO rules required the participation of at least five member nations: this level of support was not forthcoming.

The European Standard on uniaxial high-temperature tensile testing, EN 10002 : Part 5, does not contain any reference to the measurement of Young's modulus, but simply specifies the use of a class 1 extensometer for strain measurement. This implies a potential accuracy of ± 1% for the measurement of Young's modulus when testing to this standard, which is approximately one-half the accuracy achievable when testing to the previously used BS 3688. The latter standard specified the use of a grade B extensometer (equivalent to a class 0.5 device calibrated accordingly to ISO 9513 'Metallic Materials: Verification of Extensometers Used in Uni-axial Testing'). This lowering of standards and the detailed differences between the various extensometer calibration standards has been discussed in detail elsewhere (Loveday, 1991; see also Appendix 5. C. in this book), and the consequences for Young's modulus determination will be considered below.

8.2.2 Statement of uncertainty of measurement

The directives governing the preparation of ISO and EN Standards on testing specify that reported results should include a statement of the precision of the test method (Roche and Loveday, 1992). Similarly, accreditation bodies also demand an estimate of the uncertainty of measurement (e.g. NAMAS M10, 1989, Clause 8.2). For tensile testing at room temperature an estimate of the uncertainty of measurement may be obtained from the tolerances specified in the various standards for the parameters in equation 8.1, as discussed elsewhere (Loveday, 1992). In the so-called 'error budget' shown in Table 8.2, the maximum calculated errors are presented, although in practice with careful calibration the actual measurements could be more accurate.

Table 8.2 Nominal error budget for estimation of uncertainty of measurement of Young's modulus during tensile testing

	Parameters	Tolerance specified in standard	Standards	
			Testing	Calibration
a	Load	± 1%	EN 10002/1	EN 10002/2
b	Cross-sectional area	± 1%	EN 10002/1	–
c	Extensometer (displacement)	± 1% (or ± 3 μm) whichever is greater	EN 10002/1	ISO 9513 (EN 10002/4)
d	Gauge length	± 1%	EN 10002/1	ISO 9513 (EN 10002/4)
	Total uncertainty $\pm \Delta = \sqrt{(a^2 + b^2 + c^2 + d^2)}$	± 2%*		

* Excludes the influence of the ± 3 μm threshold on displacement measurement, and the influence of bending and strain rate effects: see text.

The total uncertainty value of ± 2% may generally be regarded as an acceptable value of appropriate accuracy for measurement of modulus. However, in reality at the low displacements over which modulus is generally determined, the ± 3 µm threshold for a class 1 extensometer gives an error which is greater than the ± 1% value as indicated in the classification envelope shown in Fig. 5.C.5 in this volume. The consequence of this lower threshold may be more readily appreciated from the calculated values of accuracy shown in Fig. 8.3, where it can be seen that errors in displacement measurement could be in excess of ± 20% for short-gauge-length testpieces over the strain range used for the determination of modulus.

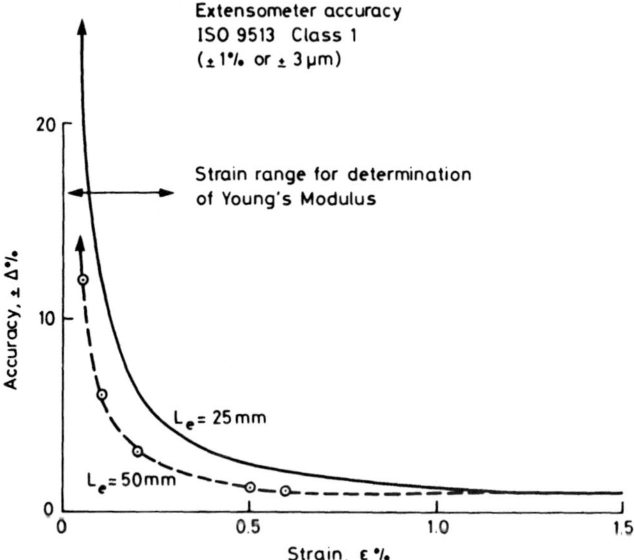

Fig. 8.3 Errors likely to be encountered in determining Young's modulus as a function of strain using a class 1 extensometer classified in accordance with ISO 9513 or BS EN 10002/4: error due to the 3 µm lower limit of bias (accuracy).

Errors in modulus measurement due to two other parameters have also not been taken into account in the error budget in Table 8.2. Firstly, the tolerance levels may not yet be rigorously specified in the testing standards, e.g. the influence of off-axis loading (bending); and, secondly, the expected accuracy is material dependent, e.g. through the influence of strain rate, or the non-homogeneity of deformation. In general in metallic materials the errors due to the strain rate are not significant since modulus does not vary with strain rate over the range specified in the tensile testing standards. However, some materials are viscoelastic or exhibit time-dependent inelastic deformation (creep) even at room temperature, e.g. 316 stainless steel, and hence the influence of strain rate effects may need to be taken into account.

The effects of off-axis loading (bending) are dependent upon factors such as testpiece geometry, the configuration of the extensometry, e.g. single-sided or dual averaging devices, and testpiece gripping arrangements. The recent analysis of experimental data in low-cycle fatigue (Kandil and Dyson, Chapter 7 in this volume) indicates that bending is a likely cause of the observed scatter in the results, and it is possible that bending may also be an explanation for scatter in Young's modulus data measured during tensile testing (see below). However, as yet a rigorous analysis has not been completed to validate this hypothesis. At present the standards for tensile testing (EN 10002/1 and EN 10001/5) merely state that the force is to be applied along the axis so as to produce minimum bending or torsion in the testpiece gauge length. Similarly, it is specified that the testpiece should be gripped so as to ensure that the force is applied as axially as possible. Thus acceptable limits of bending are not quantified, although reference is made to methods for verifying the alignment in ASTM E1012 'Practice for Verification of Specimen Alignment under Tensile Loading'. The problems of alignment in tensile testing have also been discussed by Christ and Swanson (1976) and by Bressers (1982).

Bearing in mind that the extensometer gauge length L_e is frequently less than the parallel gauge length of the testpiece, and indeed since extensometer gauge lengths as short as 10 mm may be permitted (EN 10002/5), it should be appreciated that the measurement uncertainty may well be in excess of $\sim \pm 20\%$ unless long-gauge-length testpieces with $L_e \geqslant 50$ mm and high-accuracy extensometers are specified, e.g. class 0.2 devices. However, if such extensometers are to be used, it will be necessary to employ sophisticated calibrators to provide traceability to the national measurement system for accreditation purposes. Such a calibrator, incorporating an NPL laser interferometer (Downs and Raine, 1979) is shown in Fig. 8.4; further details of the calibrator are given elsewhere (Day and Harrison, 1982; Loveday, Appendix 5.C in this volume).

8.2.3 Modulus measurements during tensile testing
at room temperature

Results of Young's modulus measurements on samples of Nimonic 75, measured using either a single-sided strain gauged clip gauge or a dual

Fig. 8.4 Calibration of high-temperature tensile extensometers using a calibrator incorporating an NPL laser interferometer.

averaging pair of side-entry high-temperature extensometers, are shown in Fig. 8.5, obtained using the NAMAS accredited tensile testing facility shown in Fig. 8.6. Statistical analysis gave a mean value of Young's modulus of $\bar{E} = 212.1 \pm 13.1$ GPa with a 95% confidence limit. This data set was obtained using testpieces with a circular cross-section with a gauge diameter of 8 mm and a parallel gauge length of 40 mm, with threaded grip ends. The measurements were carried out over a strain rate range from 2×10^{-5} up to 1 per minute, and over that range the modulus values were independent of strain rate. Similarly, the results were independent

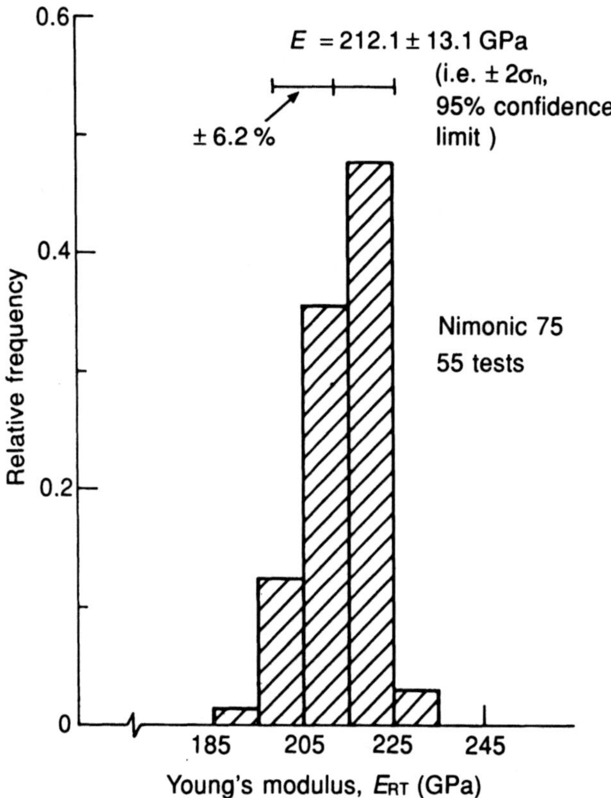

Fig. 8.5 Distribution of room temperature modulus values of Nimonic 75 determined during tensile testing.

of the type of extensometer employed for the measurements, both types being calibrated in accordance with BS 3846. The material, Nimonic 75, is a solid solution nickel-chromium alloy and was part of the bar stock certified for use as a creep reference material (Gould and Loveday, 1990), similar to that being certified for use as a room temperature tensile testing reference material (Loveday, 1992).

The scatter in the data set shown in Fig. 8.5 is not untypical of modulus values measured using tensile testing machines, but is perhaps larger than generally acknowledged. The data indicate the difficulty of obtaining precise values of modulus using a tensile testing machine. However, it should be appreciated that the data all lie within ± 13%, which is within the limits calculated for the uncertainty of measurement if allowance is made for the extensometer threshold limit at low displacements shown in Fig. 8.3 and bearing in mind that the influence of bending is unquantified. In addition, the scatter band incorporates scatter attributable to inherent material variability which is as yet unquantified, but is likely to be less than ± 2%.

Fig. 8.6 NAMAS accredited tensile testing facility used at NPL for the static determination of Young's modulus.

8.2.4 Modulus measurements prior to tensile creep testing

Creep testing carried out in accordance with BS 3500 should comply with the recommendation that the extension measured on either side of the testpiece, typically measured using extensometers of the type shown in Fig. 8.7, should not differ by more than 10%, and the mean extension observed should be consistent with the known value of Young's modulus of the material. The side-to-side differences in the transducer reading usually indicate that the testpiece is exhibiting bending. Unfortunately, the magnitude of the bending cannot be determined from measurements in a single plane if the testpiece has a circular cross-section, and at least three or more radially mounted strain measuring devices are necessary to determine the bending (Christ and Swanson, 1976; Bressers, 1982; ASTM E1012). Thus the practice of adjusting the testpiece until the difference in extension measurements on opposite sides of the testpiece is

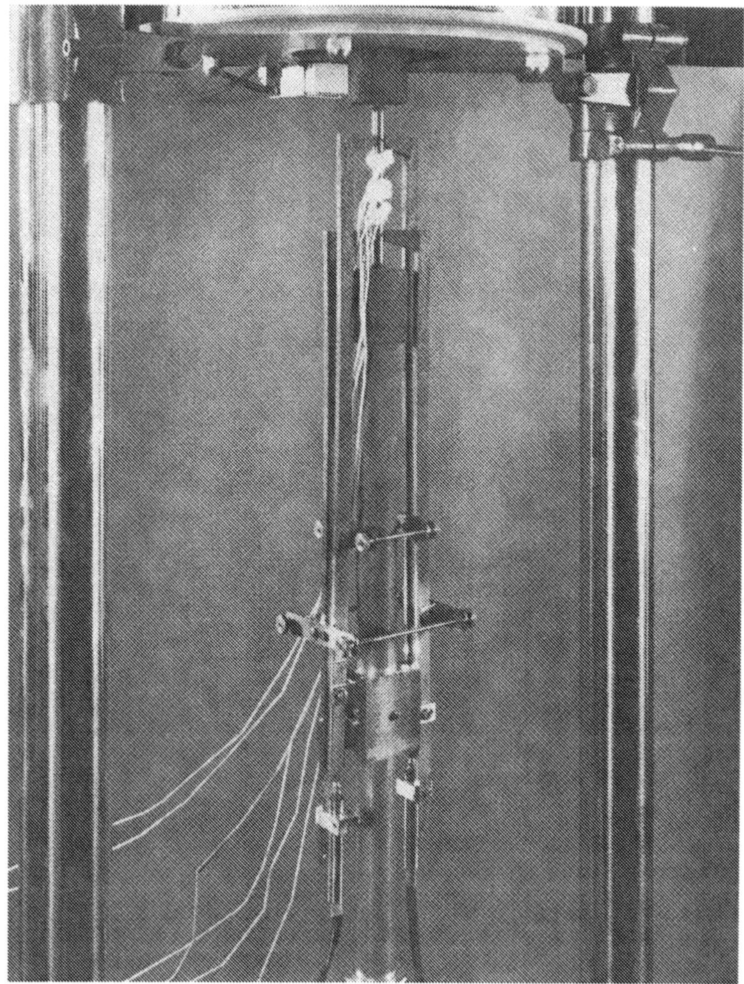

Fig. 8.7 Creep testpiece assembled with side-to-side averaging extensometer.

less than 10% is not a *sufficient* condition to ensure that the bending is below an acceptable level, but may merely indicate that the plane of maximum bending is orthogonal to the measurement plane of the extensometers (Loveday, 1986).

A typical distribution of Young's modulus values determined at room temperature prior to creep testing for Nimonic 80A is shown in Fig. 8.8. In this particular example the data comprised measurements from 132 testpieces with 50 mm gauge lengths, measurements being undertaken

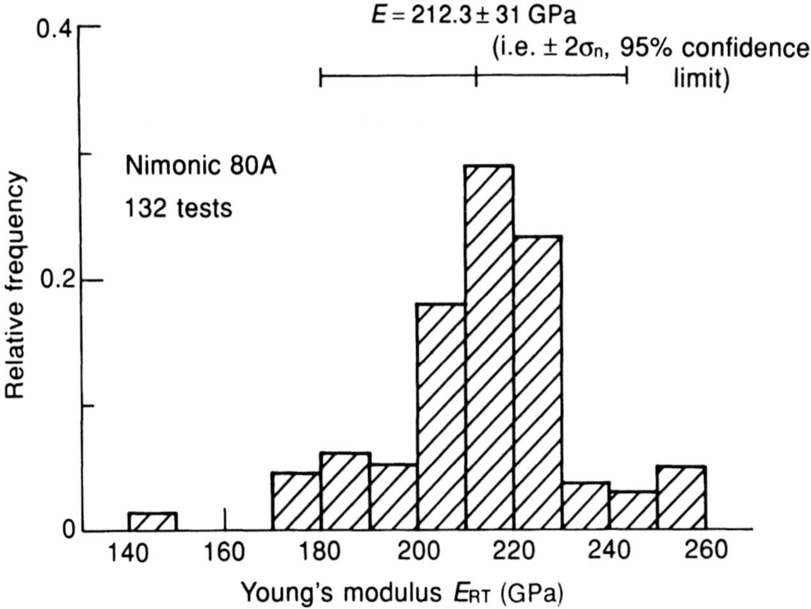

Fig. 8.8 Distribution of room temperature modulus values for nimonic 80A determined prior to creep testing.

using several machines and operators. It can be seen that values in the extreme cases ranged from 140 to 260 GPa, i.e. approximately ± 25% of the mean value. Statistical analysis of the data gave a mean value of Young's modulus of $\bar{E} = 212 \pm 31$ GPa with a 95% confidence limit (i.e. $\bar{E} \pm 2S$, where S is the standard deviation). Again, this data set exhibits a wider scatter than is generally acknowledged, but nevertheless is typical of modulus values measured prior to creep testing. The data from which the modulus values were obtained usually comprised readings recorded at five or six discrete loads, and in general the data points gave a well-defined straight line. It should also be noted that often the applied loads correspond to perhaps only 15% of the yield stress, and total displacements measured during modulus determinations are relatively small, typically ranging from 0.006 to 0.02 mm: thus high-resolution extensometry is clearly necessary if the measurement errors are to be at an acceptably low level.

The creep extensometers were classified as grade A devices (nominal accuracy $\sim \pm 0.2\%$): nevertheless the scatter in the measured modulus values measured prior to creep testing (Fig. 8.8) is considerably greater than that obtained during tensile testing (Fig. 8.5). This apparent anomaly may partially be due to differences between repeatability, i.e. same machine, same operator etc., and reproducibility, i.e. different machines, different operators etc. In addition, it may also be a consequence of the lower loads applied to the testpieces when determining modulus prior to creep testing, where typically the maximum load is only ~ 15% of the proof stress, whereas in the tensile test the upper load point used for the determination of modulus is typically ~ 50% of the proof stress. Thus the load train is not pulled into alignment in the creep test as much as in the tensile test, and hence the effects of bending are likely to be greater. It should be noted that despite the wide scatter in modulus, the subsequent creep properties were well within the accepted accuracy for creep data, clearly indicating that modulus is not a good quality assurance indicator and suggesting that testpiece bending is not very important in uniaxial creep testing of ductile materials, presumably because stress redistribution during the early stages of creep compensates for misalignment of the loading train (Loveday, 1986).

8.3 A NEW PROCEDURE FOR CALCULATING YOUNG'S MODULUS FROM UNIAXIAL TESTS

There is often some uncertainty associated with data collected at the start of a load–displacement test, possibly due to the settling in of the testpiece within the grips or to testpiece misalignment. When the material displays only a very limited range of linearity between stress and strain, this settling in may cause some difficulty in calculating Young's modulus. The uncertainty can disguise the true origin of the stress–strain curve, which in turn can affect the calculated values of proof stress and strain at failure. The linear region is confined to very small strains, for example between 0.1 and 0.2%, for certain metal matrix composites, which leads to particularly large errors in modulus determination using conventional methods. To overcome this, an analysis procedure has been developed which is supported by a software system which calculates Young's modulus to a high degree of accuracy by using both the tangent and the secant moduli to the stress–strain curve (Roebuck *et al.*, 1993). The procedure and software also allow the calculation of proportional limit, proof stress, failure strength and strain to failure.

8.3.1 Calculation of Young's modulus

Figure 8.9(a) is a schematic representation of a stress–strain curve showing a linear region (P–Q) preceded by a non-linear region (O–P). The initial problem is to calculate a reasonably accurate value of the gradient of the linear segment P–Q which can be used to determine the true origin O' of the curve. It can be shown that the traditional method of using a ruler and estimating the stress and strain intercepts with the axes can produce variations in modulus value of up to 10%, and is open to bias by the researcher (Roebuck *et al.*, 1992). Instrumented systems which use computers to collect and analyse the data use a variety of methods, most

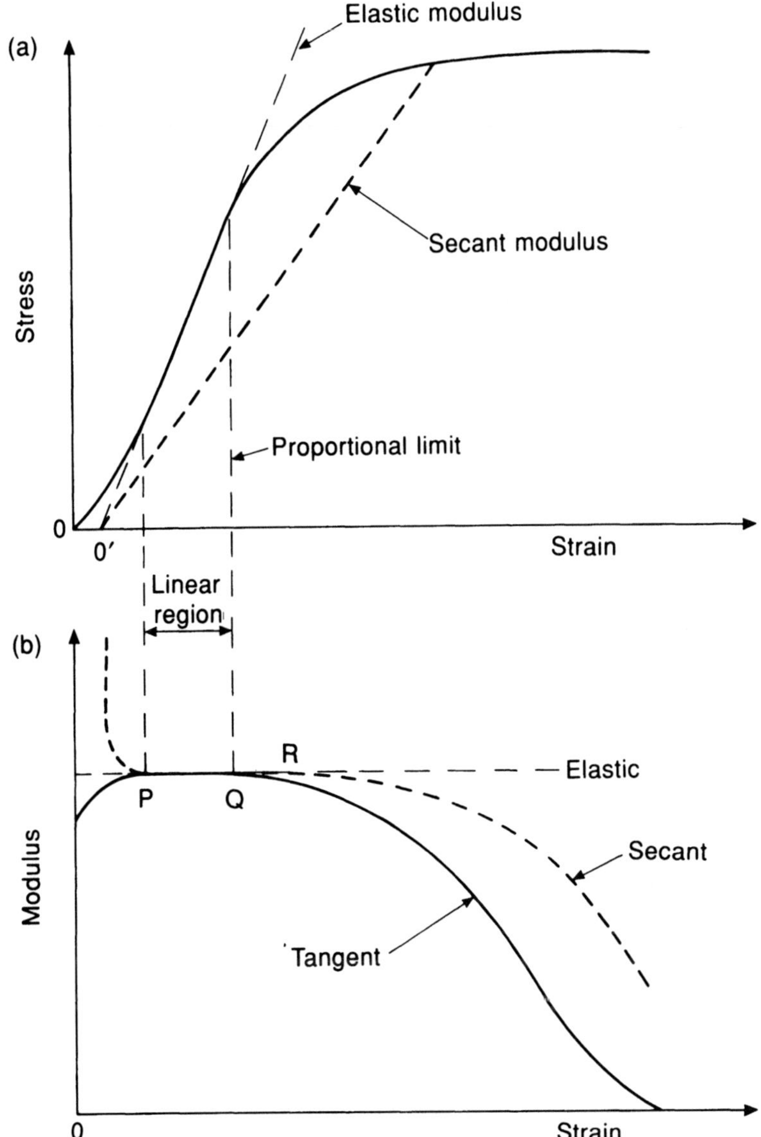

Fig. 8.9 Schematic diagram of calculation method: (a) stress–strain curve (b) modulus–strain curve.

of which invite the operator to specify the points P and Q and then calculate the gradient either of the chord P–Q from the data points P and Q or of a straight-line fit between the points P and Q. This type of calculation depends on the accurate selection of P and Q and is still open to bias by the operator. The new system includes a fully automated calculation of Young's modulus, obtained by analysing data collected during tensile or compression tests.

This procedure removes the need for operator decisions and hence minimizes operator error or bias. Even with this technique some data sets can be difficult to analyse because of noise or excessive non-linearity. Consequently the procedure includes a further option whereby a trial modulus, plotted as a horizontal line on a modulus–stress curve, can be selected by the operator and fitted to the data set.

8.3.2 Analysis procedure

The tangent modulus of a portion of the stress–strain curve represented by $n + 1$ points is calculated by performing a least-squares quadratic fit over a fixed number of points $2z + 1$ around each datum point P $(z \leqslant p < n - z)$, and differentiating the resulting equations to calculate the gradient at point P. It is assumed that the equation relating the stress y to the strain x is given by

$$y = Ax^2 + Bx + C \tag{8.2}$$

so that the sum of the squares of the residuals S_p which must be minimized to give the best fit to the curve around any point (x_p, y_p) can be defined as

$$S_p = \sum_{i=p-z}^{p+z} \left\{ A_p x_i^2 + B_p x_i + C_p - y_i \right\}^2, \quad p = z, \ldots, n - z \tag{8.3}$$

where A_p, B_p, C_p are the coefficients to be determined by the least-squares fit. The selection of the value z can be varied, but it should be noted that the first and last groups of $(z - 1)$ points in the data cannot be subjected to a least-squares fit; therefore the value z should be the smallest which gives a reasonable fit to the data. A minimum of 15 points in total is recommended, but the menu-driven procedure allows different values for z to be chosen to allow the effect of this choice to be assessed.

In order to minimize S_p the relation 8.3 must be differentiated with respect to the coefficients A_p, B_p and C_p and the resulting expression set to zero. The three equations can then be rearranged to give three simultaneous linear equations in three unknowns:

$$A_p \sum_{i=p-z}^{p+z} x_i^4 + B_p \sum_{i=p-z}^{p+z} x_i^3 + C_p \sum_{i=p-z}^{p+z} x_i^2 = \sum_{i=p-z}^{p+z} x_i^2 y_i$$

$$A_p \sum_{i=p-z}^{p+z} x_i^3 + B_p \sum_{i=p-z}^{p+z} x_i^2 + C_p \sum_{i=p-z}^{p+z} x_i = \sum_{i=p-z}^{p+z} x_i y_i \tag{8.4}$$

$$A_p \sum_{i=p-z}^{p+z} x_i^2 + B_p \sum_{i=p-z}^{p+z} x_i + C_p \sum_{i=p-z}^{p+z} 1 = \sum_{i=p-z}^{p+z} y_i$$

which can readily be solved to calculate the coefficients A_p, B_p and C_p. The coefficients can be calculated more efficiently by the following method:

$$\text{if} \quad S_p = \sum_{i=p-z}^{p+z} f_i \tag{8.5}$$

$$\text{then} \quad S_{p+1} = S_p - f_{p-z} + f_{p+z+1}$$

which is valid for all the sums required to solve equations of the type 8.4.

The tangents E_p to the curve at each point (x_p, y_p) can be calculated by differentiating the local equations of type 8.2 to give

$$E_p = 2A_p x_p + B_p, \quad p = z, \ldots, n-z \tag{8.6}$$

The results of these calculations, when plotted against strain as shown in Fig. 8.9(b), generally produce a more or less horizontal region marked P–Q which is equivalent to the region P–Q in the stress–strain curve in Fig. 8.9(a). It should be noted that the tangents are normally plotted against stress, which generally results in a longer horizontal region P–Q. They have been plotted against strain in Fig. 8.9(b) merely to demonstrate the calculation method.

The next stage in the calculation is to find automatically the modulus value of the linear region P–Q. This is easily done using a computer by finding the deviation from the mean value of tangent modulus for groups of m consecutive points throughout the tangent modulus curve shown in Fig. 8.9(b) and then selecting the group of m points with minimum deviation from the mean tangent modulus. If $m = 2q + 1$ is the number of points required to calculate the deviation from the mean, then the mean value $\bar{E}_p$ of the tangent modulus of a local set of $2q + 1$ data points centred on the point P is given by

$$\bar{E}_p = \frac{1}{2q+1} \sum_{i=p-q}^{p+q} E_i, \quad p = z+1, \ldots, n-z-q$$

and the deviation from the mean r_p is simply

$$r_p = \sum_{i=p-q}^{p+q} \left\{ \bar{E}_p - E_i \right\}^2$$

The value q defining the number of points for this procedure is again variable, but now the first and last groups of $z + q - 1$ points are excluded from the search for the best mean. In the schematic diagrams in Fig. 8.9 a very clearly defined linear region is shown, but for many materials this linear region is non-existent or at best very short. It is therefore inadvisable to select a high value of q in the search for the linear region. On the other hand, too low a value for q can lead to the selection of the wrong linear region, especially if the test data are slowly changing or very dense in a local region. The current software version uses $q = 5$ but further work is needed to increase confidence in the selection of the value $q = 5$.

The mean value of tangent modulus $\bar{E}_p$ having minimum deviation is used for the first estimate of Young's modulus. If examination of this initial estimate shows it to be inconsistent with the data when plotted as a modulus–stress (or modulus–strain) curve, then a manual selection using the computer software can be made from a plot of tangent modulus against stress displayed on the VDU. A more accurate manual selection of Young's modulus can be made from the tangent modulus curve (Fig. 8.9(b)) than from the stress–strain curve (Fig. 8.9(a)).

Once the first estimate of Young's modulus has been determined, either automatically or manually, a new origin (O' in Fig. 8.9(a)) can be calculated for the stress–strain curve. The secant modulus can be determined for every point within the data, by calculating the gradient of the secant line from the new origin O' to each data point. The secant modulus–stress (or modulus–strain) curve generally has a horizontal region (marked P–R on Fig. 8.9(b)) that is longer and often more clearly defined than that produced by the tangent modulus–stress (or modulus–strain) graph. This is because the gradient of the stress–strain curve is perturbed in the region where the curve begins to deviate from the straight line (point Q on Fig. 8.9(a)), the amount of the perturbation being dependent on the selected value of z, whereas the calculation of secant modulus requires neither curve fitting (once a true origin to the stress–stress curve has been determined) nor differentiation of the data.

The procedure described above to find the region P–Q is now repeated for the secant modulus curve, but with the value $q = 7$, to identify the region P–R. This new mean value is considered to be the definitive value of Young's modulus, and is used to redefine the origin O' of the curve and hence to recalculate the secant moduli. Failure to resolve the linear portion of P–Q in Fig. 8.9(b) where the tangent and secant moduli coincide can be a result of using an incorrect estimate of the origin O'. This can frequently arise when the length of P–Q is relatively small.

8.3.3 Calculation of proportional limit

When the best values for the origin and Young's modulus have been found it is possible to estimate a value of the proportional limit, shown in the schematic diagram of Fig. 8.10. The equation of the straight line representing Young's modulus can be written as

$$\sigma = E\varepsilon$$

where E is Young's modulus and σ and ε are stress and strain respectively (note that the straight line passes through the new origin O'). This equation is used to calculate nominal strain values $\varepsilon'_i = \sigma_i / E$ for consecutive

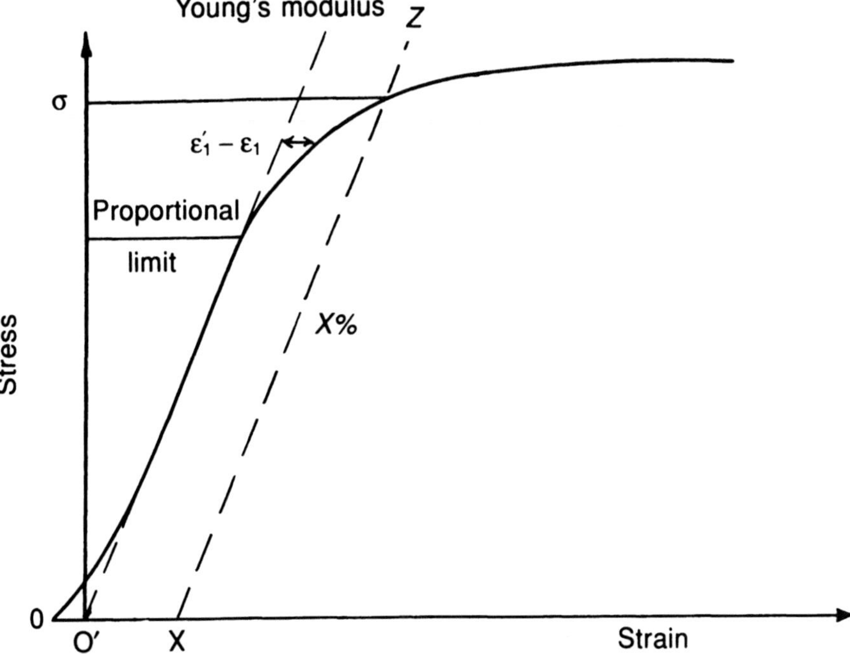

Fig. 8.10 Schematic diagram of proportional-limit/proof-stress calculations.

stress values σ_i, and the proportional limit is defined as the last point on the stress–strain curve where $(\varepsilon_i - \varepsilon_i') < 0.005\,\varepsilon_i'$. Agreement is needed on the value of the factor to use when defining the proportional limit.

8.3.4 Calculation of proof stress

To determine the $X\%$ proof stress a straight line X–Z, having a gradient equal to the Young's modulus, is fitted through the required value X on the strain axis, as shown in Fig. 8.10. If

$$y = E(x - X)$$

where E is the calculated Young's modulus value, is the equation of the line which passes through the point (X, O'), then it is possible to define the point $(\varepsilon_j, \sigma_j)$, where $E(\varepsilon_j - X) \geqslant \sigma_j$, which is the nearest point in the stress–strain data to the intersection of the proof stress line with the stress–strain curve. To find the precise value of the point of intersection a quadratic least-squares fit is carried out using 15 points of the stress–strain curve around the point $(\varepsilon_j, \sigma_j)$ in the manner described by equations 8.2–8.4, and the resulting quadratic equation is then simultaneously solved with the linear equation of the proof stress line to find the stress σ at the point of intersection as follows. If

$$\sigma = a\varepsilon^2 + b\varepsilon + c = E(\varepsilon - X)$$

defines the point of intersection of the quadratic equation and the straight line, then

$$a\varepsilon^2 + \varepsilon(b - E) + (c + EX) = 0.$$

From this, simple algebra can be used to calculate

$$\varepsilon = \frac{-(b - E) \pm \sqrt{[(b - E)^2 - 4a(c + EX)]}}{2a}$$

and hence σ, the required $X\%$ proof stress.

8.3.5 Calculation of failure strength and strain to failure

The value for failure strength is simply determined as the largest stress value in the data set which is normally, but not always, the last point collected before failure of the testpiece or when the test is stopped. Strain to failure is the final strain value of the data, after taking account of the adjustment to the new origin O'.

8.3.6 Experimental results from the new modulus analysis

An example of the use of the technique is given in Fig. 8.11, which shows the results of the analysis for a set of data collected during a tensile test on a SiC fibre reinforced aluminium metal matrix composite (MMC). The data set is part of NPL's contribution to an interlaboratory exercise, organized by the UK MMC Structural Analysis Group and currently involving six organizations, to measure the elastic constants of a fibre reinforced MMC. Further details and results are presented in Lord (1992). The material for the exercise was supplied by BP Research Centre, Sunbury, but was manufactured by Textron Specialty Materials in the USA. It is an eight-ply diffusion bonded unidirectional MMC using SCS-2 SiC fibres in a 6061 aluminium matrix. The size of the test panel was 230 mm × 110 mm × 1.4 mm with a fibre volume fraction of ~ 45%. The SiC fibres are produced by chemical vapour deposition and are typically 140 µm in diameter. Straight-sided tensile testpieces were machined from the panel using a diamond bonded slitting wheel following procedures recommended by Textron. Aluminium tabs, 40 mm long and 1.5 mm thick with a taper angle of 30°, were bonded to the ends of the testpiece prior to testing to protect the fibres and provide a mechanism for load transfer to the testpiece. All the mechanical tests at NPL were carried out on an Instron 1197 electromechanical testing machine, fitted with self-aligning wedge-action grips, at room temperature in displacement control at a crosshead speed of 1 mm min^{-1}.

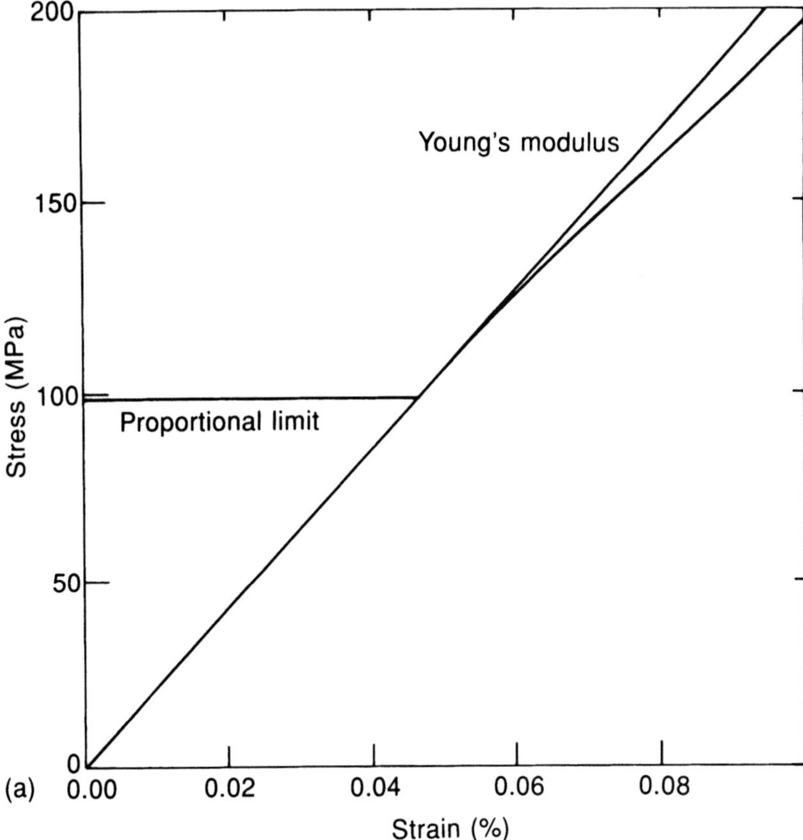

Fig. 8.11 (a) Result of tensile test on SCS-2/Al material (longitudinal strain gauges): stress–strain curve in low-stress region.

Because metal matrix composites have low strains to failure, accurate strain measurement techniques are required. The use of only one strain measuring device can lead to considerable uncertainty in the measurement of Young's modulus and therefore strain was measured on both faces of the testpiece, either with two strain gauges or with a dual averaging extensometer. Many of the problems associated with measuring the elastic properties of fibre reinforced metals (FRM) are a consequence of the low proportional limit, which in itself is a direct result of residual stresses in the material. These large residual stresses can develop during manufacture because of thermal expansion mismatch between the matrix and the fibre. The magnitude of the residual stresses depends on a number of factors including the matrix type, the type and volume fraction of reinforcement, the fabrication conditions and any subsequent heat treatment cycle. In most cases for FRM large tensile residual stresses develop in the matrix, and a consequence of this is that

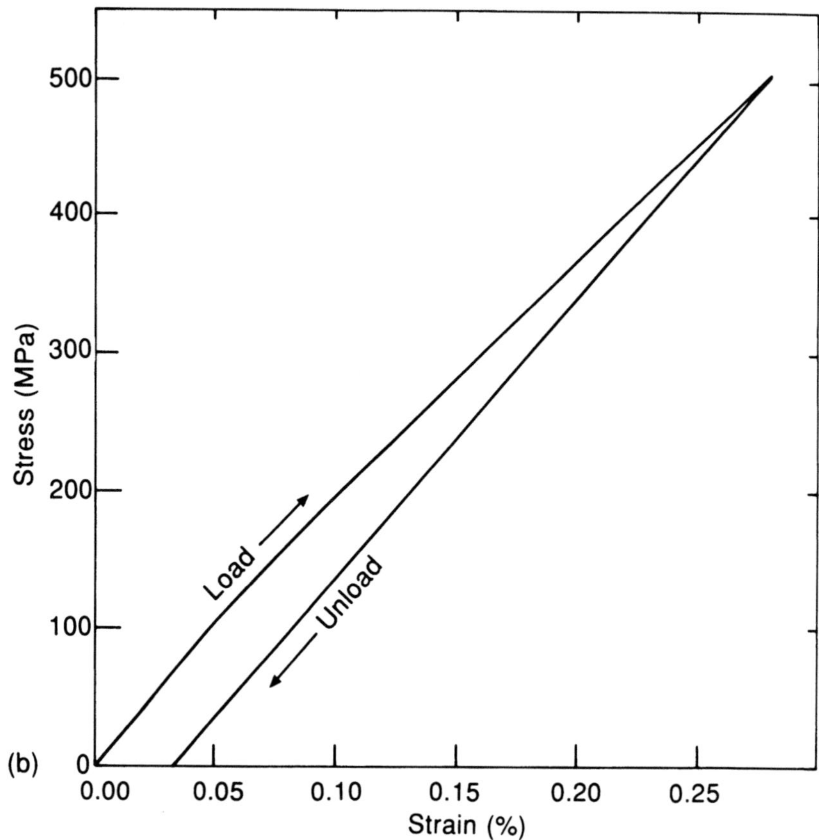

Fig. 8.11 (b) Result of tensile test on SCS-2/Al material (longitudinal strain gauges): entire stress–strain curve for loading and unloading cycles.

when a testpiece is loaded at room temperature the matrix soon yields and it becomes difficult to measure accurately the proportional limit and Young's modulus from the short linear portion of the stress–strain curve. Loading–unloading tests were used in this exercise to extend the linear portion of the stress–strain curve by work hardening the matrix. An additional advantage of this procedure was that, provided certain conditions were met, a single testpiece could be tested by each of the organizations involved. The tensile testpiece was first loaded at NPL to a stress level of 500 MPa to extend the elastic portion of the stress–strain curve and then loaded and unloaded for a further three cycles (until no hysteresis could be detected) before being passed to the next organization for testing. Part way through the exercise the testpiece was returned to NPL, new strain gauges were applied and the testpiece was retested.

The software designed to analyse the NPL data was used to produce the three graphs shown in Fig. 8.11. The early part of the stress–strain

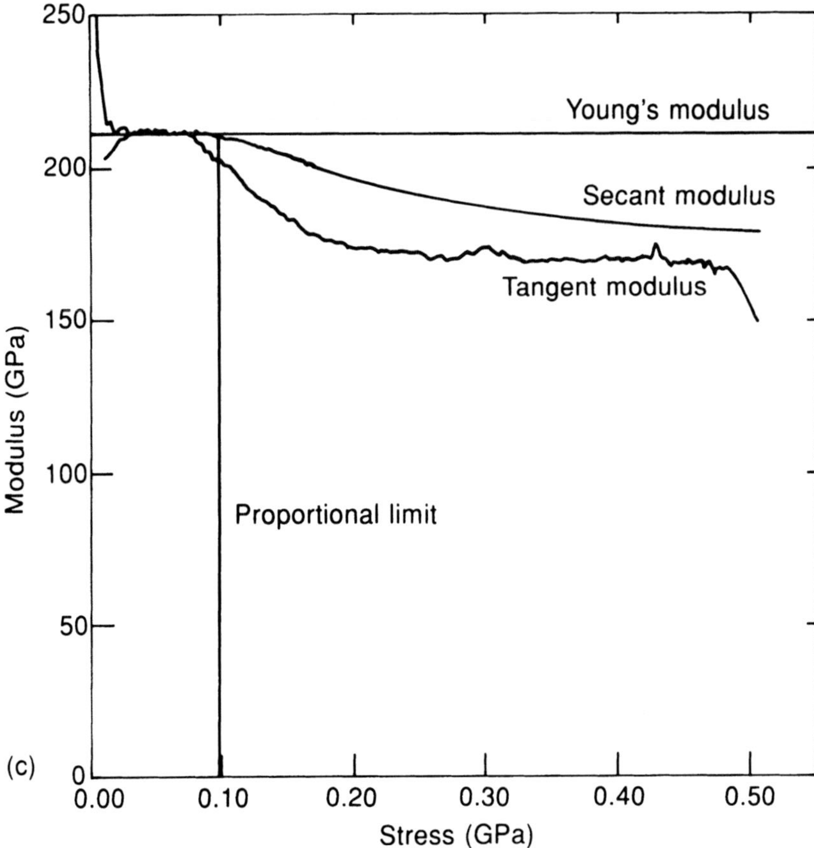

Fig. 8.11 (c) Result of tensile test on SCS-2/Al material (longitudinal strain gauges): modulus–stress curve.

curve is expanded in Fig. 8.11(a) to show Young's modulus and the proportional limit. The initial elastic portion of the curve where the fibre and matrix are both deforming elastically is very short because of residual stresses in the material. Residual stresses develop during the manufacture of fibre reinforced MMCs because of the thermal expansion mismatch between the fibre and matrix. Preliminary tests indicate that, at room temperature, the tensile residual stress in the matrix can be of the order of 200 MPa – a figure comparable to the yield strength of the unreinforced aluminium matrix alloy – and the consequence of this is that the material yields almost as soon as it is loaded.

The whole of the stress–strain curve is plotted in Fig. 8.11(b) showing, for this test, the loading and unloading curves and marking the position of the maximum stress and the strain value at which it occurred. (Note that for some tensile tests the stress decreases prior to failure of the testpiece, in which case the value and position of the maximum stress are not so clearly defined and it is important to mark them on the graph.)

Table 8.3 Tensile results for FRM longitudinal testpiece

Organization	Test number	Strain measurement device*	Modulus E_L (GPa)	Modulus mean value and SD (GPa)	Poisson's ratio
A	1	XY sg	212.0	210.4	0.238
	2	XY sg	210.0		0.241
	3	XY sg	209.2	1.2	0.242
	4	XY sg	210.2		
B	5	sg	204.3	207.1	0.240
	6	sg	208.4		0.240
	7	sg	208.5	2.4	0.240
C	8	XY sg	223.0	220.8	0.267
	9	XY sg	220.8		0.271
	10	XY sg	227.1	4.6	0.288
	11	XY sg	215.5		0.267
	12	XY sg	217.5		0.270
D	13	–	195.2	198.4	–
	14	–	205.0		–
	15	–	194.9	5.7	–
A	16	XY sg	211.7	210.9	0.245
	17	XY sg	210.6		0.244
	18	XY sg	210.6	0.5	0.245
	19	XY sg	210.7		0.244
E	20	ext	193.9	–	–

* XY sg longitudinal and transverse strain gauges
sg   longitudinal strain gauges
ext   extensometer

Figure 8.11(c) shows Young's modulus and the proportional limit plotted on a graph of secant and tangent moduli against stress. (The software also permits a graph of secant and tangent moduli against strain, although this option is not often used.) In some cases, with fibre reinforced MMC, two modulus values are quoted to characterize the whole of the stress–strain curve. In this case, however, careful examination of the modulus–stress curve shows that, beyond the proportional limit, the secant and tangent moduli continue to fall – and there is no second linear portion to the curve. The 'noise' on the tangent modulus curve could be reduced by increasing the number of data points used in the least-squares fit from 15 to say 25, but it can be seen from the tangent modulus curve that this material has a relatively short linear region and increasing the number of points could mean the loss of the tangent modulus data in the early part of the curve. In the absence of a clearly defined linear region the analysis can be forced to use a value of tangent modulus selected by the operator. The difficulties of defining an initial modulus in this material become apparent when we examine the modulus–stress curve before the proportional limit. The flat portion of the curve is not

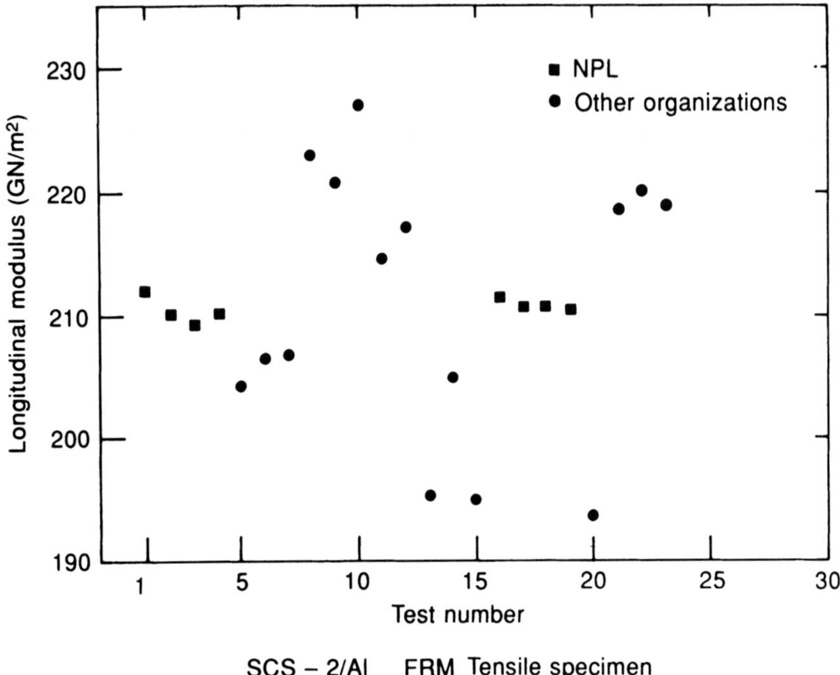

Fig. 8.12 Longitudinal Young's modulus values from an interlaboratory exercise on a fibre reinforced MMC.

clearly defined, but the actual value of Young's modulus determined from the plot of secant modulus (Fig. 8.11(c)) is a good fit to the data, as can be seen from the expanded portion of the stress–strain curve in Fig. 8.11(a).

Table 8.3 shows results obtained from all tests in the interlaboratory exercise. The mean value of longitudinal modulus E_L from tests carried out by all the organizations involved is 210.0 GPa with a standard deviation of 8.7 GPa, i.e. within $\pm$ 8% with a 95% confidence limit. The mean value from all the tests carried out at NPL was 210.6 GPa with a standard deviation of 0.9 GPa. The modulus values given in Table 8.3 are plotted in Fig. 8.12.

8.3.7 Software validation

The validation of the software used to determine modulus and tensile test properties is an important matter to address, particularly since in many modern testing machines no analogue output of the stress–strain curve is provided, and in some cases only the tensile properties at predetermined conditions are displayed, without digitized readings defining the stress–strain curve. Indeed in some production line environments the testing machine computer only indicates that the test properties are greater than previously specified values for conformance with product release requirements. Thus the operator becomes entirely reliant upon the computer for the generation of the test results.

The computer software outlined above has been validated using a theoretical approach by analysing numerical reference data to simulate the stress–strain curve and comparing the computer calculated properties with the known values (Roebuck *et al.*, 1993). In addition the software was used in parallel with the testing machine software whilst testing a material with well-established characteristics; a tensile reference material would obviously have been ideal for this task. For either of these two approaches to be undertaken on tensile testing machines it is clearly a requisite that means must be provided for feeding in digitized reference data, or that output points are provided enabling the analogue signals from the machine load and strain sensors to be fed into an independent software analysis system. The interfaces needed to achieve these requirements should in the future be standardized and specified in the testing or calibration standard.

8.4 FLEXURAL RESONANCE TECHNIQUE

Accurate values of dynamic Young's modulus E' and loss factor d_E for rigid low-loss materials can be determined by means of the flexural resonance method. This technique involves the application of a harmonic force or moment of constant amplitude at some point along a strip of the material and the monitoring of the vibration amplitude as a function of the forcing frequency (100 Hz to 10 kHz). At the so-called 'resonance' frequencies f_{rn}, maxima are observed in the vibration amplitude (Fig. 8.13) corresponding to the appearance of standing waves in the sample. Here the mode number n has values of 1, 2, 3 etc., the value 1 representing the fundamental natural vibration mode and higher numbers the successive harmonics or overtones. According to theory the storage modulus at frequency f_{rn} may be obtained from measurements of f_{rn} and the sample dimensions, and d_E is approximately equal to $(f_{2n} - f_{1n})/f_{rn}$, where $f_{2n} - f_{1n}$ is the resonance peak width at an amplitude $1/\sqrt{2}$ times the maximum amplitude (Fig. 8.13). The loss factor may alternatively be determined from the decay rate of the vibration amplitude after removal of the driving force.

The average strain amplitudes in standing-wave resonance tests are typically of order 10^{-6}, so that problems associated with the non-linearity of dynamic properties do not arise. Clamping errors may also be avoided by employing free–free vibrational modes of appropriately suspended samples. However, the resonance techniques require samples that are sufficiently rigid (E' from 1 to 200 GPa) and low loss (d_E values below about 0.2) to support their own weight and to exhibit resolved resonance peaks.

The international standard ISO 6721/3 describes the measurement of the dynamic Young's modulus and loss factor for polymeric materials by flexural resonance. Standards for ceramic materials are currently being

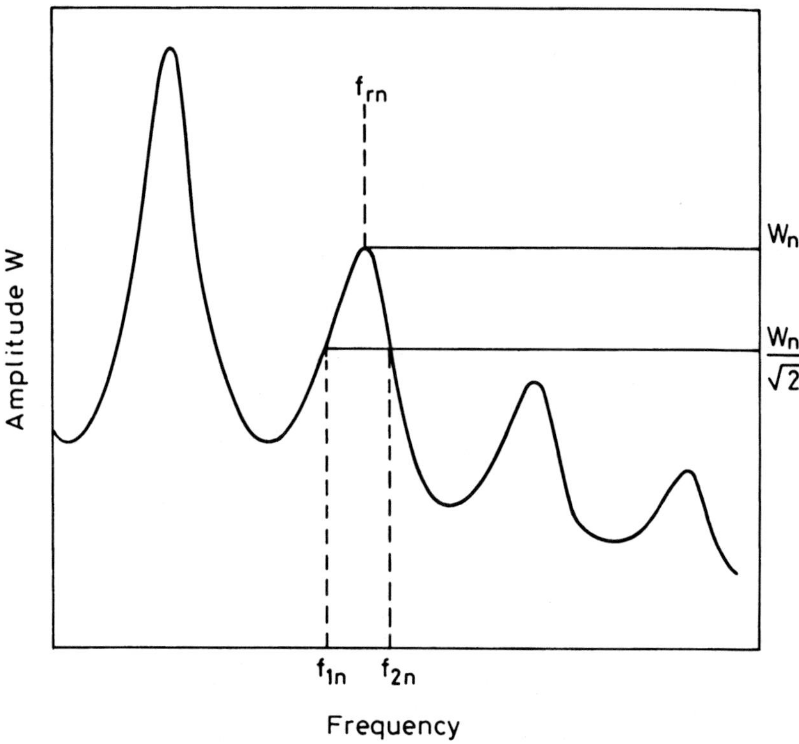

Fig. 8.13 Illustration of resonance peaks and definition of a resonance frequency f_{rn} and associated frequencies f_{2n} and f_{1n}.

developed by CEN TC 184/WG3 and by ASTM Committee E28 (Table 8.1). The acquisition of data using a resonance technique on a ceramic testpiece which is a suitable standard reference material is described by Dickson and Wachtman (1971).

8.4.1 Apparatus

The equipment used at NPL is illustrated in Fig. 8.14 and shown schematically in Fig. 8.15. A rectangular strip of material S with typical dimensions $150 \times 10 \times 2$ mm is suspended horizontally by two short loops of fine nylon filament. The separation between the loops can be varied so that they may be located at symmetrically placed nodal points for the different vibrational modes (Read and Dean, 1978). Transverse (vertical) sinusoidal forces are applied to one end of the sample by means of an electromagnetic transducer T fed by a variable frequency oscillator Os and power amplifier PA. The resulting alternating magnetic field acts upon a small piece of magnetic metal M1 bonded to the specimen. For detecting the vibration amplitude we employ the probe P of a proximitor

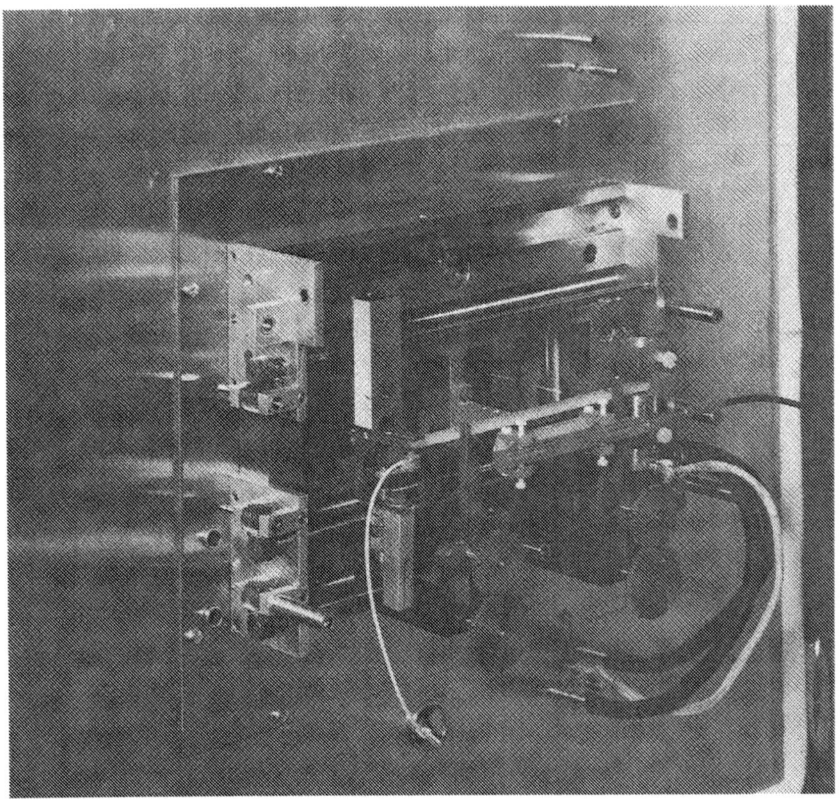

Fig. 8.14 NPL audiofrequency resonance apparatus.

which faces a small conducting mass M2 bonded to the other end of the sample. Operation of the proximitor involves the formation of eddy currents in the probe which is energized by the proximitor circuit Pr. An output voltage is obtained from Pr proportional to the gap separation between the probe tip and the conducting surface for spacings between 0.25 and 1 mm. Transverse bending vibrations in the sample are thus monitored from the cyclic variation of output voltage after amplification by a measuring amplifier MA. A narrow-band tracking filter F, tuned to the drive frequency by the oscillator, is inserted into the measuring amplifier. This eliminates extraneous noise and is essential for measurements on the higher modes owing to the low signal to noise ratios. The measuring amplifier is equipped with a meter which records the RMS value of the displacement signal on both a linear and a decibel scale. This is employed when manually tuning the oscillator to resonance (maximum amplitude) and subsequently determining the resonance peak width. The relevant frequencies are measured with a frequency counter FC connected to the oscillator output. A two-channel oscilloscope Sc may be used to monitor the drive current and displacement signals and a level recorder

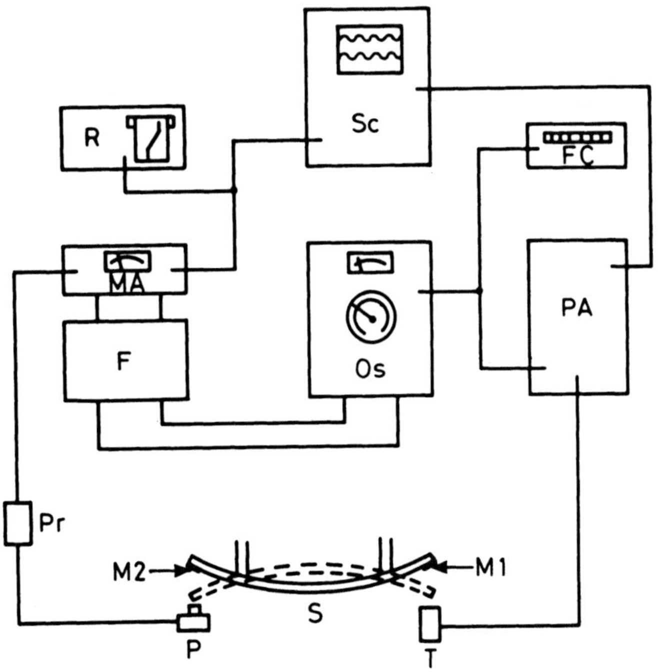

Fig. 8.15 Block diagram of audiofrequency flexural resonance equipment.

R is employed to determine d_E for very low-loss specimens by recording the free decay of vibration amplitudes after cutting the oscillator output.

The drive transducer was designed with a low inductance in order to maintain a high level of force to frequencies above 20 kHz. It comprises a 100-turn coil wound around a cylindrical steel core and is supplied by the PA with alternating currents up to 5 A. A variable DC offset up to 5 A serves to polarize the magnet and so yield undistorted harmonic forces with frequencies equal to the oscillator frequency (no frequency doubling). The transducer is liquid cooled to remove the generated heat. In order to apply corrections for added mass, it is convenient if M1 and M2 have equal dimensions and mass. For this purpose the two masses were each prepared as thin rectangular pieces from a spring-steel strip. Also note that the sample is placed in an enclosure with facilities for varying both temperature ($- 30$ to $100\,^{\circ}\mathrm{C}$) and air pressure.

8.4.2 Theoretical relationships and the sources and correction of errors

The theory underlying the flexural resonance method has been discussed by Read and Dean (1978). In the classical equation of motion, shear deformations are neglected, as are the rotatory inertial forces due to the rotational motions of beam elements. After the inclusion of a correction factor $1 + \psi_{sr}$ for these effects, and a factor $1 + \psi_m$ for the effects of added mass, the theory yields the following relationship for E' in terms of the measured f_{rn}:

$$E' = \frac{48\pi^2 M \ell^3 f_{rn}^2}{bh^3 \alpha_n^4} (1 + \psi_{sr})(1 + \psi_m) \tag{8.7}$$

In this equation, M, ℓ, b and h are the mass, length, width and thickness of the sample, respectively. The mode constants α_n are the solutions of $\cos \alpha \cosh \alpha = 1$ and have values which are given in Table 8.4.

Table 8.4 Mode constants α_n and added mass correction factors $\langle W_n^2(x)/W_n^2(0) \rangle$ for a free-free vibrating beam

n	α_n	$\langle W_n^2(x)/W_n^2(0) \rangle$			
		$\Delta/l = 0.018$	0.026	0.034	0.042
1	4.730	0.92	0.88	0.85	0.82
2	7.853	0.87	0.81	0.76	0.71
3	10.996	0.82	0.74	0.67	0.61
4	14.137	0.77	0.68	0.60	0.53
5	17.279	0.72	0.62	0.53	0.45
6	20.420	0.68	0.56	0.47	0.39
7	23.562	0.64	0.51	0.42	0.34

For isotropic materials, ψ_{sr} is given to a good approximation by

$$\psi_{sr} = \frac{\alpha_n^2 h^2}{12 \ell^2} \left[1 + \frac{E'}{\kappa_s G'} \right] \tag{8.8}$$

where G' is the dynamic shear modulus and κ_s is a constant which accounts for the non-uniformity of shear stresses on each cross-section. A value of 0.87 is generally used for κ_s. Equation 8.8 also provides a reasonable approximation for anisotropic materials providing that E' and G' are identified with the longitudinal tensile and shear moduli E_1 and G_{13}, respectively.

The following expression was derived by Read and Dean (1978) for the added mass correction factor:

$$\psi_m = \frac{4(m_1 + m_2)}{M} \left\langle \frac{W_n^2(x)}{W_n^2(0)} \right\rangle \tag{8.9}$$

where $m_1 + m_2$ is the magnitude of the two added masses and $\langle W_n^2(x)/W_n^2(0) \rangle$ is the mean square value of the ratio $W_n(x)/W_n(0)$. Here $W_n(x)$ is the transverse displacement amplitude at a distance x along the strip and $W_n(0)$ is the displacement amplitude at $x = 0$. The averages were evaluated over the small length Δ (about 5 mm) that each mass extends along the strip from the respective end. Values of these averages are included in Table 8.4 for various values of Δ/l.

Regarding the determination of the loss factor, it follows from equation 5.21 of Read and Dean (1978) that

$$d_E = \frac{f_{2n}^2 - f_{rn}^2}{f_{rn}^2} \tag{8.10}$$

where f_{2n} is the frequency above f_{rn} at which the vibration amplitude is $1/\sqrt{2}$ times the resonance amplitude, corresponding to a 3 dB amplitude drop (Fig. 8.13). If f_{1n} is the frequency below f_{rn} at which the amplitude is 3 dB below the maximum, then with the approximation $f_{1n} + f_{2n} \approx 2f_{rn}$, it can be shown that

$$d_E = \frac{f_{2n} - f_{1n}}{f_{rn}} \tag{8.11}$$

Values of d_E up to about 0.4 can be obtained using equation 8.10 or 8.11 from fundamental resonance peaks since these peaks are then well resolved. The second and higher-mode resonance peaks are much more influenced by overlap from the lower-mode, larger-magnitude peaks and with increasing loss become broadened on the low-frequency side. Since equation 8.10 involves measurements on the high-frequency side of the peak it minimizes the effects of asymmetry and is thus routinely employed at the NPL for loss factor determinations. For the higher modes, d_E values up to about 0.15 can be determined by this method.

The bandwidth method cannot be used for the accurate determination of d_E values below about 0.02 for $n = 1$ and below 0.002 for $n > 1$. In these cases measurements are made of the free decay of vibration amplitudes after cutting the drive force (Read and Dean, 1978).

Values of loss factor are not generally influenced by added mass and, in the case of isotropic materials, corrections to d_E for shear deformations are usually negligible since d_G and d_E values are approximately equal. Anisotropic materials, however, with large E'_1/G'_{13} ratios and large differences between d_G and d_E may require large shear corrections to d_E (Read and Dean, 1978).

In addition to the influence of added mass, shear and rotatory inertia, other potential sources of error with the flexural resonance technique have been discussed in detail in Read and Dean (1978). If the supporting loops are located within 0.5 mm of the nodes then errors in E' due to the added stiffness of the suspension are reduced to below 0.2% and errors in d_E due to energy dissipation through the suspension are minimized. Residual contributions to d_E up to 0.002 may still be significant for low-loss materials although empirical corrections could be applied. Contributions to d_E from air damping up to 0.001 are possible but these are usually negligible or may be corrected for when necessary. The presence

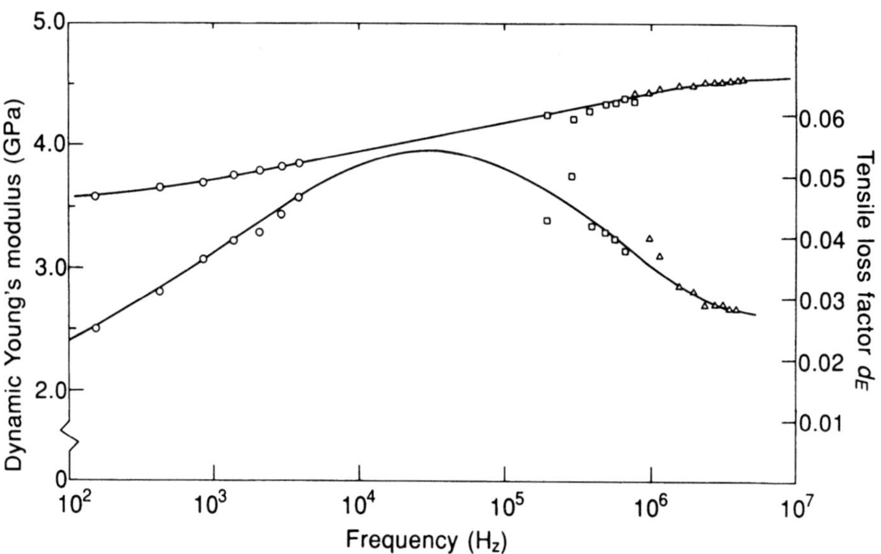

Fig. 8.16 Frequency dependence of the dynamic Young's modulus E' and loss factor d_E for PVC obtained using flexural resonance and ultrasonic wave methods: ○ by flexural resonance; □ by ultrasonics using 500 kHz transducers; △ by ultrasonics using 3 MHz transducers.

of a static magnetic field on the drive transducer can significantly affect E' and d_E values obtained from measurements in the fundamental mode. This effect has been related to the potential energy contribution from the field and is eliminated by ensuring that the spacing between the transducer and the drive mass exceeds about 5 mm for $n = 1$.

8.4.3 Some illustrative results

(a) Data for poly(vinyl chloride) (PVC)

Figure 8.16 shows values for the dynamic Young's modulus E' and the tensile loss factor d_E for PVC obtained using the flexural resonance method and the ultrasonic wave method described in section 8.5. Results cover most of the frequency range between 100 Hz and 10 MHz and coincide with the central frequency range spanned by the β-relaxation mechanism in PVC. The PVC was obtained in sheet form from ICI plc under the trade name of Darvic.

The flexural resonance measurements were made on a strip cut from the sheet and measuring $181 \times 9.84 \times 3.03$ mm. Data were obtained for modes $n = 1$ to 7 which covered the frequency range between about 100 Hz and 4 kHz. E' values were obtained from the measured f_{rn} and sample dimensions using equation 8.7. Corrections for shear and rotatory inertia were made on the basis of equation 8.8 with an E'/G' ratio of 2.7 appropriate to isotropic materials. These corrections varied from 0.2% at $n = 1$ to 5.3% at $n = 7$. Equation 8.9 was employed to correct for the steel drive and detector masses, each of which extended 5 mm along the sample at opposite ends. The added mass corrections varied from 2.2% at $n = 1$ to 1.3% at $n = 7$. Although the resonance frequencies could be determined to within $\pm 0.1\%$, the absolute accuracy in E' values is around $\pm 1\%$, determined largely by errors of about $\pm 0.2\%$ in the specimen thickness. Estimated errors of about $\pm 3\%$ in d_E are associated mainly with the measurement of resonance peak widths.

(b) Silicon carbide reinforced aluminium

In Table 8.7 are presented results for a silicon carbide reinforced aluminium composite supplied by BP Research Centre from Textron, USA (SCS2-SiC/6061). The flexural resonance measurements were made on a strip about 121 mm long by 12.7 mm wide by 1.39 mm thick, with the reinforcing fibres aligned along the specimen length. On the basis of equation 8.7, E' values were obtained for modes $n = 1$ to 5 which covered the frequency range between 830 Hz and 11 kHz. Shear and rotatory inertia corrections were made using equation 8.8 with an E'_1/G'_{13} ratio of 4.83. This ratio was based on an apparent E'_1 value (208 GPa) obtained from the fundamental resonance frequency assuming that ψ_{sr} is negligible for $n = 1$ and a G'_{13} value (43 GPa) obtained from torsional resonance measurements (Read and Dean, 1978). The corrections for shear and rotatory inertia varied from $\pm 0.16\%$ at $n = 1$ to $\pm 2.1\%$ at $n = 5$. Added mass corrections were applied according to equation 8.9 and these varied from $\pm 2.9\%$ at $n = 1$ to 1.6% at $n = 5$. The corrected E'_1 values for modes 1 to 5 varied by less than $\pm 1\%$. Since this is within the estimated accuracy of $\pm 2\%$ (determined largely by errors in thickness measurement) the value of E_1 given in Table 8.7 is an average value for the five modes investigated.

8.5 ULTRASONIC WAVE PROPAGATION METHODS

8.5.1 Introduction

The determination of elastic moduli by ultrasonic wave techniques relies on the relationships between the acoustic wave velocities in a material and the density and elastic properties of the material. At frequencies in the region of 1 MHz, acoustic wavelengths are generally smaller than testpiece dimensions and the mode of wave propagation is independent of the shape or dimensions of the testpiece. The wave is then termed a bulk wave, and the relationships between wave velocity v and elastic properties in isotropic materials take the simple form

$$C = \rho v^2 \tag{8.12}$$

where ρ is the testpiece density and C is a component of the elastic stiffness matrix for the material.

In isotropic materials, there are only two independent stiffness components C_L and C_T. In terms of other elastic properties,

$$C_L = K + (4/3)G$$
$$C_T = G \tag{8.13}$$

where K is the bulk modulus and G the shear modulus. In isotropic materials there are accordingly two modes of bulk wave propagation, one in which the displacement is in the direction of travel (the longitudinal mode) and one in which the displacement is normal to the direction of propagation (the transverse mode). Assigning velocities v_L and v_T respectively to these modes, equation 8.12 becomes

$$C_L = \rho v_L^2$$
$$C_T = \rho v_T^2 \tag{8.14}$$

In anisotropic materials, these relationships are less simple. The number of stiffness components increases and depends on the elastic symmetry of the material. Furthermore, in an arbitrary direction, there are three bulk modes that can be propagated: one is quasi-longitudinal and the other two are quasi-transverse, implying that, in general, wave displacements are neither along nor normal to the wave direction. The measurement of wave velocities along a range of different directions enables all the elastic constants of anisotropic materials to be determined. Apparatus is described in section 8.5.2 whereby these measurements may be made rapidly and accurately and, in many situations, on a single testpiece.

Ultrasonic wave methods have been used successfully to characterize the elastic anisotropy of fibre reinforced composite materials, (Dean and Turner, 1973; Zimmer and Cost, 1970; Dean and Lockett, 1973; Dean, 1974). Where the fibres are uniaxially aligned and the packing arrangement in the plane normal to this axis is random, the material is transversely isotropic and the elastic symmetry is hexagonal. The relationships for determining elastic properties for this class of material are explained in section 8.5.3. In section 8.5.5, the acquisition of data is described for a carbon fibre reinforced polymer and a silicon carbide reinforced metal.

Where the material is viscoelastic, the moduli are complex quantities and the amplitude of the ultrasonic wave is attenuated through the dissipation of wave energy by the cyclic deformation of the material during wave propagation. The attenuation is related to the imaginary components of the complex moduli. These relationships are described in section 8.5.3, and their use is illustrated in section 8.5.5 for the determination of the components of the complex tensile and shear moduli of poly(vinyl chloride) over a range of frequencies.

8.5.2 Measurement of bulk wave velocity and attenuation

The standard ASTM E494 describes a method for determining the velocity of ultrasonic bulk waves in materials. Based on this, ASTM Committee E28 is preparing a standard for measuring the elastic moduli of ceramics using ultrasonics. A European Standard is also under preparation in TC 184/WG3 (see Table 8.1).

(a) Apparatus

Apparatus suitable for determining the velocity of ultrasonic waves along different directions in a testpiece is shown schematically in Fig. 8.17. Measurements involve timing the transit of ultrasonic pulses through a known path length in the testpiece (Hartmann and Jarzynski, 1974; Read and Dean, 1978; Read, Dean and Duncan, 1991). Different

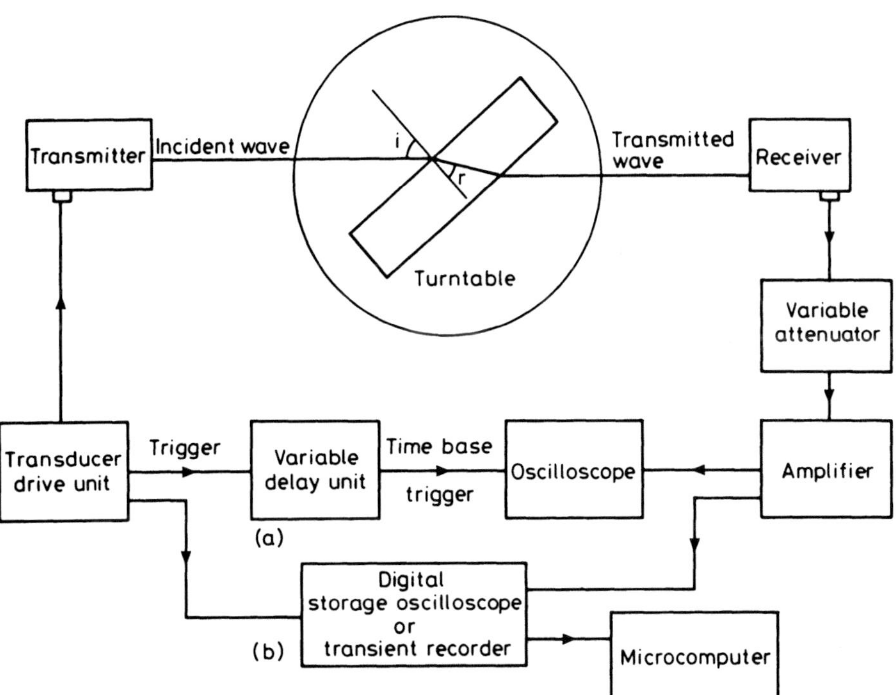

Fig. 8.17 Schematic diagram of ultrasonic pulse apparatus suitable for the measurement of wave velocity and attenuation.

modes of propagation and paths in the testpiece are achieved through refraction of an incident beam of longitudinal wave pulses. The apparatus consists of two coaxial ultrasonic transducers immersed in a water bath whose temperature may be accurately controlled to within ± 0.1 °C. One acts as a transmitter, the other as a receiver. For isotropic materials, the testpiece requires one pair of parallel faces only. For anisotropic samples, measurements along a larger number of directions are needed and the testpiece should take the form of a rectangular parallelepiped prepared ideally so that the edges coincide with symmetry directions in the material. The testpiece is mounted vertically on a turntable between the transducers so as to intercept the beam. The angle of incidence can be varied by a stepper motor which operates in steps of 0.1°. At normal incidence, only longitudinal waves are generated in the testpiece. At arbitrary angles, two refracted waves may be launched, one being quasi-longitudinal (pure longitudinal in isotropic materials) and the other quasi-transverse (pure transverse in the isotropic case). These waves are polarized in the plane containing the incident and refracted wave directions. For reliable measurements of absorption coefficient using refracted waves, it is necessary to select a receiving transducer whose diameter is greater than the beam diameter from the transmitter in order that the receiver intercepts the whole beam after the beam has been displaced laterally following refraction by the testpiece.

An alternative experimental arrangement allows direct contact between the transducers and the faces of the testpiece, as in Fig. 8.18. The water bath and turntable are then not needed. The transducer separation can be varied to accommodate samples of different thicknesses, and both longitudinal and shear wave transducer pairs are required. This arrangement is suitable for isotropic specimens but is not appropriate for determining all the stiffness components of anisotropic materials because it requires waves to be launched in a range of directions. It does, however, provide useful support for the determination of shear moduli in those anisotropic materials for which transverse wave velocities can be obtained with only limited accuracy by the immersion method (see section 8.5.5).

The transmitting transducer is powered by a series of high-voltage, short-duration pulses from the transducer drive unit (typically 300 V for 10 ns). A pulse repetition interval of about 1 ms is satisfactory. The pulses are detected by the receiving transducer and amplified via a circuit with an accurate attenuator. A useful facility for setting normal incidence enables the transmitting transducer to be used also as a receiver to detect the pulse reflected from the front surface of the specimen. An example of a pulse waveform is given in Fig. 8.19, the frequency and length being determined by the transducer design.

The determination of wave velocity requires measurement of the difference in the transit times of pulses with and without the testpiece in the beam. This may be achieved using an accurate variable delay unit as shown in Fig. 8.17(a). The amplified pulse waveform from the receiver transducer is displayed on an oscilloscope whose time base is triggered by the transmitter pulse generator through the delay unit. This unit should be capable of inserting delay times greater than the pulse transit time between the transducers and should have a resolution in the region of 1 ns. This enables the arrival time of a pulse to be determined by adjusting the delay until some characteristic feature of the pulse coincides with the centre of the graticule on the oscilloscope screen. An early, zero-voltage cross-over is typically used as the timing feature as indicated in Fig. 8.19. When a testpiece is inserted in the beam, a new arrival time is determined and the transit time is deduced from the difference in arrival times.

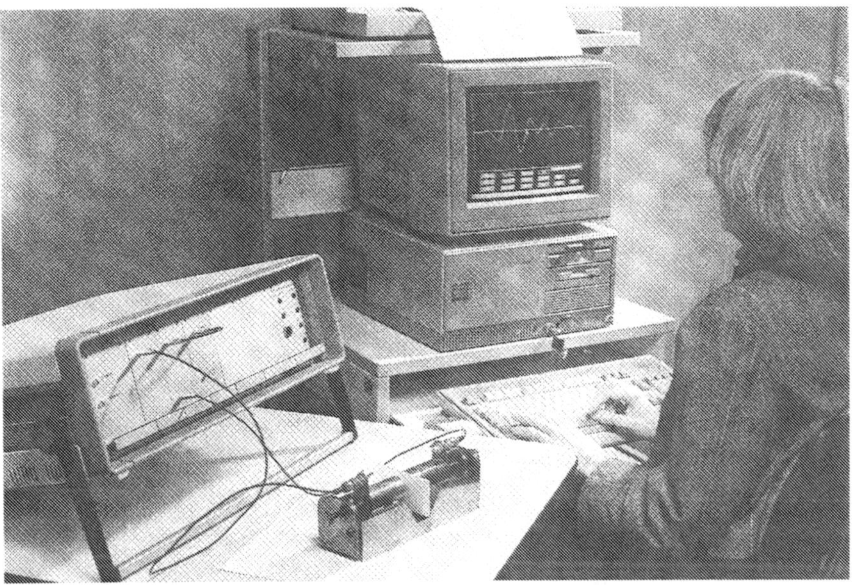

Fig. 8.18 Apparatus used for modulus measurement using direct contact of ultrasonic transducers with the end faces of a testpiece.

An alternative method for measuring transit times is shown in Fig. 8.17(b). Here a transient recorder is used to obtain digital records of pulses with and without the testpiece in the beam. From computer analysis of these records, the transit time can be derived. Very high-frequency sampling rates are needed for high accuracy in the determination of transit times. Digital records of pulses also allow Fourier transforms of each pulse waveform to be carried out (Read, Dean and Duncan, 1991). From the sine transform, phase information of discrete frequency components of the pulse is obtained, from which velocities

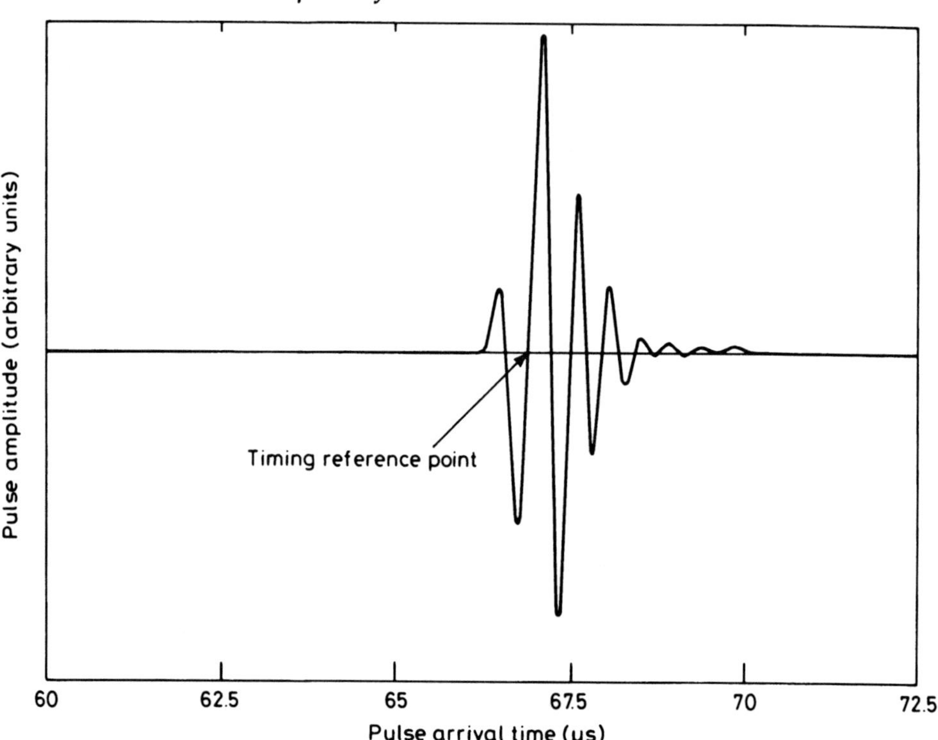

Fig. 8.19 Trace of the waveform of a typical pulse through water. The resonance frequency of the transducer is 2 MHz.

can be deduced. The cosine transform yields values for the relative amplitudes of these frequency components which can be used to determine absorption coefficients for viscoelastic materials.

(b) Determination of velocities and absorption coefficients

If v is the velocity in the testpiece and v_w the velocity in water, then the angles of incidence i and refraction r are related by the equation

$$\frac{\sin i}{v_w} = \frac{\sin r}{v} \tag{8.15}$$

Values for v_w at different temperatures are given in references containing properties of materials, (e.g. Bradfield, 1964). The transit time τ in a testpiece of dimension b in the direction normal to the incidence face is then given by

$$\tau = \frac{b}{\cos r}\left[\frac{\cos (r - i)}{v_w} - \frac{1}{v}\right] \tag{8.16}$$

The velocity v can then be deduced from equations 8.15 and 8.16. Longitudinal wave velocities are more conveniently obtained at normal incidence (see section 8.5.3), where from equation 8.16

$$v_L = \frac{b v_w}{b - \tau v_w} \tag{8.17}$$

The determination of absorption coefficients requires measurements of wave amplitude, which are most accurately obtained from Fourier transforms of pulses (Read, Dean and Duncan, 1991). The reduction in wave amplitude on transmission through a testpiece contains contributions from absorption by the material and from reflection losses at the front and back surfaces of the testpiece. The contribution from reflection losses can be eliminated by comparing transmitted amplitudes through testpieces of different thicknesses at the same angle of incidence. If the transmitted amplitudes of a frequency component f of a pulse are $A_1(f)$ and $A_2(f)$ through testpieces of thickness b_1 and b_2 and r is the common angle of refraction, then the absorption coefficient at f is (Read, Dean and Duncan, 1991)

$$\alpha(f) = \frac{\cos r}{(b_2 - b_1)} \left[0.115a + \log_e \frac{A_1(f)}{A_2(f)} \right] \tag{8.18}$$

where a is the difference in attenuator settings in decibels for the two thicknesses.

8.5.3 Relationships between wave velocity and material properties

(a) Anisotropic elastic materials

The theory of bulk wave propagation in anisotropic materials has been developed by Musgrave (1970). The application of the theory to transversely isotropic materials such as fibre composites has been described in Dean (1974), Zimmer and Cost (1970) and Read and Dean (1978). The theory relates bulk wave velocities to the elastic stiffness components C_{pq} for the material and its density. For transversely isotropic materials there are five independent components C_{pq} and, assuming the 1-axis coincides with the unique axis (the direction of fibre orientation), these are C_{11}, $C_{22} = C_{33}$, C_{44}, $C_{55} = C_{66}$, $C_{12} = C_{13}$ and $C_{23} = C_{22} - 2C_{44}$. Once all of these components have been determined, Young's and shear moduli and Poisson's ratios can be calculated using the equations in section 8.5.5(b).

Along an arbitrary direction in an anisotropic material, three bulk waves can be propagated. These have different polarizations and, in general, different velocities. In the apparatus described in section 8.5.2(a), the boundary conditions at the water/specimen interface allow only two of these waves to be generated. If the wave direction makes an angle θ with the unique axis, then the velocities of these waves are given by the roots of the equation

$$[(C_{66}\cos^2\theta + C_{22}\sin^2\theta - \rho v^2)\,(C_{11}\cos^2\theta + C_{66}\sin^2\theta - \rho v^2)$$

$$- \sin^2\theta\,\cos^2\theta(C_{12} + C_{66})^2] = 0 \qquad (8.19)$$

For propagation along the unique axis, $\theta = 0°$ and equation 8.19 has roots

$$v = (C_{11}/\rho)^{1/2}$$
$$v = (C_{66}/\rho)^{1/2} \qquad (8.20)$$

The first root represents a pure longitudinal wave and the second a pure shear wave. The former can be launched using the apparatus shown in Fig. 8.17 when the testpiece is oriented such that the ultrasonic beam strikes the face orthogonal to the axis at normal incidence. The shear wave cannot be launched in this configuration. Similarly, when $\theta = 90°$, the roots to equation 8.19 are

$$v = (C_{22}/\rho)^{1/2}$$

$$v = (C_{66}/\rho)^{1/2} \qquad (8.21)$$

These give the velocities of the longitudinal and shear waves, respectively, propagating in a direction orthogonal to the unique axis, the shear wave being polarized in the direction of the unique axis. The longitudinal wave velocity in equation 8.21 can be determined experimentally when the ultrasonic beam is incident normal to a face containing the unique axis.

If the testpiece is now oriented so that the unique axis is vertical in Fig. 8.17, then wave velocities are determined in the plane of isotropy. These velocities are independent of direction and take values

$$v = (C_{22}/\rho)^{1/2}$$
$$v = (C_{44}/\rho)^{1/2} \qquad (8.22)$$

Both can be determined experimentally, the former at normal incidence and the latter at any angle of incidence at which the shear wave is generated by refraction. The remaining components C_{66} and C_{12} must be determined from measurements of the quasi-longitudinal or quasi-transverse wave velocities in a plane containing the unique axis. These are roots of equation 8.19 from which values for C_{66} and C_{12} can be determined by optimization if velocities have been measured at three or more values for the angle θ. For highly anisotropic materials, a graphical method may alternatively be used with reasonable accuracy to obtain these components (Dean and Lockett, 1973; Read and Dean, 1978).

(b) Isotropic elastic materials

For isotropic materials,

$$C_{11} = C_{22} = C_{33} = L$$

$$C_{44} = C_{55} = C_{66} = G$$

$$C_{12} = C_{13} = C_{23} = L - 2G \tag{8.23}$$

where L and G are the constrained tensile modulus and the shear modulus respectively. It follows from equations 8.14 and 8.20 that

$$C_L = L = \rho v_L^2$$

$$C_T = G = \rho v_T^2 \tag{8.24}$$

(c) Isotropic viscoelastic materials

With viscoelastic materials, stiffness components L and G in equation 8.23 are complex quantities $L' + iL''$ and $G' + iG''$. Equation 8.24 therefore implies that the wave velocities are also complex, and it is the imaginary components of the velocities that are responsible for an exponential attenuation of wave amplitudes with distance travelled in a viscoelastic material (Read and Dean, 1978; Read, Dean and Duncan, 1991). An absorption coefficient α can therefore be defined equal to the inverse of the distance travelled in the material which causes a reduction in the wave amplitude to a factor $1/e$. If α_L is the absorption coefficient for longitudinal waves and the longitudinal loss factor $d_L = L''/L'$ of the material is $\leqslant 0.2$, then

$$d_L = \frac{2\alpha_L v_L}{\omega} \tag{8.25}$$

where $\omega = 2\pi f$ is the angular frequency of the wave. Here v_L is the wave velocity and, for materials with $d_L \leqslant 0.2$, is effectively the real part of the complex velocity. Thus

$$L' = \rho v_L^2 \tag{8.26}$$

Similarly, if α_T is the absorption coefficient of the shear wave, the shear loss factor d_G is given by

$$d_G = \frac{2\alpha_T v_T}{\omega} \tag{8.27}$$

Also

$$G' = \rho v_T^2 \tag{8.28}$$

8.5.4 Factors limiting accuracy

Wave velocities can be measured with high accuracy in those materials which cause negligible distortion of the transmitted pulse. This leads to an associated high accuracy in the derivation of stiffness components

C_{11}, C_{22}, C_{66} and C_{44}. Since wave velocities are less sensitive to the magnitude of C_{12}, this component is determined with a lower accuracy.

The accuracy in the measurement of wave velocities can be dependent upon testpiece dimensions. If the dimension b in equation 8.16 is small (in the region of 1 mm), the accuracy that can be achieved in the ratio τ/b can limit the accuracy in the determination of the velocity. Furthermore, when one of the dimensions of the testpiece is comparable with, or smaller than, the wavelength of a wave travelling in the plane normal to this dimension, the assumption that a bulk wave is propagating may not be valid. Thus the close proximity of the surface of the testpiece may influence the character and hence the velocity of the wave.

The accuracy in the measurement of the arrival time of a pulse depends on the sensitivity and accuracy of the delay circuit in Fig. 8.17(a) or the sampling interval of the transient recorder in Fig. 8.17(b). In the apparatus developed by the National Physical Laboratory, the maximum sensitivity of both of these pieces of equipment is 1 ns.

With viscoelastic materials, distortion of the transmitted pulse arises through the frequency dependence of the velocity and attenuation of each frequency component of the pulse. The resulting change in pulse shape introduces an uncertainty in locating the reference point on the transmitted pulse and hence an error in the determination of transit time. This can be reduced by comparing phase information on discrete frequency components obtained from Fourier transforms of the pulse before and after transmission through the testpiece.

Pulse distortion can also be significant in composite materials where the reinforcing phase or included voids have a size which gives rise to significant scattering of the frequency components of the pulse. An uncertain error is introduced in velocity measurements which can be reduced by using ultrasonic transducers of lower frequency. In this context, we note that negligible distortion is observed in 3 MHz pulses after transmission through void-free testpieces of carbon fibre reinforced plastics. The fibre diameter and spacing in these materials is in the region of 10 μm and the wavelength is in the region of 1 mm.

8.5.5 Some results for anisotropic and viscoelastic materials

(a) An isotropic viscoelastic material

Figure 8.16 compares values for the dynamic Young's modulus E' and the tensile loss factor d_E of PVC obtained by the ultrasonic wave method with the results described in section 8.4.3(a) using flexural resonance. The ultrasonic results were obtained with two transducers having frequencies of 500 kHz and 3 MHz using two testpiece thicknesses of 4.00 mm and 9.01 mm. Digital records of pulses were stored at sampling intervals of 10 ns for the 500 kHz pulses and 5 ns for 3 MHz pulses. Sine and cosine fast Fourier transforms were obtained of longitudinal and transverse wave pulses in the polymer, from which longitudinal and shear wave velocities and absorption coefficients were derived at discrete frequencies defined by the transform range (1024) and the sampling interval.

Values for L' and G' were obtained using equations 8.26 and 8.28 and the dynamic Young's modulus E' may then be deduced from

$$E' = \frac{G'(3L' - 4G')}{L' - G'}$$ (8.29)

Values for the absorption coefficient of the transverse wave were obtained from the average of measurements at angles of incidence of $\pm 55°$ where the amplitude of the refracted transverse wave was a maximum. The absorption coefficient of the longitudinal wave was measured at normal incidence. Loss factors d_L and d_G were calculated using equations 8.18, 8.25 and 8.27 and the Young's loss factor was deduced from

$$d_E = d_G - \frac{G'L'(d_G - d_L)}{(L' - G')(3L' - 4G')}$$ (8.30)

(b) Anisotropic elastic materials

Table 8.5 Elastic properties of a carbon fibre reinforced PEEK material*

C_{11}	$C_{22} = C_{33}$	C_{44}	$C_{55} = C_{66}$	$C_{12} = C_{13}$	C_{23}	ρ
125	13.5	3.3	5.4	7.0	6.9	1586
E_1	$E_2 = E_3$	G_{23}	$G_{12} = G_{13}$	ν_{21}	ν_{23}	ν_{12}
120	10.0	3.3	5.4	0.34	0.49	0.03

* The 1-axis is in the direction of fibre alignment. Units are GPa for moduli and kg m^{-3} for density.

Tables 8.5 and 8.6 list elastic moduli for two uniaxial fibre composite materials. The longitudinal and transverse Young's and shear moduli, E_1, E_2, G_{12} and G_{23} respectively, and Poisson's ratios were determined from measured stiffness components using the following equations:

$$E_1 = \frac{C_{11}(C_{22} + C_{23}) - 2C_{12}^2}{C_{22} + C_{23}}$$

$$E_2 = \frac{(C_{22} - C_{23})(C_{11}C_{22} + C_{23}C_{11} - 2C_{12}^2)}{C_{11}C_{22} - C_{12}^2}$$

$$G_{12} = C_{66}$$

$$G_{23} = C_{44}$$

$$\nu_{21} = \frac{C_{12}}{C_{22} + C_{23}}$$

$$\nu_{23} = \frac{C_{23}C_{11} - C_{12}^2}{C_{22}C_{11} - C_{12}^2}$$

$$\nu_{12} = \frac{C_{12}(C_{22} - C_{23})}{C_{22}C_{11} - C_{12}^2}$$

Poisson's ratios ν_{ab} are defined here as the ratio of the lateral strain along the a-axis to the longitudinal strain along the b-axis for uniaxial loading along the b-axis.

Data for a carbon fibre reinforced poly(etheretherketone) (PEEK) are given in Table 8.5. This material was supplied by ICI plc. A single testpiece of dimensions $70 \times 15 \times 5.4$ mm was used, the fibre axis coinciding with the longest dimension. Measurements of longitudinal wave velocity with direction in a plane normal to the fibre axis revealed that the testpiece was accurately transversely isotropic. The testpiece caused negligible distortion of transmitted pulses, which gave confidence in the accuracy of the recorded data.

Table 8.6 Elastic properties of a silicon carbide fibre reinforced aluminium material

C_{11}	$C_{22} = C_{33}$	C_{44}	$C_{55} = C_{66}$	$C_{12} = C_{13}$	C_{23}	ρ
248	144	35	44	64	74	2810
E_1	$E_2 = E_3$	G_{23}	$G_{12} = G_{13}$	ν_{21}	ν_{23}	ν_{12}
210	102	35	44	0.29	0.45	0.14

* The 1-axis is in the direction of fibre alignment. Units are GPa for moduli and kg m^{-3} for density.

In Table 8.6, results are shown for a silicon carbide reinforced aluminium composite supplied by BP Research Centre. Measurements of wave velocities were made on two rectangular testpieces which had been used for modulus determinations by conventional tensile testing (section 8.3.6) and flexural resonance (section 8.4.3(b)). Both testpieces were about 120 mm long by 13 mm wide by 1.4 mm thick. In one, the fibres were aligned along the length of the testpiece and in the other they lay in the direction of the width. The transverse testpiece was used to obtain the components C_{11}, C_{44}, C_{22} and C_{23} whilst the longitudinal testpiece gave values for C_{22}, C_{66} and C_{12}. Substantial attenuation and distortion of transmitted pulses were found at 3 MHz. The relatively large diameter of the fibres in comparison with the wavelength (approximately $100\,\mu$m and 3 mm respectively) is believed to be mainly responsible for the attenuation and distortion by scattering, although a contribution from voids in the aluminium matrix cannot be discounted. The accuracies in velocity measurements are accordingly uncertain, and this has prevented any reliable assessment of the isotropy of properties normal to the fibre direction. Transverse isotropy has been assumed.

8.6 COMPARISON OF MODULI MEASURED USING VARIOUS TECHNIQUES

8.6.1 Comparison of results on a metal matrix composite material

Values for the longitudinal Young's modulus of a silicon carbide reinforced aluminium material supplied by BP Research from Textron USA (SCS2–SiC/6061 Al alloy) have been determined by the different test methods described in sections 8.3, 8.4 and 8.5. These are compared in Table 8.7 which also gives values for the longitudinal shear modulus obtained from the resonance and wave propagation methods. There is good agreement to better than $\pm 1\%$.

Table 8.7 Comparison of values for the longitudinal Young's and shear moduli, E_1 and G_{12} respectively, of the BP Metal Composites silicon carbide reinforced aluminium material obtained by different test methods

E_1(GPa)	G_{12}(GPa)	*Method*
210.6	—	Tensile test
210	43	Resonance
210	44	Wave propagation

It is possibly not valid to interpret the close agreement between the Young's modulus values obtained by the flexural resonance and ultrasonic wave methods as an indication of the accuracy of these results. With the resonance test, the uncertainty in the determination of specimen thickness of testpieces whose surfaces have not been accurately lapped and machined parallel can lead to errors of 1 to 3% as reported in section 8.4. This source of error leads to a measured value which is below the true value. Regarding the ultrasonic value, section 8.5.5 has emphasized that pulse distortion caused by enhanced scattering of the higher-frequency components of acoustic pulses is probably responsible for the main source of uncertainty in the accuracy of the ultrasonic wave measurement. This source of error also generally gives rise to a measured modulus which is below the true value.

These results and comments represent the first stage of a programme which aims to establish test methods and procedures for data analysis for determining the elastic properties of composite materials with a high and quantifiable accuracy. They illustrate that an evaluation of sources of error in any method and the reliability of property values generated by that method benefit from the availability of comparable data obtained by other methods.

8.6.2 Comparison of results for Nimonic 75

Table 8.8 compares values of Young's modulus for Nimonic 75 found using the same three techniques as in section 8.6.1. It can be seen that there is good agreement between the three methods of determining modulus, although clearly the dispersion in the results is appreciably less in the case of the ultrasonic and flexural resonance methods in comparison with the tensile test.

Table 8.8 Comparison of Nimonic 75 room temperature modulus values measured using various techniques (Dawson, Loveday and Varma, 1992, unpublished work)

Testing technique	Modulus E (GPa)	Standard deviation (SD)	95% confidence limit = ± 2SD	Number of measurements
Tensile test	212.1	6.56	± 13.1 (= ± 6.2%)	55
Flexural resonance	213.7	1.86	± 3.7 (± 1.7%)	5
Ultrasonic wave propagation	213.5	1.87	± 3.74 (± 1.7%)	6

8.7 CONCLUSIONS

The measurement of modulus by dynamic instrumental techniques, such as the resonance and ultrasonic methods, has reached a mature stage of development. These methods are in principle applicable to a wide range of material types and are essential for measurements on anisotropic materials where more than two moduli are needed to relate stress and strain components. The most important contribution to the uncertainty in moduli obtained using these techniques is the measurement of dimensions and manufacture of the testpiece to controlled tolerances. Typically this results in uncertainties of about ± 1% in the measurement of modulus, but with care these could possibly be reduced to ± 0.5%.

By contrast it was shown in section 8.2 that measurements of Young's modulus by existing static methods are severely limited, principally by the uncertainties associated with measurement of strain using extensometers, and uncertainties of typically ± 10% are to be expected if existing standards and test methods are used.

Thus there is considerable scope for developing better static methods to reduce the uncertainties in measurement of modulus. In engineering design with ductile materials there is a need to know not just a value for Young's modulus, but also the form of the stress–strain curve including proportional limit and work hardening behaviour. In section 8.3 it was shown that by the use of an improved method of data analysis, through the use of measurements of tangent and secant modulus, uncertainties can be minimized to ± 1% and perhaps even better (Table 8.3) provided accurate strain measurement methods are used which allow for bending.

These potential advances in technique could point the way forward for the introduction of improved static methods.

There is also scope for systematic work in comparison exercises where carefully calibrated measurements are made by a range of static and dynamic techniques on a number of materials in order to validate the various methods. Several exercises of limited scope were reported in sections 8.2, 8.4 and 8.6 but inconclusive results were obtained, probably because the uncertainties associated with material homogeneity were difficult to separate from those due to the test method. Further more detailed comparison work, therefore, would be very helpful in illuminating the choices in method that face the technologist when deciding which technique is the most appropriate to adopt for measuring the moduli of a particular material.

ACKNOWLEDGEMENTS

Dr R. Morrell and Dr G. D. Sims are thanked for supplying information concerning testing standards for ceramics and fibre reinforced plastics respectively. Dr J. D. Lord and Dr L. N. McCartney are acknowledged for their significant contributions to the work described in section 8.3.

APPENDIX 8.A   SELECTION OF EQUATIONS FOR ELASTIC MODULUS MEASUREMENT OF ADVANCED TECHNICAL CERAMICS BY RESONANCE AND IMPULSE EXCITATION METHODS

R. Morrell

Notation

A_i correction factor, where i is f for flexure and t for torsion
b testpiece width perpendicular to direction of flexure
d testpiece diameter
E Young's modulus
f_l longitudinal resonance frequency
f_f flexural resonance frequency
f_t torsional resonance frequency
G shear modulus
h testpiece height in direction of flexure
l testpiece length
m testpiece mass
n mode of vibration (= 1 for fundamental mode)
x h/l

X correction factor for longitudinal resonance due to Love (1964)
y h/b
δ $d/2\lambda$
λ wavelength of vibration
v Poisson's ratio
ρ density of testpiece

8.A.1 Introduction

The objective of this appendix is to define the accuracy limitations when using various equations for calculating elastic properties from resonant frequency or impulse excitation frequency of test bars. In reviewing these equations for incorporation into a CEN Standard for testing advanced ceramics, the question of the most appropriate choice of equation arose. A study of levels of error was made, and from this the justification for the choice of equation form was obtained.

8.A.2 Equations

Different literature sources give various equations relating elastic properties to modes of vibration in bar testpieces. In each of these equations, various simplifying assumptions or correction factors have been made, giving varying levels of accuracy in calculating elastic moduli. The error levels in ignoring or truncating correction factors are calculated for each equation as functions of geometry and Poisson's ratio, and are compared with possible errors due to production and dimensional measurement of testpieces.

It should be noted that in the original derivation of the correction factors there is inevitably some error. The magnitude of the error is sometimes given in the original references, and in some cases upper limits are as large as the differences in truncation errors that are considered below. In this appendix these potential errors are ignored.

(a) Longitudinal resonance

The basic equation for longitudinal vibration of a cylindrical testpiece is

$$E = \left\{ \frac{16\,ml f_{\text{ln}}^2}{\pi d^2 n^2} \right\} \tag{8.A.1}$$

for the nth mode of vibration. For a rectangular section bar, $\pi d^2/4$ is replaced by bh, the cross-sectional area. Strictly this equation is valid only for a long thin rod or bar, and a correction term has been proposed (Love, 1964; Glandus, 1981) for the case where the rod cannot be considered long compared with its diameter. This factor is given by

$$X = 1 - \pi^2 v^2 \delta^2 \tag{8.A.2}$$

where $\delta = d / 2\lambda = dn / 4l$ for the nth mode, and modifies equation 8.A.1 to

$$E = \left\{ \frac{16 m l f_{\text{ln}}^2}{\pi d^2 n^2 X^2} \right\} \tag{8.A.3}$$

Inspection of the correction factor shows that the value of X^{-2} increases with increasing diameter to length ratio, and with increasing Poisson's ratio. The magnitude of this correction is given in Table 8.A.1 assuming Poisson's ratio is 0.3, a value higher than for most advanced technical ceramics in order to obtain a maximum likely value for the correction factor.

Table 8.A.1 Correction factor X^{-2} for longitudinal vibration

d / l		0.01	0.02	0.05	0.10	0.20
X^{-2}	$n = 1$	1.00001	1.00004	1.00028	1.00111	1.00446
	$n = 2$	1.00004	1.00028	1.00111	1.00446	1.01802

The results of the evaluation show that failure to use the correction results in underestimation of Young's modulus by a small amount. If it is assumed that the accuracy target for a standard test is 0.5%, it is desirable to keep the calculation error below 0.1%. The correction factor can therefore be ignored for the fundamental frequency if $d / l \leqslant 0.09$, or for the second mode if $d / l \leqslant 0.04$. For $n = 1$, this condition can be readily met and is an appropriate figure to use for a standard.

Table 8.A.2 Dimensional errors in longitudinal resonance

Source of error	Accuracy of measurement proposed	Magnitude of parameter	Effect of error using equation (8.A.1) (case (a))	Effect of error using separate density value (case (b))
Mass (g)	Weigh to 1 mg	10 g	< 0.01%	—
Length (mm)	Measure to 0.05 mm (calipers)	> 60 mm	< 0.1%	< 0.2%
Diameter (mm)	Measure to 0.01 mm (micrometer)	>3 mm	< 0.7%	—
Frequency (Hz)	Measure to fourth place	1000 to 50000 Hz	< 0.2%	< 0.2%
Density (Mg m^{-3})	Measure to Mg m^{-3} by immersion	> 2.5 Mg m^{-3}		< 0.4%

By comparison, the errors of measurement of dimensions can be calculated as shown in Table 8.A.2 for two cases: (a) when the equation is used in the form of equation 8.A.1 based on mass and dimensions, and (b) when the mass and dimensions are replaced by density measured independently by substituting ρ for $4m/\pi d^2 l$ (the equivalent substitution can be made for rectangular section bars). The minimum sizes chosen for the testpiece give a ratio $d/l = 0.05$. Non-uniformity of diameter is assumed to be less than $d/100$ for rod testpieces, but this is only important if calculation (a) is used.

Table 8.A.2 demonstrates that the largest error is in measuring rod diameter (or the corresponding parameters for a rectangular section bar) and that, for the d/h ratio chosen, an independent measure of density to 0.01 Mg m^{-3} will give a more accurate result for small-diameter bars, but remains the largest error. Thus the density form of the equation might be used with advantage, especially for testpieces which do not have particularly uniform diameter along their lengths, or which do not have accurately parallel faces. These results vindicate ignoring the Love correction unless the testpiece has $d/l > 0.1$ and the density is measured more accurately than to 0.01 Mg m^{-3} (assuming it is practically uniform to this level within the testpiece).

(b) Flexural resonance

There is a variety of equations used in the literature, with varying complexity depending on the correction functions employed. For a rectangular section bar using SI units, they all have the form

$$E = 0.9465 \left\{ \frac{mf^2 l^3}{bh^3} \right\} A_f \tag{8.A.4}$$

where A_f is a correction factor to account for inertia effects and shearing. This correction factor has terms in Poisson's ratio, and the depth to length ratio $h/l = x$. Goens (1931) was the first to evaluate this fully, followed by modification by Pickett (1945) and consolidation into a single form by Spinner, Reichard and Tefft (1960) to the following form, which is considered to be accurate to better than 1% for a wide range of specimen shapes, but much more accurate for slender bars:

$$A_f = 1 + 6.5850(1 + 0.752v + 0.8109v^2)x^2$$
$$- \frac{100.083(1 + 0.203v + 2.173v^2)x^4}{12 + 76.06(1 + 0.1408v + 1.536v^2)x^2} - 0.86806x^4 \tag{8.A.5}$$

where $x = h/l$. In the draft ASTM Standard for ceramics (Committee C28), the numerical terms of this equation are rounded to give

$$A_f = 1 + 6.58(1 + 0.08v + 0.8v^2)x^2 - \frac{100(1 + 0.02v + 2.2v^2)x^4}{12 + 76(1 + 0.14v + 1.54v^2)x^2} - 0.87x^4 \tag{8.A.6}$$

Inspection of this correction factor shows that its value is highly sensitive to the h/l ratio, being negligible for h/l small and increasing with increasing h/l.

In the analysis of an international round robin, Wolfenden *et al.* (1989) ignored this term completely. In existing ASTM Standards, C623 for glasses and C848 for ceramic whitewares, it is tabulated as a function h/l for various values of Poisson's ratio. In the proposed new ASTM Committee C28 Standard for advanced ceramics, the correction factor is used in the form shown in equation 8.A.6 to adjust iteratively the values of E and v from measurements of flexural and torsional vibration, although its use has now been dropped for bars of $h/l \leqslant 0.05$ when only the first two terms are required. In the Japanese Standard for ceramics, JIS R1602, it is truncated to the first two terms. There is therefore a considerable variety of approaches already employed in various standards, and in order to determine the most appropriate form to employ for the CEN draft standard, it was necessary to evaluate the correction function. It is clearly desirable to avoid having to employ the full equation 8.A.6 if a valid approximation can be developed that is accurate to within, say, 0.1%, which is smaller than the errors due to measurement of geometrical parameters.

Equation 8.A.6 can be rewritten in the form

$$A_f = 1 + A_1 x^2 + \frac{A_2 x^4}{A_3 + A_4 x^2} + A_5 x^4 \tag{8.A.7}$$

The relative values of the terms of the sum for $v = 0.3$ are evaluated in Table 8.A.3. This value of Poisson's ratio is possibly the highest likely to be determined for ceramics at ambient temperature.

Table 8.A.3 Evaluation of the terms of equation 8.A.6 for $v = 0.3$

$x = h/l$	0.01	0.02	0.05	0.1	0.2	0.5
Term 1	1	1	1	1	1	
Term 2	7.212×10^{-4}	2.885×10^{-3}	1.803×10^{-2}	0.0721	0.288	1.803
Term 3	1.047×10^{-7}	1.672×10^{-6}	6.432×10^{-5}	9.75×10^{-4}	0.0166	0.283
Term 4	8.7×10^{-9}	1.392×10^{-7}	5.43×10^{-6}	8.7×10^{-5}	1.3×10^{-3}	0.054
Total A_f	1.000722	1.00289	1.0181	1.0733	1.306	3.14

If the correction factor were to be completely ignored, in order to keep the accuracy of the Young's modulus calculation to better than 0.1% it would be necessary to have h/l less than 0.01. Such a slender specimen would be difficult to control and would need rather low frequencies for driving. It is therefore necessary to employ the correction factor to at least the second term, and this would require keeping h/l to less than about 0.05, which is much more reasonable. Under this condition, the third and fourth terms of equation 8.A.7 can be dropped since their maximum contribution is only 7×10^{-5} for $h/l = 0.05$. For bars with greater h/l, the full correction factor becomes necessary to retain 0.1% accuracy.

The evaluation of A_f in Table 8.A.3 is for $v = 0.3$. To check the evaluation for other values of Poisson's ratio, the value of A_f to the first two terms only is given in Table 8.A.4. It can be seen that the value of A_f (and hence the value of Young's modulus) is sensitive to Poisson's ratio to a level of about 0.1% over the range of expected values for the majority of advanced technical ceramics. The level of error might thus be significant if a Poisson's ratio factor is not included in the correction.

Table 8.A.4 Evaluation of A_f for varying Poisson's ratio and $h/l = 0.05$

v	0.1	0.2	0.3	0.4	0.5
Total A_f	1.01671	1.01724	1.01803	1.01908	1.02040

In conclusion, without loss of accuracy greater than 0.1%, we can write

$$A_f = 1 + 6.58(1 + 0.08v + 0.8v^2)(h/l)^2 = 1 + B(h/l)^2 \qquad (8.A.8)$$

with an upper limit of error of 0.1% for h/l less than or equal to 0.05. For $v = 0.3$ this reduces to

$$A_f = 1 + 7.21(h/l)^2 \qquad (8.A.9)$$

with an error of the order of 0.1% for values of Poisson's ratio higher or lower by 0.1. The formula given in JIS R1602 is slightly different. It assumes no particular value of Poisson's ratio, and is given as

$$A_f = 1 + 6.59(h/l)^2 \qquad (8.A.10)$$

In other words, this formula has used the assumption that $v = 0$, giving a systematic error understimating E by about 0.15%, as shown by Table 8.A.5. (Note that the factor 6.59 is obtained by upward rounding of equation 8.A.5, whereas equation 8.A.8 results from downward rounding.)

Table 8.A.5 Evaluation of B in $A_f = 1 + B(h/l)^2$ (equation (8.A.8))

v	0	0.1	0.2	0.3	0.4	0.5
B (eqn 8.A.8)	6.580	6.685	6.689	7.212	7.633	8.159
A_f for $h/l = 0.05$	1.0165	1.0167	1.0172	1.0180	1.0191	1.0204
Difference:						
$A_f(v = 0.3) - A_f$	0.15%	0.13%	0.08%	0	-0.11%	-0.24%

These errors must be compared with errors from other sources in making the measurement. These are summarized in Table 8.A.6, where it can be seen that by far the largest possible error is in the measurement of thickness, and in the parallelism of these faces. The physical limits of parallelism from surface grinding the testpieces is probably similar to the measurement accuracy, so it is difficult to reduce the error by setting the parallelism to better than $h/100$ for $h = 1$ mm. Ideally, the error can only be reduced by increasing the beam thickness to, say, 5 mm, giving 0.6% error in E. To reduce this error further down to the same significance as other errors would require a beam thickness of 10 mm, and with a ratio h/l of <0.05 this would require a testpiece of the somewhat impractical length of 200 mm (for advanced ceramics).

Table 8.A.6 Measurement errors in flexural resonance

Source of error	Accuracy of measurement proposed	Magnitude of parameter	Percentage error	
			in measurement	in formula
Mass m	Weigh to 0.001 g	>2 g	$<0.05\%$	$<0.05\%$
Length l	Measure to 0.05 mm (calipers)	>50 mm	$<0.1\%$	$<0.3\%$
Thickness h	Measure to 0.01 mm (micrometer)	1 to 2 mm	$<1\%$	$<3\%$
Width b	Measure to 0.01 mm (micrometer)	3 to 10 mm	$<0.3\%$	$<0.3\%$
Frequency (Hz)	Measure to fourth figure	100 to 20 000 Hz	$<0.1\%$	$<0.2\%$

In conclusion, therefore, if equation 8.A.8 is employed, the errors in the correction factor are insignificant, and the area for most attention is the precision of machining of the testpiece, its roughness and parallelism, and the measurement of its thickness in the direction of flexure. (Since this appendix was prepared, it has been agreed that for the purposes of a CEN Standard, equation 8.A.10 will be used to be in accordance with the ASTM Standard and JIS R1602, subject to the restriction that the length to thickness ratio is greater than or equal to 20.)

(c) Torsional resonance

There is closer agreement in the literature over the form of the equations to be employed for torsional resonance than for flexural resonance. In this case the relationship for a rectangular section is given by

$$G = \left(\frac{4lmf_t^2}{bh} \right) A_t \tag{8.A.11}$$

where the geometrical factor A_t has been calculated by Thompson (1939) and Spinner and Tefft (1961), and reviewed by Pickett (1945) and Spinner and Valore (1958). The Thompson version as used in the draft ASTM C28 Standard is

$$A_t = \frac{(h/b) + (b/h)}{4(h/b) - 2.52(h/b)^2 + 0.21(h/b)^6} \tag{8.A.12}$$

assuming $h < b$. Inspection of equation 8.A.12 shows that it is not insignificant for any value of h/b, and thus must always be employed. Table 8.A.7 shows the effect of the possible deletion of the sixth-power term. This illustrates that the error is greater than 0.1% when h/b is greater than about 0.3. Although from the point of view of encouraging torsional resonance it is desirable to have h/b less than 0.2, to restrict the validity of the equation in this way may be inappropriate, especially for the impulse excitation technique. In addition, the term does not add greatly to the complexity of the correction term, unlike the flexural case.

Table 8.A.7 Torsional resonance correction term, deletion of sixth-power term

h/b	0.05	0.1	0.2	0.5	1.0
A_t	103.51	26.948	7.437	1.8204	1.1834
A_t ignoring sixth-power term	103.51	26.948	7.437	1.8248	1.3514
Error (%)	0	0	0	$+0.24$	$+14.2$

The errors due to inaccuracy of dimensional measurements can be treated in a similar way to the longitudinal method. In this case, to estimate the errors in A_t due to errors in h and b, it is necessary to differentiate equation 8.A.12 with respect to h and b. Using the substitution $y = h/b$ equation 8.A.12 becomes

$$A_t = \frac{y + 1/y}{4y - 2.52y^2 + 0.21y^6} \tag{8.A.13}$$

Differentiating with respect to y gives

$$\frac{\mathrm{d}A_t}{A_t} = \left\{ \frac{1 - 1/y^2}{1 + 1/y} - \frac{4 - 5.04y + 1.26y^5}{4y - 2.52y^2 + 0.21y^6} \right\} y \left\{ \frac{\mathrm{d}h}{h} - \frac{\mathrm{d}b}{b} \right\} \tag{8.A.14}$$

in which $\mathrm{d}A_t$ is the error in A_t derived from errors $\mathrm{d}h$ in h and $\mathrm{d}b$ in b. This can be rewritten as

$$\frac{\mathrm{d}A_t}{A_t} = F(y) \left\{ \frac{\mathrm{d}h}{h} - \frac{\mathrm{d}h}{b} \right\} \tag{8.A.15}$$

Table 8.A.8 Torsion resonance: determination of errors due to dimensions

h/b	0.1	0.2	0.3	0.4	0.5	0.8	1.0
$F(y) = F(h/b)$	-1.832	-1.656	-1.467	-1.266	-1.051	-0.384	-0.103
Error $\mathrm{d}A_t/A_t$ ($\mathrm{d}h/h = 0.01$ $\mathrm{d}b/b = 0$)	-0.0183	-0.0166	-0.0147	-0.0127	-0.0105	-0.0038	-0.0010
Total RMS error*	0.0259	0.0234	0.0207	0.0179	0.0149	0.0054	0.0014

*Assuming the measurement error is random and the total is the root of the sum of the squares of the individual equal contributions from $\mathrm{d}h/h$ and $\mathrm{d}b/b$.

In Table 8.A.8, values of $F(y)$ and $\mathrm{d}A_t/A_t$ are tabulated in the second and third lines respectively for various h/b ratios for $\mathrm{d}h/h = \pm0.01$, the value set as the required accuracy of measurement of the test bar thickness in the draft CEN Standard. If the error db/b is set to the same level, the total RMS error (line 4 of Table 8.A.8) will be $\sqrt{2}$ times the error due to dh/h alone. In practice, the error in db/b is likely to be significantly less than dh/h in proportion to the shape of the testpiece, and for h/b less than 0.03 becomes insignificant. The data for the total error should thus be treated as pessimistic, and those in line 3 are probably more realistic.

It is clear from this table that the error in A_t is strongly dependent on the ratio h/b, ranging from 2.6% for $h/b = 0.1$ to 0.1% at $h/b = 1$. It should be noted that the errors scale linearly with choice of dh/h and db/b. In Table 8.A.9, the magnitude of this error is compared with other errors. Unlike the longitudinal case, Table 8.A.9 shows that the consequent error in G will depend principally on the measurement of h (and b if not significantly greater than h) whether or not dimensions or density are used in equation 8.A.11, and will be of the order of 1 to 2% for accuracies of measurement $h/100$ and $b/100$, but increasing with decreasing h/b. Improvement in the accuracy of G will depend on improving the accuracy of measurement of h and b (and probably also the uniformity of h and b). If a conventional micrometer is specified for measurement of thickness, the accuracy will be limited to 0.01 mm, giving a minimum specimen thickness of 1 mm (100dh). Since the error in A_t falls in proportion with that of h, the error can only be made insignificant relative to other errors if the specimen thickness increases to 5 mm or more. As with the longitudinal and flexural cases above, the limitation on accuracy for small ceramic test bars remains controlled by specimen dimensions.

Table 8.A.9 Errors in measurement of torsional modulus

Source of error	*Accuracy of measurement proposed*	*Magnitude of parameter*	*Percentage error*	
			in measurement	*in formula*
Mass m	Weigh to 0.001 g	> 2g	< 0.05%	< 0.05%
Length l	Measure to 0.05 mm (calipers)	> 50 mm	< 0.1%	< 0.1%
Thickness h	Measure to 0.01 mm (micrometer)	1 to 2 mm	< 1%	< 1%
Width b	Measure to 0.01 mm (micrometer)	3 to 10 mm	< 0.3%	< 0.3%
Frequency (Hz)	Measure to fourth figure	200 to 50 000 Hz	< 0.1%	< 0.2%
A_t	See Table 8.A.8	1 to 100	up to 2.6%	up to 2.6%

8.A.3 Conclusions

In general, the achieved parallelism and measurement accuracy of the testpiece are the principal sources of inaccuracy in making measurements of elastic moduli by resonance or impulse excitation techniques.

The magnitude of these errors on small test bars generally outweighs any questions of the accuracy of the calculation equations.

If a target accuracy of the formula to be used is set at 0.1%, which is substantially smaller than most other errors of measurement, then the following conclusions may be drawn concerning the equations to be employed:

- In longitudinal resonance, the correction term is unnecessary for length to diameter ratios of less than 20.
- In flexural resonance, the complex correction factors cited in various standards or drafts can be safely truncated or ignored for simplicity of calculation. This situation is achieved when the length to thickness ratio is greater than 20. It is recommended that equation 8.A.8 is used.
- In torsional resonance, while it is desirable to have a wide thin specimen for easy driving of resonance, curtailing the high-power correction term is not especially useful, and would restrict the validity of the equations to wide thin beams. This may be inconvenient for the impulse excitation method.

This analysis allows the safe use of selected forms of the equations, and indicates preferred ranges for the geometrical shape of testpieces.

REFERENCES

Adams, R. (1989) Measurement of elastic constants, in *Physical and Elastic Characterisation* (ed. M. McLean), *Characterisation of High Temperature Materials*, vol. 4, Institute of Metals, Chapter 1, pp. 1–23.
Bradfield, G. (1964) *Use in Industry of Elasticity Measurements in Metals with the Help of Mechanical Vibrations*, NPL Notes on Applied Science 3, HMSO, London.
Bressers, J. (1982) Axiality of loading, in *Measurement of High Temperature Mechanical Properties of Materials* (eds M. S. Loveday, B. F. Dyson and M. F. Day), HMSO, London, Chapter 17, pp. 278–95.
Broughton, W. R., Kumosa, M. and Hull, D. (1990) Analysis of the Iosipescu shear test as applied to unidirectional carbon-fibre reinforced composites. *Composites Science and Technology*, **38**, 299–325.
Christ, B. W. and Swanson, S. R. (1976) Alignment problems in the tensile test. *Journal of Testing and Evaluation*, **4**, 405–17.
Dawson, D., Loveday, M. S. and Varma, R. K. (1992) Young's Modulus: A Comparison of Ultrasonic, Resonance and Tensile Testing Techniques. Private Communication.
Day, M. F. and Harrison, G. F. (1982) Design and calibration of extensometers and transducers, in *Measurement of High Temperature Mechanical Properties of Materials* (eds M. S. Loveday, M. F. Day and B. F. Dyson), HMSO, London, Chapter 14, pp. 225–40.
Dean, G. D. (1974) Characterisation of fibre composites using ultrasonics, in *Composites – Standards Testing and Design*, IPC Science and Technology Press, pp. 126–30.
Dean, G. D. and Lockett, J. F. (1973) Determination of the mechanical properties of fibre composites by an ultrasonic technique, in *ASTM Special Technical Publication* **521**, pp. 326–46.

Dean, G. D. and Turner, P. (1973) The elastic properties of carbon fibres and their composites. *Composites*, **4**, 174–80.

Dickson, R. W. and Wachtman, J. B. (1971) An alumina standard reference material for resonance frequency and dynamic moduli measurement. I: For use at 25 °C. *Journal of Research of the National Bureau of Standards. Annals of Physics and Chemistry*, **75A** (3), May–June, 155–62.

Downs, M. J. and Raine, K. W. (1979) An unmodulated bi-directional fringe counting interferometer system for measuring displacement. *Precision Engineering*, **1**, 85–8.

Gittus, J. (1975) *Creep, Viscoelasticity and Creep Fracture in Solids*, Wiley (Applied Science), Chapter 9.

Glandus, J. C. (1981) Rupture fragile et résistance aux chocs thermiques de céramiques à usage méchaniques. Chapter 2: Modules élastiques et anélastiques des céramiques. University of Limoges Thesis.

Goens, E. (1931) Über die Bestimmung des Elästizitätsmoduls von Stüben mit Hilfe von Biegungschwingungen. *Annalen der Physik B*, **11**, 649–78.

Gould, D. and Loveday, M. S. (1990) *The Certification of Nimonic 75 Alloy as a Creep Reference Material*, CRM 42, BCR Euroreport EUR 13076 EN.

Hartmann, B. and Jarzynski, J. (1974) Immersion apparatus for ultrasonic measurements in polymers. *Journal of the Acoustical Society of America*, **56**, 1469.

Iosipescu, N. (1967) New accurate procedure for single shear testing of metals, *Journal of Materials*, **2** (3), 537–66.

Lord, J. D. (1992) *Intercomparison Exercise to Measure the Elastic Properties of a Fibre Reinforced Metal Matrix Composite*, NPL report DMM(A)83.

Love, A. E. H. (1964) *Mathematical Theory of Elasticity*, Cambridge University Press.

Loveday, M. S. (1986) High temperature axial extensometry: standards, calibration and usage, in *Strain Measurement at High Temperature* (ed. E. C. Hurst), Elsevier Applied Science, pp. 31–47.

Loveday, M. S. (1991) *Standards for the Calibration of Extensometers*, NPL report DMM(A)41).

Loveday, M. S. (1992) Towards a tensile reference material, in *Harmonisation of Testing Practice for High Temperature Materials* (eds M. S. Loveday and T. B. Gibbons), Elsevier Applied Science, Chapter 7.

Musgrave, M. J. P. (1970) *Crystal Acoustics*, Holden-Day, San Francisco.

NAMAS (1989) *The Criteria for Accreditation of Calibration or Testing Laboratories*. NAMAS M10 Accreditation Standard.

Pickett, G. (1945) Equations for computing elastic constants from flexural and torsional resonant frequencies of vibration of prisms and cylinders. *Proceedings of the ASTM*, **45**, 846–65.

Read, B. E. and Dean, G. D., (1978) *The Determination of Dynamic Properties of Polymers and Composites*, Adam Hilger, Bristol.

Read, B. E., Dean, G. D. and Duncan, J. C. (1991) Determination of dynamic moduli and loss factors, in *Physical Methods in Chemistry*, vol. 7, Wiley, Chapter 1, pp. 1–70.

Roche, R. and Loveday, M. S. (1992) Harmonisation and improvements in European Standards, in *Harmonisation of Testing Practice for High Temperature Materials* (eds M. S. Loveday and T. B. Gibbons), Elsevier Applied Science, Chapter 3.

Roebuck, B., Lord, J. D., Cooper, P. M. and McCartney, L. N. (1993) Data acquisition and analysis of tensile properties for MMC. *Journal of Testing and Evaluation*, **22** (1), 63–69.

Roebuck, B., McCartney, L. N., Cooper, P. M., Bennett, E. G., Lord, J. D. and Orkney, L. P. (1992) *UK Interlaboratory Tensile Tests on an Al alloy/SiC Particulate MMC*, NPL report DMM(A)77.

Spinner, S., Reichard, T. W. and Tefft, W. E. (1960) *Journal of Research of the National Bureau of Standards*, **64A** (2), 147–55.
Spinner, S. and Tefft, W. E. (1961) A method for determining mechanical resonant frequencies and for calculating elastic moduli from these frequencies. *Proceedings of the ASTM*, **61**, 1221–38.
Spinner, S. and Valore, R. C. (1958) Comparison of theoretical and empirical relations between the shear modulus and torsional resonant frequencies for bars and rectangular cross sections. *Journal of Research of the National Bureau of Standards*, **60** (5), 459–64.
Thompson, W. T. (1939) Effect of rotary and lateral inertia on flexural vibration of prismatic bars. *Journal of the Acoustical Society of America*, **11**, 198.
Weissberg, V. and Arcan, M. (1988) A uniform pure shear testing specimen for adhesive characterisation, in *Adhesively Bonded Joints: Testing, Analysis and Design*, ASTM STP981, pp. 28–38.
Wolfenden, A. (1990) *Dynamic Elastic Modulus Measurements in Materials*, 16 papers presented at a seminar in Kansas City, Missouri, May 1989, ASTM STP1045.
Wolfenden, W. T. *et al.* (1989) Dynamic Young's modulus measurement in metallic materials: results of an interlaboratory testing programme. *Journal of Testing and Evaluation*, **17** (1), 2–13.
Zimmer, J. E. and Cost, J. R. (1970) Determination of the elastic constants of a unidirectional fibre composite using ultrasonic velocity measurements. *Journal of the Acoustical Society of America*, **47**, 795.

STANDARDS

ASTM Standards

ASTM E28-92	Standard Test Methods for Room Temperature Dynamic Determination of Modulus of Elasticity, Shear Modulus, and Poisson's Ratio for Elastic Materials. Committee E28. Draft document, submitted for balloting.
ASTM E111-82 (88)	Standard Test Method for Young's Modulus, Tangent Modulus and Chord Modulus.
ASTM E494-92	Standard Practice for Measuring Ultrasonic Velocity in Materials.
ASTM C623-92	Test Methods for Young's Modulus, Shear Modulus and Poisson's Ratio for Glass and Glass-Ceramics by Resonance.
ASTM C747-74 (88)	Test Methods for Moduli of Elasticity and Fundamental Frequencies of Carbon and Graphite Materials by Sonic Resonance.
ASTM C848-88	Test Methods for Young's Modulus, Shear Modulus, and Poisson's Ratio for Ceramic Whitewares by Resonance.
ASTM E1012-93	Practice for Verification of Specimen Alignment under Tensile Loading.
ASTM C1198-91	Standard Test Method for Dynamic Young's Modulus, Shear Modulus and Poisson's Ratio for Advanced Ceramics by Sonic Resonance.
ASTM D3039-93	Standard Test Method for Tensile Properties of Polymer Matrix Composite Materials.

British Standards

BS 18 : 1987	Methods for Tensile Testing of Metals.
BS 2782 : Part 3	Methods of Testing Plastics. Part 3: Mechanical Properties. Methods 320A to 320F. Tensile Strength, Elongation and Elastic Modulus.
BS 3500 : Part 3 : 1987	Methods for Creep and Rupture Testing of Metals. Part 3: Tensile Creep Testing.
BS 3688 : 1989	Methods for Mechanical Testing of Materials at Elevated Temperature. Part 2: Tensile Testing.
BS 3846 : 1985	Methods for Calibration and Grading of Extensometers for Testing Metals.

EN Standards

ENV 843 : 1994	Methods of Test for Advanced Monolithic Technical Ceramics. Mechanical Properties at Room Temperature. Part 2: Determination of Elastic Moduli.
EN 2561	Aerospace Series. Carbon Thermosetting Resin Unidirectional Laminates: Tensile Test Parallel to Fibre Direction.
EN 2562	Aerospace Series. Carbon Thermosetting Resin. Unidirectional Laminates. Flexural Test.
EN 10002 : Part 1 : 1990	Tensile Testing of Metallic Materials. Part 1: Method of Test at Ambient Temperature.
EN 10002 : Part 2 : 1992	Tensile Testing of Metallic Materials. Part 2: Verification of Tensile Testing Machines.
EN 10002 : Part 4 : 1993	Tensile Testing of Metallic Materials. Part 4: Verification of Extensometers Used in Uniaxial Testing.
EN 10002 : Part 5 : 1992	Tensile Testing of Metallic Materials. Part 5: Method of Test at Elevated Temperatures.

ISO Standards

ISO 178 : 1992	Plastics. Determination of Flexural Properties.
ISO 527 : 1992	Plastics. Determination of Tensile Properties. Part 1: General Principles. Part 2: Test Conditions for Moulding and Extrusion Plastics.
ISO 5725 : Part 1 : 1991	Accuracy (Trueness and Precision) of Measurement Methods and Results. Part 1: General Principles and Definitions.
ISO 6721 : 1992	Plastics. Determination of Dynamic Mechanical Properties. Part 1: General Principles. Part 2: Torsion Pendulum. Part 3: Bending Vibration – Resonance Curve.
ISO 9513	Metallic Materials. Verification of Extensometers Used in Uniaxial Testing.

CHAPTER 9

Metrology for engineering materials

M. K. Hossain and I. R. Sced

9.1 INTRODUCTION

In recent years rapid advances have been made in materials technology, and the trend is set to continue with substantial investments being made by countries world-wide in materials development, processing and applications. The use of modern materials pervades all of industry and has a strong influence on its competitiveness. However, to convert these materials into competitive engineering products, the designer requires access to appropriate design methodologies that specify material property requirements. The need for reliable design data generated using validated and widely recognized measurement methods then becomes crucial. Furthermore, quantitative assessment of materials behaviour and performance is essential to achieve quality and reliability of products. Materials measurements have an important and widespread impact. This is illustrated in Fig. 9.1, which shows that the demand for the development of new or improved measurement methods originates from materials producers to enable them to provide reliable data that the market needs. There is also demand for data from users of materials so that products can be made and supplied to the market with a reliable expectation of performance.

Measurements methods, once developed, would then be used on a range of materials in order to prove and improve the method, leading to a number of activities:

- assessment of the quality of the measurements (accreditation and traceability);
- validation of the method;
- harmonization;
- prediction (short cuts to avoid numerous and long-term tests).

The output of the work leads to standards as a key deliverable, with codes of practice and reference materials being developed either before standardization or as a consequence of the standard. The beneficiaries ultimately of this type of work are trade, industry and government.

In this chapter we discuss some of the key issues in materials metrology and the relationships of the national measurement system. The term 'materials metrology' is probably an unfamiliar one. It can be defined simply using the normal meaning of the word 'metrology' as 'the science of measurement as applied to materials'. Materials metrology is a convenient way of bringing together an increasing awareness of the need for a rigorous approach to the measurement of materials properties; for an improved understanding of the sources of variability; for the consequent development of improved measurement methods; and for their eventual harmonization for international acceptance. All this is aimed to improve the quality and reliability of materials data and to generate a recognized framework for materials measurements.

9.2 COMPARISON OF MATERIALS METROLOGY WITH PHYSICAL METROLOGY

9.2.1 Physical metrology

It is instructive to consider briefly the measurement system that is well established internationally for physical quantities such as length, mass, temperature, force, pressure, time and voltage – all of fundamental importance to manufacturing industry and to society in general. The central theme of such a system is that all measurements, wherever they may be made, should be traceable through a hierarchical system directly to the national measurement standards, which in turn are coordinated with those of the other trading partners. The measurement standards system can be envisaged as having those basic standards with the highest accuracy at the top, with an unbroken chain of traceable calibrations down to different working levels. In the measurement of length, for example, the shop floor standards can, in principle, be traced through calibrations to primary standards, with the accuracy improving at each level. Furthermore, the primary standard is applicable, albeit indirectly, to the most wide-ranging usage in an industrial society.

9.2.2 Materials metrology

The measurement infrastructure for materials, on the other hand, is much more complex and, as a result, there is a lack of internationally recognized compatibility for the measurement of key properties based on a sound understanding of the science underpinning materials behaviour.

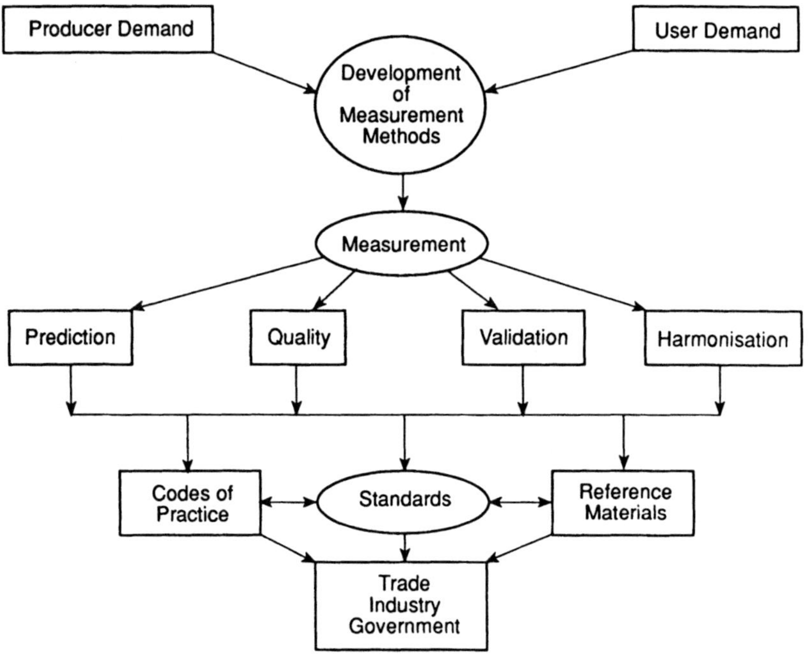

Fig. 9.1 The impact of materials metrology in a wider context.

Consider three sets of measurements related to an industrial product or a component:

- its dimensions;
- the strength and stiffness (modulus) of the constituent material;
- the resistance of the material to corrosion and wear.

To a manufacturer it is essential that a product can be made to known dimensions and checked by measurements; however, this on its own is not sufficient for the market-place which dictates that the product must also be fit for the purpose for which it was built. Immediately, one can see the requirements of a reliable databank of measured materials properties such as strength, modulus and corrosion resistance.

For any measurement there are four successive requirements:

1. definition of the measurand, i.e. the quantity to be measured;
2. definition of conditions under which the measured value is to be applied in practice;
3. development of an adequate measurement method giving reproducible results; and
4. establishment of accuracy and traceability.

In the case of dimensional measurements, and indeed the measurement of other basic physical properties, all four requirements mentioned above can be generally met, notwithstanding the fact that research is needed to develop new and improved methods with higher accuracies for complex measurement needs, driven by increasing sophistication of industrial products and manufacturing technology.

For the measurement of strength and modulus of the constituent materials of a product, the concepts are readily understood and defined so that these properties can be determined for established materials and applications using standardized methods with potentially known accuracy and traceability. However, for newer (composite) materials or for new or established materials exhibiting non-linear, viscoelastic and viscoplastic behaviour, validated measurement methods are not always available, with the consequence that requirements 3 and 4 above cannot be met satisfactorily at present. Indeed, as more materials with tailored properties are developed, and materials are used near their limits of performance in complex and demanding engineering structures, there is an increasing concern with the validity and accuracy of materials testing and analysis.

When one considers the corrosion and wear properties of materials, the issues become more complex and even the precise definition of the quantity to be measured is often debatable. From a practical point of view, measurements tend to be more application and environment specific and thus it becomes difficult to develop a simple hierarchical measurement framework. Detailed treatments of the state of the art in these areas are given in Chapters 11 and 12.

9.3 EXAMPLES OF MATERIALS MEASUREMENT PROBLEMS

9.3.1 Modulus measurement of metal matrix composites (MMCs)

In recent years, the advent of quality ceramic fibres and particulate reinforcements has led to considerable development of metal matrix composites using light metals such as Al, Ti or Mg as the matrix. These materials offer high strength, stiffness and the potential for operating at elevated temperatures, with properties that can be tailored for a particular application by a suitable choice of matrix and reinforcement. Likely applications are in diverse fields such as airframes, reciprocating parts in automobiles, substrates for electronics assemblies and leisure goods.

At present there is only one standard developed specifically for MMC. This standard, ASTM D3552, relates to the measurement of tensile properties of fibre reinforced MMCs. For isotropic MMCs, such as particulate composites, conventional metal standards need to be assessed, but there are limitations to the degree to which these standards can be

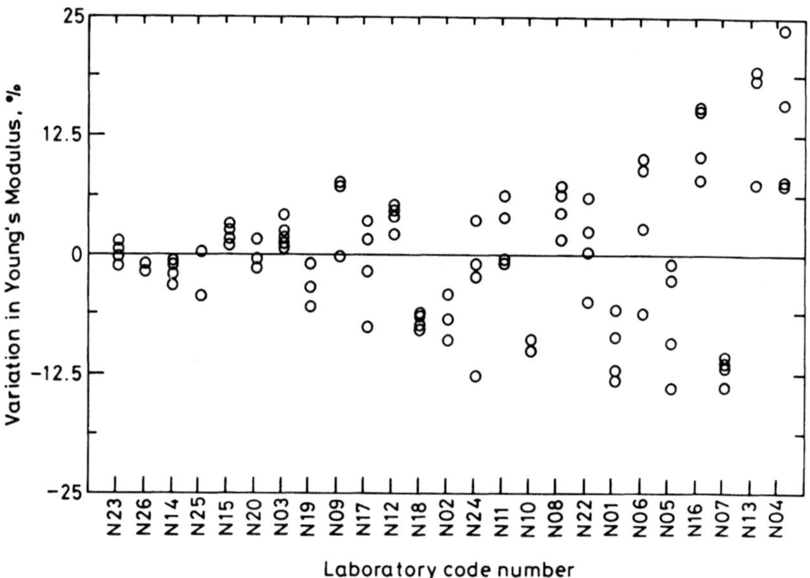

Fig. 9.2 Results from round robin exercise on measurement of Young's modulus of an aluminium MMC showing percentage variation from a mean value of 98 kN mm^{-2} (Roebuck and Lord, 1991).

applied. Hence the accurate determination of even a basic property such as modulus poses problems.

Figure 9.2 shows the results of a round robin exercise carried out by the NPL on the measurement of Young's modulus for a silicon carbide particulate reinforced aluminium alloy: the observed variation is clearly unsatisfactory. The results show that the variation is not due to materials scatter but is related to metrological problems and a poor understanding of their importance. The three most significant factors that emerged were as follows (Roebuck *et al.*, 1993; Chapter 8 in this book).

- The importance of taking an average of readings from two strain measurement devices to measure longitudinal strain. In this case, flat testpieces were being used; from other work there are indications that for some materials, when symmetrical testpieces are used, four orthogonally positioned strain measuring devices may be required.
- The need for a reliable technique to calculate the elastic modulus from experimental data.
- The importance of correctly calibrated extensometry.

Thus, for these newer high-performance materials, measurement and analyses of data require a much more rigorous treatment than for conventional materials.

9.3.2 Fracture toughness testing for MMCs

The tolerance of a material to the existence of cracks arising from manufacture or service, known as fracture toughness, is a very important engineering parameter. The plane strain fracture toughness K_{IC} is used by engineers and designers in structural integrity calculations to determine the relationship between defect size and applied stress. Fracture toughness testing of particulate MMCs (Roebuck and Lord, 1991) has indicated that it can be difficult to obtain plane strain fracture toughness values that are valid according to criteria specified in standards such as ASTM E399 and BS 5447 for metallic materials. Results can be invalid because of excess crack curvature, possibly caused by macroscopic residual stresses, and excessive nonlinearity of the load–deflection curves. The latter is thought to be affected by crack closure and/or crack bridging mechanisms.

A rigorous metrological approach is helping to elucidate these mechanisms and will lead to either new testing techniques or new acceptance criteria that will enable reliable fracture toughness testing (Roebuck and Lord, 1991; Roebuck, 1992).

9.3.3 Creep of polymeric materials

In order to characterize the creep of a polymer, a minimum of some 25 sets of measurements is required (ISO/DIS/11403/1). To extend the data to a useful temperature range increases this to several hundred measurements. When one considers that a single basic polymer can exist on the market in many hundreds of different grades, the magnitude of the testing task is formidable.

An alternative approach is to develop a validated model, and to do this it is necessary to understand the effects of physical ageing in polymers. With increasing elapsed time t after processing, continuous increases in density are observed for most solid polymers owing to molecular rearrangements associated with the approach to structural equilibrium. This physical ageing process yields pronounced decreases in molecular mobility and related properties such as creep rate, impact strength, dielectric constant and loss factor.

Figure 9.3 shows the variation of the compliance $D(t)$ with log t at 23 °C for polypropylene samples of different ages (Read, Dean and Tomlins, 1988). Here $D(t)$ is defined as the time-dependent longitudinal strain divided by the constant uniaxial applied stress. At long times, the experimental compliances (point symbols) are considerably lower than the values obtained by extrapolations (broken lines) of short-term data

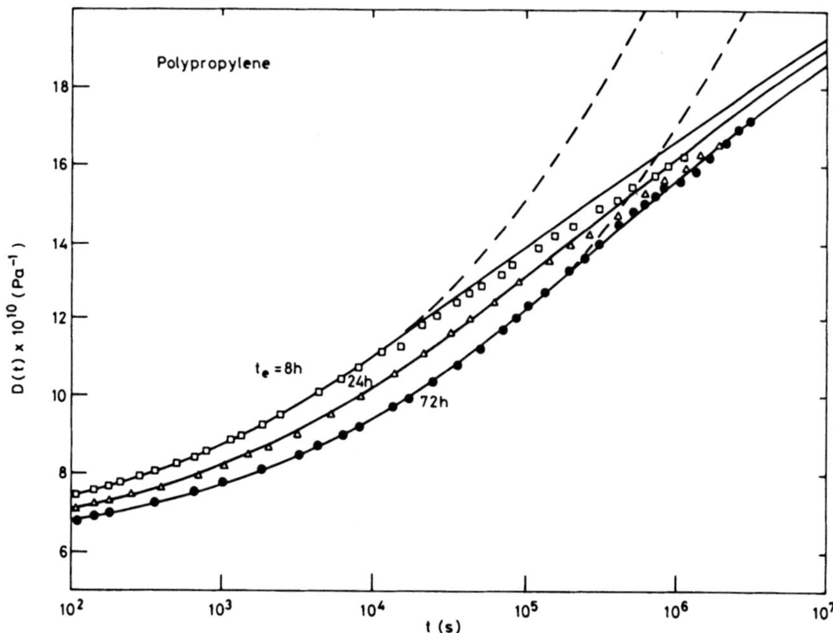

Fig. 9.3 The effect of physical ageing on the prediction of creep compliance for polypropylene (Read, Dean and Tomlins, 1988).

which have neglected further ageing during the test period. However, if account is taken of polymer ageing, the predicted compliances (continuous curves) are within 3% of the observed values for two or three decades beyond the short-term region. If, therefore, a polymeric component is subjected to a constant load one month after processing, its deformation may be confidently predicted for times greater than 10 years from tests of about one week's duration. A model based on known physical principles associated with accurate traceable measurements can thus provide a cost-effective answer to the measurement problem in this case.

9.3.4 Tensile testing

When the application of metrological principles to materials measurement is considered, it soon becomes apparent that many of the problems are generic and apply to some extent to many materials. Their relative importance will, of course, often be material specific. Typical generic problem areas in tensile testing are, for example, the testpiece alignment, the means of stress transfer and the effects of residual stresses and stress concentrations. It will be immediately appreciated that these do not act in isolation but more often interact with each other. Testpiece alignment, for example, becomes more critical with brittle materials where bending stresses can significantly influence the result, and this effect would interact clearly with residual stresses and stress concentrations. (Amaral and Pollock, 1989). Alignment is also important when determining certain properties in established engineering materials and must be addressed if good pedigree databases are to be produced (see Chapter 7).

Another example is that of stress transfer into the testpiece. Machine grips may typically be hydraulic, screw clamp or wedge box, or the testpiece may be threaded or have a shouldered design. Dependent upon the grip system, the gripping stresses may be relatively constant during a test, or may change with load. Combine this with stress concentration effects, particularly for parallel testpieces such as those used for aligned fibre composites (in which for the material to function correctly the load must be transferred to the fibres), and it is found that the test result can depend critically on the gripping system (Gray and McCombe, 1992; Gray and Sharp, 1988).

9.4 AN INFRASTRUCTURE FOR MATERIALS METROLOGY

The fundamental need for materials metrology stems from the fact that users must have reliable data on materials properties and performance for the gamut of materials potentially of interest to them. Data and testing are required in support of quality control of raw materials, evaluation of new and improved materials, materials selection, preliminary product design, process selection, detailed product design, quality assurance during manufacture, product assessment, monitoring of in-service performance, residual life and so on. These wide-ranging activities require information on a range of material parameters. For structural applications of engineering materials, data may be required on:

- mechanical properties e.g. stiffness, strength, creep, toughness, impact, fatigue and wear;
- environmental properties, e.g. resistance to heat, light, humidity and corrosive gases;
- surface properties, e.g. composition, structure and adhesion;
- process-related properties, e.g. viscosity, surface tension and heat transfer.

Functional materials may require measurements of electrical, optical, magnetic or a combination of these properties. Given the wide range of materials now available, the field of material measurement, testing and analysis is very large. What complicates matters further is the fact that

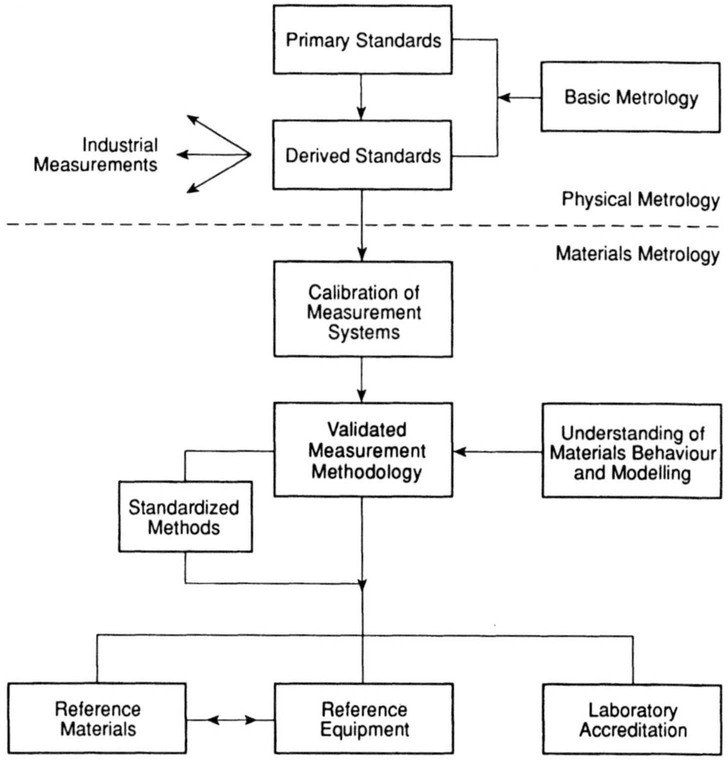

Fig. 9.4 The relation between physical metrology and materials metrology.

different material classes often exhibit special features of behaviour which must be taken into account in a measurement method if the results are to be valid and of practical value.

How does one ensure quality in measurement? A key consideration in this regard is traceability, but this is more difficult to achieve in materials property measurements than for physical measurements and creates a significant challenge. It seems that a materials measurement infrastructure cannot just be hierarchical in nature, but requires a multi-pronged approach involving reference (standardized) methods, reference machines, reference materials and laboratory accreditation. This concept can be described by Fig. 9.4. The link between physical metrology and materials metrology is established through the traceable calibration of the equipment used for the measurement of materials properties. Primary standards of mass and length, for example, provide the basis for calibration of load and displacements via secondary or derived standards. To develop measurement methods that are reliable, validation is required, and this can only be achieved through a proper scientific understanding of materials behaviour. Modelling is often necessary to enable interpolations and extrapolations of measurements to be carried out, as well as to identify the correct measurement parameters and the conditions under which these measurements should be made.

Once a validated method has been established, standards can be developed and a combination of reference materials, reference equipment and laboratory accreditation may be used to make traceable measurements of known uncertainties as discussed below. The four key strands of a measurement infrastructure for materials are: standardized measurement methods; reference metrological machines; reference materials; and laboratory accreditation. These are now discussed in more detail.

9.4.1 Standardized measurement methods

In order to generate materials property data of high pedigree, the measurement methods must be clearly understood and the effects of all the significant variables quantified. Documented measurement methods can then be produced so that prescribed levels of repeatability and reproducibility can be achieved using material exhibiting 'zero' or low scatter. Once the measurement method has been validated to enable uncertainties to be stated, the method can be standardized and harmonized on a global basis. Statements of measurement uncertainty are only now becoming mandatory in testing standards.

9.4.2 Reference metrological machines

Another key component of the materials measurement infrastructure should be metrological machines for certain measurements. This will take the form of a state-of-the-art machine employing best known practice. Such a machine can then be used to explore rigorously the effects of all the significant testing variables on the measurement results. A clear understanding of the effects of these variables may then be used to define the real limits of practical testing. It is by using this approach that some researchers are now beginning to find that certain long-established and accepted methods of material testing have not been clearly understood or properly validated (Chapters 7 and 8).

An interesting example of an early approach to the reference machine concept is that of a Charpy pendulum impact testing machine that was installed at the NPL on behalf of the Bureau Communitaire de Référence (BCR) in 1981. This machine is maintained to higher metrological tolerances for its critical parts than those required by any current standard. It has been used in various intercomparison exercises over a number of years and been shown to have good repeatability and reproducibility using impact reference materials (Marchandise, 1992). As a consequence, it has been selected as one of a limited number of European machines for use on certification of impact toughness V-notch Charpy testpieces. The key point was the decision to maintain a dedicated, high-precision machine purely for investigative and certification work.

Brief mention has been made here of the certification of reference testpieces, and that conveniently brings us to consider the role of such reference materials.

9.4.3 Reference materials

The certified reference material (CRM) is a further necessary component of materials metrology. A CRM has been defined by the International Standards Organization (ISO) as a material or a substance of which one or more physical or chemical properties are sufficiently well established to be used for the calibration of an apparatus or for the verification of a measurement method. The material is accompanied by, or traceable to, a certificate stating the property value of concern and is issued by an organization (frequently recognized nationally or internationally) which is generally accepted as technically competent.

Such materials are well characterized with *defined* repeatable and reproducible properties. In certain circumstances, these can provide a means of calibration for materials testing, although this approach should only be adopted when calibrations cannot be achieved through the normal measurement standards system. Such is the situation, for example, in the case of hardness measurement where it becomes necessary to check standardized metrological machines against standard hardness blocks. These could be regarded as transfer standards.

Although such a system works satisfactorily for conventional materials, the determination of hardness in engineering ceramics and other very hard materials, which are in the advanced materials category, require refinement in the measuring equipment and techniques as well as the analysis of results (Butterfield, Clinton and Morrell, 1989). Hence an internationally acceptable framework does not yet exist for this purpose.

For advanced materials in general, there will be a need for numerous reference materials since each will have a relatively narrow application range depending on the material type and usage. NPL has been active in developing reference materials for mechanical testing and surface analysis. For example, with support from BCR, NPL has recently produced a reference material of nimonic 75 for creep testing at 600 °C and 160 MPa (Gould and Loveday, 1992). The material was certified for creep strain measurement using results from a round robin exercise in which 12 laboratories from six nations participated, and the measurements from over 60 creep tests, each lasting in excess of 600 hours, were used to determine the certified properties. By using robust procedures and careful statistical analyses of experimental results, the uncertainty in the certified value of the minimum creep rate has been reduced to a low level of about ± 11%, with

95% confidence. Part of the same batch of the solid solution nickel-base nimonic 75 is now in the process of being certified for use as a room temperature tensile CRM. Such reference materials are becoming increasingly vital for accreditation purposes for laboratories generating materials data that are needed to design competitive plants and products.

9.4.4 Laboratory accreditation

This is the fourth element in the materials measurement infrastructure, and signifies recognition of a testing laboratory's competence to perform test methods in specified fields. It provides confidence in the overall quality of operation of a laboratory with regard to its staff, facilities, equipment, calibration, test methods and procedures, recording of results, and production of reports. Accreditation, however, does not guarantee that an individual test result or datum point is correct or valid; it merely judges the competence of a laboratory. Chapter 10 deals with this aspect of materials metrology in detail.

9.5 MEASUREMENT QUALITY

Two aspects of materials data require further discussion: accuracy and traceability. When referring to 'accuracy' one has to distinguish between the precision achievable by the measuring instruments or methods, and the accuracy meaning how close the measured value is to the real value. Usually for material properties the real value cannot be calculated from theoretical considerations, making it difficult to define the accuracy of results. In the measurement of materials properties it is best to refer only to repeatability, defined as the variability of results obtained on a single site by a single operator using a single measuring instrument, and to reproducibility, which refers to the variability between sites with different measuring equipment and different operators. Precision of measurement, often loosely referred to as 'accuracy', is used in documented standards for material testing (ISO 5725).

In materials testing, measurement uncertainties are not regularly quoted, and this is a matter of some concern. The situation is beginning to change, however; the drive is coming partly from the standards bodies and partly from laboratory accreditation bodies. For instance, it is the declared policy of ISO and CEN/CENELEC/ECISS that all new standards concerned with testing techniques will contain a 'statement of uncertainty' or a method of calculating the 'accuracy' of the test method based upon tolerances specified in the relevant standard. Examples of recent work in this area can be found for tensile testing (Loveday, 1992).

As regards traceability, it is important not only that the data should be generated using methods and equipment that have been traceably calibrated, but that the data once generated are themselves traceable to their source. In this way the data will acquire a pedigree that will serve as a definition of the quality of the data. A further important point that must also be considered relates to the increasing use of built-in or custom-written software for data capture and processing in materials testing. Here traceability must exist from the original measurements through the data processing procedure to the final results.

9.6 ORGANIZATIONAL STRUCTURE FOR MATERIALS METROLOGY

In physical metrology there is a wide and long-standing international structure including many national standards laboratories. This is not the case with materials metrology. In the developed countries, with the exception of a few laboratories like the USA (National Institute of Science and Technology), the UK (National Physical Laboratory), Germany (Bundesanstalt für Materialforschung und -prüfung) and Japan (Ministry of International Trade and Industry, Science and Technology Agency Laboratories), materials metrology work does not have a strong focus and involves a combination of many government, private sector and academic institutions. This means international agreements and harmonization of the measurement framework are more difficult to achieve in the materials field than for physical metrology.

9.6.1 Development of standards

Standards for test and measurement methods have a very important role in materials metrology. However, rapid advances in the technology pose problems for the setting of standards. Although continual progress is being made in relating materials characteristics to performance, the quantitative basis from which widely acceptable standards for test and measurement methods can be developed are not available in many areas (Hossain, 1992a). Without such standards, there are often disagreements not only on results but sometimes on what is actually being measured.

World-wide, there is now a drive from industry to develop internationally harmonized standards for testing, and there is strong and growing activity in Europe, Japan and the USA. This is not surprising because design data generated using standard test methods enable engineers to establish whether a new product made from an advanced material or process is likely to perform better in service than an existing product made from conventional materials. Therefore, they help to stimulate the market for advanced materials and products made from them. Also, it is widely accepted that standards support:

- materials specification;
- quality control;
- consumer protection;
- health and safety.

Who has a role in materials metrology? Clearly, standardization bodies since they have the responsibility for the generation of documented measurement methods that laboratories must work to. These standards will in turn need measurement and testing methods which must be developed and validated before they can be used for standards testing. It is here that laboratories such as the NPL play a major role.

Measurement methods must be developed against a background of sound materials knowledge. Wherever appropriate and possible, metrological machines should be used to define clearly the limits on testing variables that contribute to scatter in the measurement result. The active participation of test laboratories and industrial laboratories is important at all stages. Finally, the national accreditation bodies – who have a role in facilitating documented standardized measurement methods and traceable calibration – are an integral part of the generation of materials data.

9.6.2 The role of pre-standards research

In pre-standards research, which helps to generate the technical basis from which commonly accepted standards and specifications can be prepared, international collaboration is vital. A mechanism already exists through a body known as VAMAS (Hossain, 1992b) which operates under a Memorandum of Understanding signed in 1987 by the G7 countries and the Commission of the European Communities. Many research groups from government, industry and academic laboratories are working together in 14 technical areas covering a wide range of advanced materials (see Appendix to this chapter). Measurement methods are being developed for thermal, mechanical, electrical and environmental properties of materials. Based on VAMAS results, three standards have been completed so far and 11 more are in prospect.

The work of VAMAS may be described by using the example of Technical Working Area 2 (TWA2), which is dealing with surface chemical analysis techniques such as Auger electron spectroscopy (AES), X-ray photoelectron spectroscopy (XPS) and secondary ion mass spectroscopy (SIMS), which are being widely used in industry. The objective of the VAMAS programme in this area is to produce, by an internationally coordinated effort, the reference procedures, reference data and reference materials necessary to establish standards for surface chemical analysis. The area is led by NPL with nearly 100 laboratories participating from about 10 countries to cover 30 separate projects.

Already the VAMAS work has produced reference materials, a standard data transfer format for spectral data, and recommendations for standardized energy and intensity scale calibration methods for AES. These methods have full traceability to the relevant measurement standards in physical metrology. The work of TWA2 has been considered so important that a new ISO Technical Committee has been set up which in future will accelerate the transfer of the VAMAS TWA results into international standards.

9.6.3 Future

To build on this type of work, what is needed is the identification of key laboratories in various countries dealing with materials metrology. With a wider international network it should be possible to expand the activity and also make better use of the limited resources available for this type of work. Also important to recognize here is that such a network should help to accelerate the generation of internationally harmonized test and measurement standards for materials properties, thereby promoting international trade in advanced materials and products using them.

9.7 CONCLUSIONS

Compared with physical metrology, materials metrology is quite complex and less hierarchical. However, it is of crucial importance for ensuring reliability and quality of industrial products.

Key aspects of metrology such as repeatability, reproducibility and traceability require the development of validated measurement methods based on sound physical understanding of materials characteristics and behaviour. With increasing application of IT in manufacturing technology and the push towards the limit of performance of modern materials, the need for more accurate properties data is growing rapidly.

To aid international trade in advanced materials and their products, standardized test and measurement methods developed with good metrological bases are required. There is an international drive to produce a range of essential standards for advanced materials.

The organizational structure for materials metrology is not as well focused as in physical metrology. However, a number of key laboratories such as NPL (UK), NIST (USA), BAM (Germany) and MITI/STA Laboratories in Japan are providing the nucleus of a wider international network operating under VAMAS to develop a widely recognized infrastructure for materials measurement. The position should be strengthened through wider participation from industry and government.

ACKNOWLEDGEMENT

The chapter has been prepared as part of the Materials Measurements Programme of underpinning research financed by the United Kingdom Department of Trade and Industry.

APPENDIX 9.A   VAMAS TECHNICAL WORKING AREAS

TWA no.	*Title*	*Materials under study*	*Chair*
TWA1	Wear Test Methods	Alumina, silicon nitride, AISI 52100 steel	USA, Germany
TWA2	Surface Chemical Analysis	Wide-ranging reference materials, metallic and non-metallic	UK
TWA3	Ceramics	Alumina, zirconia-alumina	USA
TWA4	Polymer Blends	Polycarbonate/ polyethylene blend, Orgalloy R-6000 commercial blend	UK
TWA5	Polymer Composites	Glass and carbon fibre reinforced resins	UK
TWA6	Superconducting and Cryogenic Structural Materials	Niobium-tin and niobium-titanium filaments, cryogenic steels	Japan
TWA7	Bioengineering Materials	Hydroxyapatite, alumina, zirconia	Japan
TWA8	Hot-Salt Corrosion Resistance	Rene 80 and IN738, nickel-base superalloys, protective coatings	UK
TWA9 (completed)	Weld Characteristics	304 and 316 stainless steels	UK
TWA10	Material Databanks	Creep and fatigue data from low- and high-alloy steels	EC, USA
TWA11 (completed)	Creep Crack Growth	Chromium/molybdenum/ vanadium ferritic steels	UK
TWA12	Efficient Test Procedures for Polymer Properties	Wide range of polymers	UK
TWA13	Low-Cycle Fatigue	IN718 and nimonic 101 nickel-base alloys, 316L and 9Cr/1Mo steel	EC
TWA14	The Technical Basis for a Unified Classification System	Engineering ceramics	USA
TWA15	Metal Matrix Composites	Discontinuous SiC reinforced aluminium	USA

REFERENCES

Amaral, J. E. and Pollock, C. N. (1989) Machine design requirements for uniaxial testing of ceramic materials, in *Mechanical Testing of Engineering Ceramics at High Temperatures* (eds B. F. Dyson, R. D. Lohr and R. Morrell), Elsevier Applied Science, p. 51.

Butterfield, D. M., Clinton, D. J. and Morrell, R. (1989) *The VAMAS Hardness Tests Round-Robin on Ceramic Materials*, VAMAS Technical Report 3.

Gould, D. and Loveday, M. S. (1992) A reference material for creep testing, in *Harmonisation of Testing Practice for High Temperature Materials* (eds M. S. Loveday and T. B. Gibbons), Elsevier Applied Science, Chapter 6, pp. 85–109.

Gray, T. G. F. and McCombe, A. (1992) Influence of specimen dimension and grip in tensile testing steel to EN 10002. *Ironmaking and Steelmaking*, **19**, 402–8.

Gray, T. G. F. and Sharp, J. (1988) Influence of machine type and strain-rate interaction in tensile testing, in *ASTM Symposium on Precision of Mechanical Tests*, STP1025.

Hossain, M. K. (1992a) Standardisation for advanced materials: experience and strategies for the future. *Bulletin of Materials Science*, **15**(1), 77–89.

Hossain, M. K. (1992b) VAMAS – current status and future trends. *VAMAS Bulletin*, no. 15, 2–5.

ISO DIS 11403/1 : 1992   Plastics – Acquisition and Presentation of Comparable Multipoint Data. Part 1: Mechanical Properties.

ISO 5725 : 1986   Accuracy (Trueness and Precision) of Measurement Methods and Results. Part 1: General Principles and Definitions.

Loveday, M. S. (1992) Towards a tensile reference material, in *Harmonisation of Testing Practice for High Temperature Materials* (eds M. S. Loveday and T. B. Gibbons), Elsevier Applied Science, Chapter 7, pp. 111–53.

Marchandise, M. (1992) The calibration and standardisation of the impact toughness test, in *Harmonisation of Testing Practice for High Temperature Materials* (eds M. S. Loveday and T. B. Gibbons), Elsevier Applied Science, Chapter 5, pp. 67–83.

Read, B. E., Dean, G. D. and Tomlins, P. E. (1988) Effects of physical ageing of creep in polypropylene. *Polymer*, **29**, 2159.

Roebuck, B. (1992) Parabolic curved crack fronts in fracture toughness specimens of particulate reinforced Al alloy MMC. *Fatigue and Fracture of Engineering Materials and Structures*, **15**(1), 13–22.

Roebuck, B. and Lord, J. D. (1991) Fracture toughness testing for metal matrix composites, in *Proceedings of the Conference on Test Techniques for MMC*, November 1990, London, Institute of Physics Short Meetings Series 28, pp. 110–28.

Roebuck, B., Lord, J. D., Cooper, P. M. and McCartney, L. N. (1993) Tensile properties, data acquisition and analysis for particulate and fibre reinforced MMC, in *ASTM Symposium on Accuracy of Load and Strain Measurements*, 17–19 November 1992, Miami: *Journal of Testing and Evaluation*, **22**(1), 63–9, January 1994.

CHAPTER 10

Accredited testing and reference materials

W. T. K. Henderson and G. B. Thomas

10.1 INTRODUCTION

The accreditation of laboratories for carrying out calibration and testing is increasingly recognized internationally as an effective means of raising the quality of performance of such laboratories and, through the medium of accreditation standards, a means of international intercomparison of testing and calibration competence. The majority of developed countries have established national accreditation services and many others are in the throes of setting them up. In the United Kingdom the National Measurement Accreditation Service (NAMAS) was established in 1985 by the amalgamation of the National Testing Laboratory Accreditation Service set up in 1981, and the British Calibration Service set up in 1966. To date, NAMAS has handled some 1500 accreditations. This experience, long and extensive by international standards, has put NAMAS in a pre-eminent position in the international league and the NAMAS practices have been used as a model by many other accreditation services.

Internationally, the need to reduce barriers to trade places considerable importance on interactions between national accreditation services. These take place through bilateral contacts and through participation in international organizations, in particular the International Laboratory Accreditation Conference (ILAC), the Western European Calibration Cooperation (WECC) and the Western European Laboratory Accreditation Cooperation (WELAC). These interactions have made, and continue to make, significant contributions towards an increasing commonality in approach to accreditation world-wide and to a closer harmonization of accreditation services within Europe (Leemput, 1992).

National accreditation services aim to:

- provide a national unified laboratory accreditation service which establishes widespread recognition of the competence of accredited calibration and testing laboratories;

- improve the authority and standard of calibration and testing within a country and thereby enhance the quality and reputation of its goods in markets both at home and overseas;
- eliminate multiple assessment of calibration and testing laboratories;
- negotiate agreements on mutual recognition with other national schemes and thereby obtain international acceptance of the competence of accredited laboratories;
- provide publicity for accredited laboratories and a service to their users through the publication of a directory of accredited laboratories.

10.2 THE ACCREDITATION PROCESS

Accreditation is voluntary and is open to any laboratory carrying out objective calibrations or tests. These may include not only independent commercial laboratories but also those in government, higher education institutes and other parts of the public sector and, increasingly, laboratories that are part of manufacturing organizations. Accreditation is granted as a result of assessment of a laboratory's documented procedures and technical competence against stringent published criteria.

To illustrate the accreditation process, the procedures adopted in the UK under the auspices of NAMAS will be considered in detail. NAMAS has played a major part in the development of the international acceptance of accreditation criteria, and the NAMAS Accreditation Standard (NAMAS, 1989) is fully compatible with the European Standard EN 45001 and is virtually identical to ISO IEC Guide 25 (1990).

Peer assessment by independent assessors is a key principle of the NAMAS system. Great care is taken in the selection and training of specialist assessors, who must have attended the NAMAS four-day assessor training course before appointment and use on assessments.

A laboratory applying for accreditation does so in terms of the scope of calibrations or tests for which accreditation is sought and sets out the material or products to be calibrated or tested, the type of test or measurement and the method or specification to be used. Wherever possible, the method is defined by reference to a British, European or International Standard or, if necessary, to a fully documented and validated internal procedure.

The quality manual for the organization and the associated procedures for the areas for which accreditation is sought must be examined by the assessor. The assessment process entails a detailed examination of the quality documentation and a thorough examination of the laboratory including observation of calibrations or tests being performed. The assessment covers all aspects of the laboratory's operation including staff qualifications and training, maintenance of equipment, sample handling and identification, the working environment, the recording and reporting of data and also the arrangements for internal audit of performance and for review. All aspects of assessment relate to criteria set out in the NAMAS Accreditation Standard: any departures are recorded as non-compliances. To be granted accreditation, a laboratory must address and resolve all non-compliances to the satisfaction of the assessor(s) and NAMAS.

The grant of accreditation includes an accreditation certificate and a schedule defining the scope of measurements or tests accredited and the methods used. An accredited laboratory is permitted to use the NAMAS logo on reports and certificates. The names of accredited laboratories and details of their accreditation are published in the NAMAS *Directory of Accredited Laboratories.*

10.2.1 International aspects

NAMAS accreditation is a widely recognized assurance of quality in calibration and testing, particularly with respect to the creation of the Single Market in Europe. The NAMAS criteria have been used to fulfil the UK's need to meet EC Article 100 Directives. In the UK, where New Approach Directives require testing as an important element for compliance, UK government departments often require that such testing be carried out in laboratories accredited by NAMAS.

The conclusion of international mutual recognition agreements ensures the acceptability of NAMAS certificates and reports overseas. At present NAMAS is party to three bilateral agreements and to two European multilateral agreements covering calibration and testing, as shown in the Appendix to this chapter. The European agreements were made under the auspices of the Western European Calibration Cooperation (WECC) and the Western European Laboratory Accreditation Cooperation (WELAC).

10.2.2 Maintaining accreditation

The grant of accreditation is however not the end of the story. It is only the end of the beginning. To retain accredited status a laboratory must continue to demonstrate that the required standards are maintained. Its customers need to be confident that it will perform calibrations and tests consistently from one day to the next, and have a right also to expect that the same measurement or test carried out at different accredited laboratories to a standard specification will produce similar results. Surveillance is thus a major ongoing function of the NAMAS Executive, and a structured programme of visits is an important element. To assist maintenance of standards in some fields, where the concept of traceability to national standards of measurement is not applicable, the accreditation standard permits the Executive to make participation in recognized interlaboratory comparison schemes or proficiency testing a mandatory requirement. The effectiveness of such schemes often depends on the availability of suitable reference materials or standards.

10.3 REFERENCE MATERIALS

A reference material is defined as a material or substance of which one or more properties are sufficiently well established to be used for the calibration of an apparatus, for the assessment of a measurement method, or for assigning values to materials (ISO Guide 30, 1992).

A special case of a reference material is the certified reference material (CRM). This is defined as a reference material of which one or more property values are certified by a technically valid procedure, accompanied by or traceable to a certificate or other documentation which is issued by a certifying body (ISO Guide 30).

10.3.1 Uses of reference materials in accreditation

From the accreditation point of view, reference materials are considered useful in two ways: first, as a source of reliable samples for carrying out verification and check tests; and second, as material for use in proficiency testing programmes and interlaboratory comparisons. Both procedures fall under the general requirement for a laboratory to ensure the quality of the results it provides to clients. For certain types of testing, calibration of equipment may not be a practicable means of establishing traceability of measurement. This would apply particularly to cases where the measurement is of a characteristic or property that is not a simple physical quantity, or where several interdependent variables may be involved in determining the result: examples are the identification and quantification of chemical composition and the identification and counting of features such as asbestos fibres. In such cases, the use of reference materials is essential for a laboratory to provide evidence of the adequacy of its procedures.

The selection of the appropriate type of reference material for the intended use needs consideration. Certified reference materials are often relatively expensive, reflecting the time-consuming, meticulous nature of the preparation and characterization processes. In many cases they are also in short supply. In such cases it may be more appropriate to use a CRM to characterize sub-reference materials or working standards rather than to use it directly. However, such a step is not to be undertaken lightly: the homogeneity of the proposed working standard will need to be established and the costs of this, together with the costs of material, use of facilities and expert manpower to determine the property value(s) may be considerable. It should also be noted that, simply because of the additional measurement step involved in the comparison of the sub-reference material with the CRM, it will have a greater uncertainty in respect of its property value(s) than the CRM. The sub-reference material may also possess a greater degree of heterogeneity, further increasing the uncertainty associated with its property value.

10.3.2 Verification and check testing

(a) Verification

In a number of cases, reference materials are used within a laboratory for verification purposes. Some testing standards require the use of reference materials to verify that a piece of apparatus is operating correctly and giving results within specified limits. An example in the materials testing field is the requirement contained in certain standards for the verification of machines used for the impact testing of metals. (Such standards include BS131 Part 6, ASTM E23 and BS EN 10045/2.) In addition to the requirements for the machine to meet specified dimensional and alignment limits, a verification of the machine's performance is required. This is achieved by testing reference testpieces under specified conditions and establishing that the results are within defined limits.

The standards for testing the hardness of metals also require that the machine's performance be verified against standard reference blocks. From one point of view, therefore, hardness test blocks could be considered to be the most widely used type of mechanical test reference material, although their role is more correctly regarded as that of a transfer standard. (Since 1988, the standard hardness scales for the UK are those realized by IMGC in Turin.)

A need for verification using reference materials may still be considered to exist even though it is not specifically demanded in a standard. For example, the use of 'black box' testing equipment that cannot be checked or calibrated readily in a conventional manner is growing; computer-based methods are increasingly used for controlling tests, recording data and evaluating results; and technical staff with less experience and fewer qualifications are increasingly employed to operate such equipment. These factors indicate significant growth in the demand for reference materials to monitor performance.

(b) Check testing

Regular check tests carried out for the purpose of in-house quality control would involve the use of CRM or sub-reference materials. This function may also be performed – albeit at a reduced level of precision – by testing samples retained from a batch previously tested. The lower level of precision in this case is associated with the extent to which the homogeneity of the selected batch is known. Furthermore, although this method is convenient if material is readily available, the stock of retained samples in a given batch tends to be low, as customers generally minimize the amount of material allocated for testing. However, where reference materials are not available, this approach is better than nothing and should provide assurance that no significant problems remain undetected for long periods.

Problems that can arise may include: (1) deterioration of testing equipment, either over a period since the previous calibration, or suddenly owing to malfunction of part of the apparatus (including software); and (2) unauthorized changes in procedure (e.g. short cuts that may appear valid to an operator at first sight). Sometimes events of this sort will show up immediately as results that are very obviously wrong, but more subtle errors could go unnoticed for a significant period and may lead to a requirement for expensive retesting. The extent of the damage will be limited if effective check tests are implemented at regular intervals, selected according to the amount of testing undertaken on the equipment in question.

An important advantage of using reference materials for regular check testing is that the uncertainty of the property value (or values), as determined using a defined test procedure, has been established. Where the same procedure is used by the laboratory the precision of the measurement process can be checked at any time by using the reference material. Also, provided that the uncertainty associated with the property value of the reference material is sufficiently small, the material can be used to determine the uncertainty of the measurement process used by the laboratory.

10.3.3 Intercomparison testing

The second way in which reference materials are relevant to the accreditation process concerns proficiency testing or interlaboratory comparison. These exercises not only serve to satisfy the requirement for a laboratory to exercise control over the quality of its results, but also enable a laboratory to demonstrate its proficiency in relation to others in respect of a specific test (or tests). Not every test (or even every testing field) is subject to proficiency testing. Indeed, in comparison with the number of individual tests for which laboratories seek accreditation, the extent and availability of proficiency testing schemes is relatively low at present, particularly so in the engineering materials area.

Nevertheless, accreditation bodies not only encourage participation in schemes where they exist but, in many cases, insist upon it as a condition of gaining and maintaining accredited status. While NAMAS use of mandatory proficiency testing at present is limited to the scheme concerning counting of asbestos and similar fibres (regular interlaboratory counting exchanges (RICE)) it is expected to increase. There are several other proficiency testing schemes, mainly in the chemical and microbiological testing fields, in which NAMAS strongly recommends participation. Those laboratories that do not participate are required to demonstrate by some other means that they have the necessary competence to perform these tests.

10.3.4 Availability and cost of reference materials

Clearly, significant quantities of reference materials will be required in the operation of proficiency testing schemes, particularly in cases where the test is destructive. This is the case with the large majority of tests used to establish mechanical properties of engineering materials. Rather less material will be required within any one laboratory for the purposes of regular in-house check testing. However, the total demand for reference materials is potentially large. The additional cost of implementing in-house check tests using certified reference materials is, in many cases, limited to the purchase of the material, since the requirement to carry out such tests exists in any case. The costs of testpiece preparation and testing would therefore be incurred whatever material was selected for this purpose.

On the other hand, a proficiency testing programme incurs costs for material, preparing and issuing samples, analysing results, communicating with participants and general administration of the scheme. In the absence of certified reference materials, the cost of materials for a laboratory inter-comparison exercise, for example, would be replaced by those required to identify, produce and characterize a suitable reference material.

In general, there are very few certified reference materials available in the materials testing field. They include the impact testpieces referred to above (available from BCR and NIST), a reference material for creep testing of metals (from BCR and NPL) and rubber samples for comparison of stress–strain properties (from NIST).

10.3.5 Future requirements for reference materials

There are requirements for additional certified reference materials, suitable for other types of material properties test. A priority appears to be for a reference material for tensile testing of metals, but other types of test such as fracture toughness K_{IC}, fatigue crack growth and high strain fatigue on a range of engineering materials do not have suitable, widely available reference materials.

An effective reference material for mechanical testing should be well characterized, homogeneous, and certified by well-documented, standard procedures. It should be readily available to a controlled specification, at economic cost and with an assurance of long-term supply from a QA approved supplier. Measurement of the certified value(s) should be carried out in technically competent laboratories, thoroughly familiar with the material and the relevant testing, with a quality system ensuring traceability of measurement, where appropriate, to national or international standards. Experienced operators in laboratories accredited by a recognized accreditation scheme would meet these requirements.

In most cases the property values of certified reference materials are determined from the results of tests carried out by several laboratories, rather than being based on only one source of testing. This approach enables the influence of different types of equipment, operators, testpiece or sample preparation procedures and similar factors to be incorporated in the limits of uncertainty for the defined property value(s). These considerations are particularly important in, for example, the determination of chemical composition. They are also relevant to many of the tests used to determine the mechanical properties of materials. However, in this case there is the prospect of developing 'standardizing machines' that could provide a source of traceable measurement transferred through a reference material to other testing machines. Such an approach is comparable with that used in the case of hardness testing of metals.

10.4 CONCLUSIONS

Laboratory accreditation is assuming an ever higher profile. This is true both in countries which have well-established accreditation schemes, such as NAMAS, and in countries where schemes may not yet exist but are being actively developed – in some cases with urgency. A primary reason for this is the pressure that arises from international efforts to reduce barriers to trade while continuing to ensure the quality and safety of goods supplied through requirements for testing and certification.

Testing to harmonized international standards will become increasingly common, and direct comparisons between laboratories across frontiers will become more frequent. These circumstances will enhance the importance of reference materials and CRMs for the purposes of quality control and interlaboratory comparisons. The demand for such materials appears set to increase.

APPENDIX 10.A MUTUAL RECOGNITION AGREEMENTS

UK bilateral mutual recognition agreements

Country	*Accreditation body*	*Year of agreement*	
		Calibration	*Testing*
Australia	(NATA)	1985	1985
New Zealand	(TELARC)	1985	1985
Hong Kong	(HOKLAS)	1989	1989

Multilateral mutual recognition agreements

Country	Accreditation body	Year of agreement	
		WECC calibration	WELAC testing
Denmark	(DANAK)	1990	1992
Finland	(CMA)	1989	
Finland	(FINAS)		1993
France	(BNM-FRETAC)	1989	
France	(RNE)		1992
Germany	(DKD)	1989	
Ireland	(ILAB)	1993	1993
Italy	(SIT)	1989	
Italy	(SINAL)		1993
Netherlands	(NKO)	1989	
Netherlands	(STERLAB)		1992
Norway	(NKT)	1992	
Norway	(NA)		1993
Spain	(RELE)		1992
Sweden	(SWEDAC)	1989	1992
Switzerland	(SCS)	1989	1993
UK	(NAMAS)	1989	1992

REFERENCES

ASTM E23 : 1993 Standard Test Methods for Notched Bar Impact Testing of Metallic Materials.

BS 131: Part 6: 1989 Method for Precision Determinations of Charpy V-Notch Impact Energies for Metals.

BS EN 10045: Part 2: 1993 Charpy Impact Test on Metallic Materials. Part 2: Method for the Verification of Impact Testing Machines.

EN 45001: 1989 General Criteria for the Operation of Testing Laboratories.

ISO (1992) *Terms and Definitions Used in Connection with Reference Materials*, Guide 30.

ISO/IEC (1990) *General Requirements for the Competence of Testing Laboratories*, Guide 25.

Leemput, P. J. van de (1992) Laboratory accreditation: standards, reference materials and proficiency testing, in *Harmonisation of Testing Practice for High Temperature Materials* (eds M. S. Loveday and T. B. Gibbons), Elsevier Applied Science, Chapter 4.

NAMAS (1989) *General Criteria of Competence for Calibration and Testing Laboratories*. NAMAS M10 Accreditation Standard.

CHAPTER 11

Metrology of wear

H. Czichos

11.1 DEFINITION OF WEAR AND CATEGORIES OF WEAR MEASURING AND TESTING

Wear is defined as follows (OECD, 1989): 'the progressive loss of substance from the operating surface of a body occurring as a result of relative motion at the surface'. Wear is usually connected with friction: 'the resisting force tangential to the common boundary between two bodies when, under the action of an external force, one body moves or tends to move relative to the surface of the other'.

Because of the great variety of friction and wear phenomena (or broadly tribological phenomena) in industrial applications, mechanical engineering equipment and laboratory studies, there are different types of wear measurements and tests (or broadly tribotests) which can be classified according to German Standard DIN 50 322 : 1984 into the following main categories (Fig. 11.1):

I.   machinery field tests: testing of actual machinery under practical operating conditions;

II.   machinery bench tests: testing of actual machinery under practice-oriented (simplified, simulated or accelerated) operating conditions;

III.   systems bench tests: testing of specific systems under practice-oriented operating conditions;

IV.   components bench tests: testing of specific components under practice-oriented operating conditions;

V.   model tests: testing of model testpieces under practice-oriented operating conditions;

VI.   laboratory tests: testing of arbitrary testpieces under laboratory operating conditions.

The categories of tribotesting have a different scope, and results obtained in one category of tribotesting cannot be simply transferred to another category. This chapter reviews the metrology of wear for the category of laboratory tests.

Cate-gory	Type of tests	Symbol
I	Machinery field tests	
II	Machinery bench tests	
III	Systems bench tests	
IV	Components bench tests	
V	Model tests	
VI	Laboratory tests	

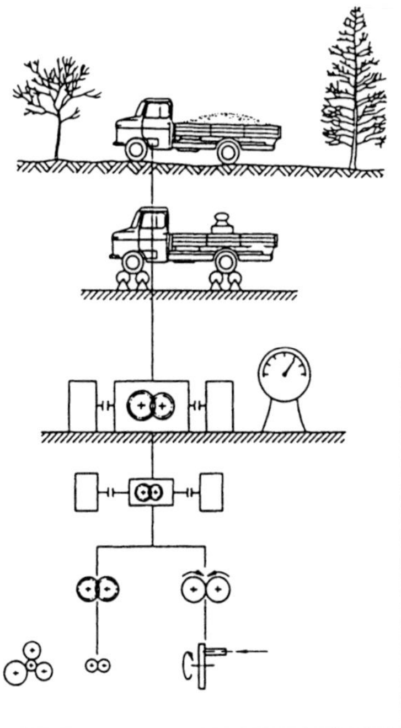

Fig. 11.1 Categories of tribotesting.

11.2 FUNDAMENTALS OF THE METROLOGY OF WEAR

In the metrology of wear, it must be recognized that wear – as well as friction – must be related to the whole system of interacting components, taking into account various processes, parameters and influencing factors (Czichos, 1978). This is obvious from a comparison between the test conditions to obtain strength data on the one hand and friction and wear data on the other (Fig. 11.2).

In strength tests (Fig. 11.2(a)) the deformation or fracture resistance of a material testpiece in a given environment is determined under the action of a certain stress mode, like tension, compression, shear, bending or torsion. The resulting strength data (in terms of force per cross-section, or energy) are considered as intrinsic material properties.

In a friction or wear test (Fig. 11.2(b)) the resistance against motion (i.e. friction) or the resistance against surface damage (i.e. wear) of a material/material pair (dry system) or a material/lubricant/material combination (lubricated system) in a given environment is determined

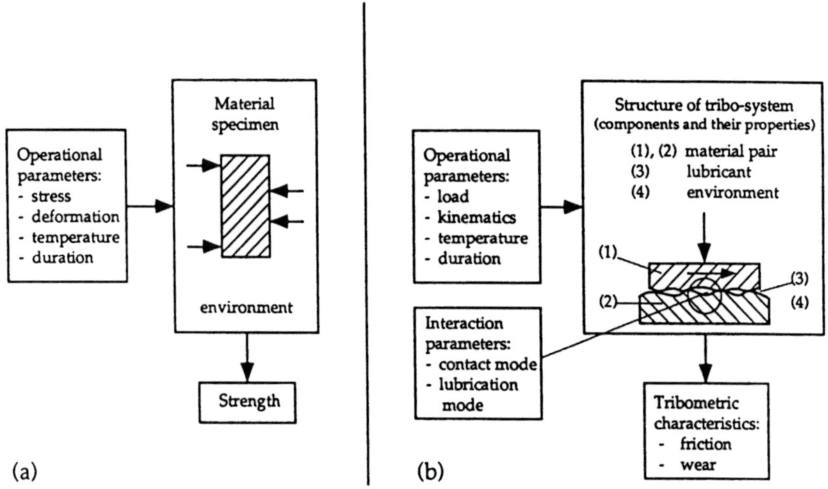

Fig. 11.2 Characteristics and parameters of (a) strength tests (b) friction and wear tests.

under the action of a certain type of motion, like sliding or rolling. The resulting tribometric characteristics, in particular the friction or wear data, must be understood as 'tribological systems characteristics' connected basically with the following group of parameters:

- structural parameters, defining the components (materials, lubricant, environment) of the test and their physical, chemical and technological properties;
- operational parameters, i.e. the loading, kinematic and temperature conditions and their functional duration;
- interaction parameters, characterizing, in particular, the action of the operating parameters on the structural components of the tribological system and defining its contact and lubrication mode, and the interfacial wear processes.

In the following, these parameters, essential to the metrology of wear, will be briefly reviewed in a systematic manner.

11.3 PARAMETERS AND CONDITIONS
RELEVANT TO THE METROLOGY OF WEAR

11.3.1 Structural parameters

Figure 11.3 illustrates typical examples of tribosystems subject to friction and wear together with corresponding simplified test configurations

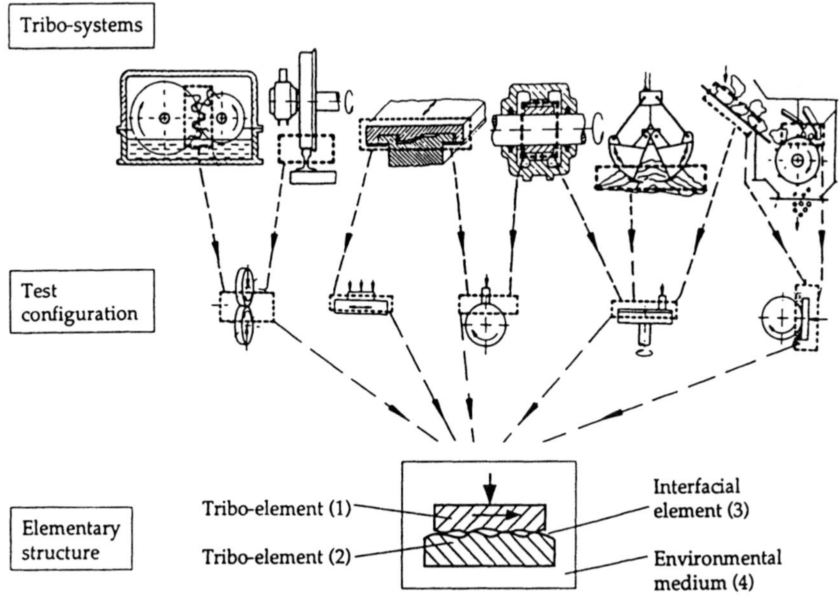

Fig. 11.3 Examples of engineering tribosystems, test configurations and their elementary structure.

and their elementary structure. In any friction and wear situation, four tribocomponents are involved:

1. first triboelement;
2. second triboelement;
3. interfacial element, e.g. lubricant, dust particles;
4. environmental medium, e.g. air, corrosive atmosphere.

The friction and wear data of tribosystems depend on various properties of their structural components 1 to 4, which can be classified in most cases into the following groups:

Triboelements 1 and 2
- geometric parameters: geometry, dimensions, surface topography etc.
- chemical parameters: volume composition, surface composition etc.
- physical parameters: thermal conductivity etc.
- mechanical parameters: elastic modulus, hardness, fracture toughness etc.
- microstructural parameters: grain size, dislocation density, stacking fault energy etc.

Interfacial (fluid) element 3, environmental (gaseous) medium 4
- chemical parameters: composition, additive content, acidity, humidity etc.
- physical parameters: density, thermal conductivity, flash and fire point etc.
- mechanical parameters: viscosity, viscosity–temperature characteristics, viscosity–pressure characteristics etc.

In any wear test, the basic properties of the elements 1 to 4 must be properly defined. In dry wear tests it is recommended to perform the tests in a closed test chamber with an environmental medium of controlled chemical composition and humidity.

In addition to these basic structural elements of wear tests, detrimental elements like dirt, dust and moisture may also be present and must be recognized in the analysis of structural parameters.

11.3.2 Operational parameters

Operational parameters can be considered (with the exception of friction-induced temperatures) as independent variables which may be varied in tribotesting to obtain the experimental friction and wear data depending on them. The basic operational parameters are as follows.

(a) *Type of motion*

Type of motion refers to the kinematics of triboelements 1 and 2, to be classified in terms of sliding, rolling, spin, impact and their possible superpositions. The kinematics may be continuous, intermittent, reverse or oscillating.

(b) *Load*

Load F_N is defined as the total force (including weight) which acts perpendicular to the contact area between triboelements 1 and 2. The load can be applied by various means to the test system, e.g. as a mass (dead load), by a spring, by hydraulic means or by electromagnetic means. In Fig. 11.4 the different possibilities for the application of load are shown schematically for the example of a pin-on-cylinder test system.

In static loading, the different principles of load application lead to the same value of the load force F_N. In dynamic loading, however, owing to the different mass-spring-damper combinations inevitably involved in the different loading principles, different load–time behaviour may result. Therefore the dynamic $F_{N(t)}$ behaviour may be different from the initial static load F_N.

For the measurement of the load, force transducers based on strain gauges, inductive elements or pressure-sensitive detectors evaporated

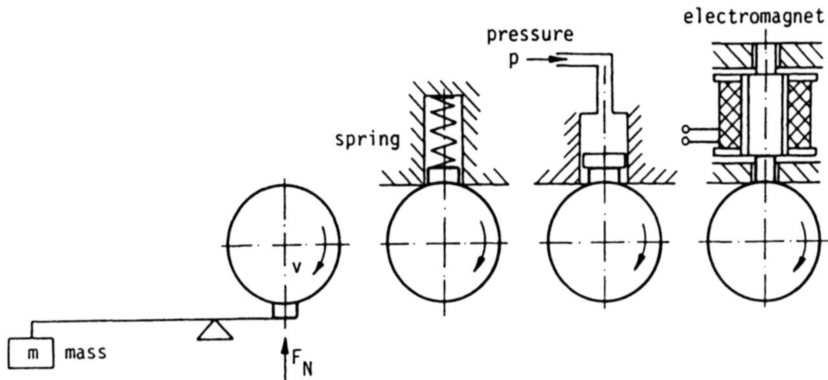

Fig. 11.4 Load application methods in tribotesting.

on the surface of the testpiece in the contact zone are available. The measurement of the load by means of an electromechanical force transducer has the great advantage that the F_N signal together with the signal of the friction force F_F can be fed to an electronic divider which calculates directly the actual value of the friction coefficient $f = F_F / F_N$ during the test.

The load F_N applied to the test system determines, in connection with the geometric area of contact A, the nominal contact pressure $p = F_N / A$. Since the value of A may change due to wear during the test, the value of p may also undergo some changes as a function of time.

(c) Velocity

Velocity v is specified with respect to the vector components and the absolute values of the individual motions of triboelements 1 and 2. According to Fig. 11.5 a distinction must be made between the relative velocity v_r (relevant for friction-induced temperature rises), the sum velocity v_s (relevant in lubricated tribosystems for the formation of an elastohydrodynamic (EHD) film) and the slide to roll ratio, or slip s.

(d) Temperature

The temperature T of the structural components depends on location and time. There is an initial (steady state) temperature. The friction-induced temperature rise (average temperature rise and flash temperatures) is estimated on the basis of friction heating calculations (Winer

Metrology of wear

Type of motion		Velocities u, v and slip s				$s = 2\left[\dfrac{u_1 - u_2}{u_1 + u_2}\right]$
		u_1	u_2	$v_r = \|u_1 - u_2\|$	$v_s = \|u_1 + u_2\|$	
Simple sliding	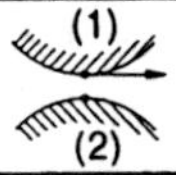	$u_1 > 0$	$u_2 = 0$	$v_r = u_1$	$v_s = u_1$	$s = 2$
Sliding		$u_1 > 0$ $\|u_1\| > \|u_2\|$	$u_2 < 0$	$v_r > u_1$	$v_s < u_1$	$2 < s < \infty$
Pure sliding		$u_1 > 0$	$u_2 = -u_1$	$v_r = 2u_1$	$v_s = 0$	$s \pm \infty$
Sliding		$u_1 > 0$ $\|u_1\| < \|u_2\|$	$u_2 < 0$	$v_r > \|u_2\|$	$v_s < \|u_2\|$	$-\infty < s - 2$
Simple sliding	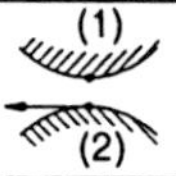	$u_1 = 0$	$u_2 < 0$	$v_r = \|u_2\|$	$v_s = \|u_2\|$	$s = -2$

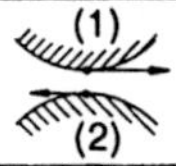

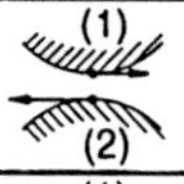

(a)

Type of motion		Velocities u, v and slip s				$s = 2\left[\dfrac{u_1 - u_2}{u_1 + u_2}\right]$
		u_1	u_2	$v_r = \|u_1 - u_2\|$	$v_s = \|u_1 + u_2\|$	
Simple sliding	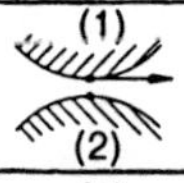	$u_1 > 0$	$u_2 = 0$	$v_r = u_1$	$v_s = u_1$	$s = 2$
Rolling with slip	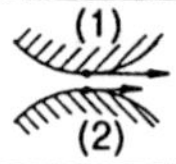	$u_1 > 0$ $u_2 = u_1$	$u_2 > 0$	$v_r < u_1$	$v_s > u_1$	$0 < s < 2$
Pure rolling		$u_1 > 0$	$u_2 = u_1$	$v_r = 0$	$v_s = 2u_1$	$s = 0$
Rolling with slip	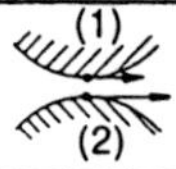	$u_1 > 0$ $u_1 < u_2$	$u_2 > 0$	$v_r < u_2$	$v_s = u_2$	$-2 < s < 0$
Simple sliding		$u_1 = 0$	$u_2 > 0$	$v_r = u_2$	$v_s = u_2$	$s = -2$

(b)

Fig. 11.5 Kinematics of tribosystems: (a) sliding (b) sliding and rolling.

and Cheng, 1980) or by experimental techniques. The main techniques for the measurement of friction-induced temperatures are illustrated in Fig. 11.6. In most cases thermocouples of different types or optical infrared pyrometer techniques are utilized. The techniques may be difficult to handle or the results difficult to interpret, so that it is still not easy to perform accurate and meaningful measurements of friction-induced temperatures. The thermoelements commercially avaiable have time responses from 10 s to 10 ms, depending on size and the thermocouple materials. In using thermocouples two methods may be applied:

- One (or both) of the actual testpieces forms part of the thermocouple (intrinsic thermocouple).
- One or more complete thermocouples are inserted in one or both of the testpieces.

With pyrometric techniques, either the thermal radiation emitted from the surroundings of the frictional interface of the optically non-transparent testpieces is detected, or one partner has to be made optically transparent in the infrared region. In both instances the optical emissivity as a function of temperature and wavelength of the radiation is the crucial point.

(e) Time dependence

The set of operational parameters (F_N, v, T) is time dependent, e.g. in load cycles, or heating and cooling intervals.

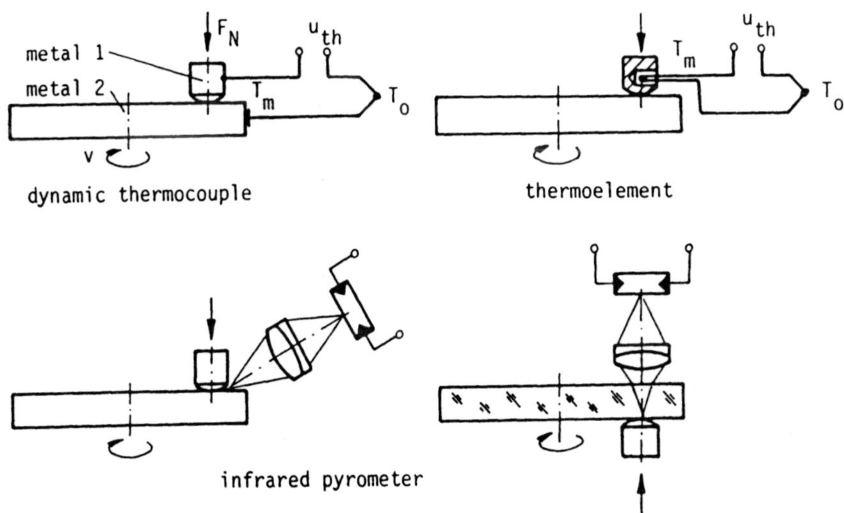

Fig. 11.6 Temperature measuring methods in tribotesting.

(*f*) Duration

The duration t of the operation, performance or test must be known. In addition to these functional operational parameters, disturbances like external vibrations and radiation may also need to be taken into consideration.

11.3.3 Interaction parameters

Interaction parameters characterize the action of the operational parameters on the structural components of the test system. These parameters define in particular the contact mode, the lubrication mode and the interfacial wear processes of a system with a given material/material or material/lubricant/material structure.

The formation of a contact is a necessary condition for all friction and wear processes which occur, by definition, in the momentary contact area A_c. The kinematics and time-dependent variation of the contact area leads to wear tracks (A_{w1}, A_{w2}) which may be different for the tribocomponents 1 and 2 of a given tribosystem. As illustrated in Fig. 11.7 for the example of a pin-on-disc tribotesting system, a tribocontact parameter ε can be defined, also named the contact area to wear track ratio (Czichos, 1978).

If for a given tribosystem the tribocontact parameters of tribocomponent 1 and tribocomponent 2 are unequal, i.e. $\varepsilon_1 \neq \varepsilon_2$, the two tribocomponents are subject to different tribological actions with respect to the kinematics, stresses, friction-induced heating and material–atmosphere interactions, as summarized in Fig. 11.7. It follows that in investigations of tribosystems, the response of *both* tribocomponents to the interfacial tribological processes must be carefully analysed. For example, a material for a certain triboengineering component should only be tested in a test configuration where the material testpieces have the same tribocontact parameter as in the real triboengineering system.

11.4 CHARACTERIZATION AND MEASUREMENT OF WEAR

The tribological interactions leading to materials deterioration and loose wear particles can be broadly divided into two classes (Fig. 11.8):

- Stress interactions: these are due to the combined action of load forces and frictional forces, and lead to wear processes described broadly as surface fatigue and abrasion.
- Materials interactions: these are due to intermolecular forces either between the interacting solid bodies or between the interacting solid bodies and the environmental atmosphere (and/or the interfacial

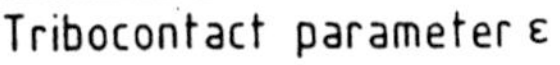

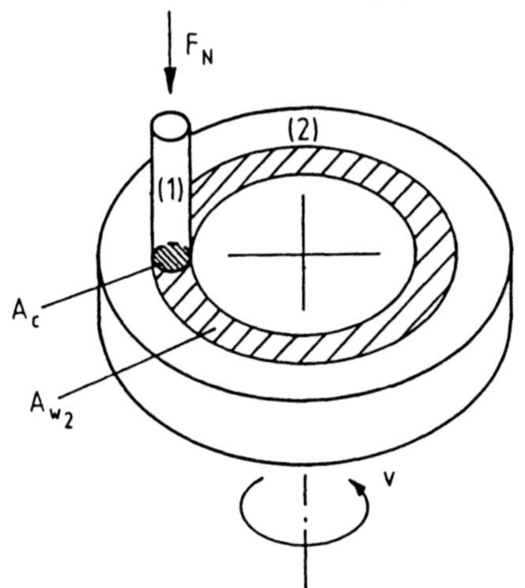

ε : ratio $\dfrac{\text{contact area}}{\text{wear track on specimen}}$

$$\varepsilon_1 = \frac{A_c}{A_{w_1}} \; ; \; \varepsilon_2 = \frac{A_c}{A_{w_2}}$$

e.g.: pin-on-disc-configuration:

pin: $\varepsilon_1 = 1$
disc: $\varepsilon_2 \ll 1$

Characteristics of tribo-components for which $\varepsilon = 1$	Characteristics of tribo-components for which $\varepsilon < 1$
permanent contact	intermittent contact
no macroscopic cyclic stressing	cyclic stressing
permanent friction heating	intermittent friction heating
reduced material-atmosphere interactions	direct material-atmosphere interactions on $A_w - A_c$

Fig. 11.7 The tribocontact parameter: an important interaction characteristic for the components of tribosystems.

medium), and lead to wear processes described broadly as tribo-chemical reactions and adhesion.

For the determination of wear with a test system of defined structural, operational and interaction parameters, different measuring techniques may be utilized. According to German Standard DIN 50 321 : 1979 'Wear Measuring Quantities', wear may be detected by measuring:

1. direct wear quantities, such as
   (a) changes of geometry of the testpieces (see Fig. 11.9), namely

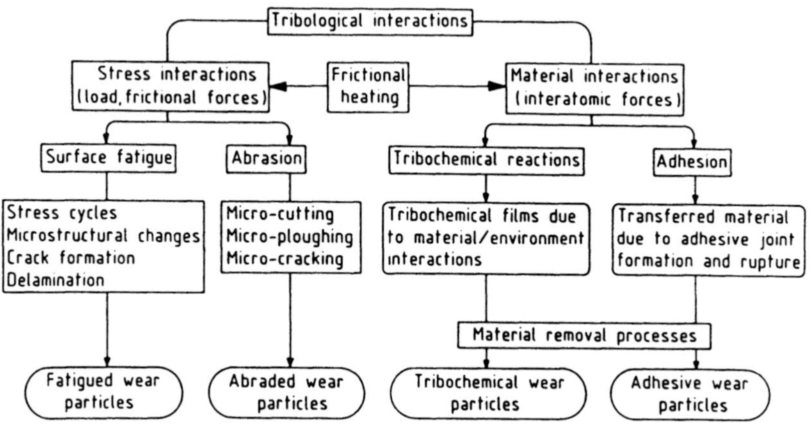

Fig. 11.8 Tribological interactions and wear mechanisms.

    (i)   changes in linear dimensions (wear length W_1)
    (ii) changes in cross-sections (wear area W_q)
    (iii) changes in volume (wear volume W_v);
  (b) changes of the mass of the testpieces;
  (c) the amount of worn material loss; or
2. relative wear quantities, i.e. wear rates such as
  (a) the wear to time ratio (wear velocity);
  (b) the wear to distance ratio; or
3. the wear coefficient, defined as

$$K' = \frac{\text{wear volume (mm}^3)}{\text{load} \times \text{distance (N m)}}; \text{ or}$$

4. the dimensionless Archard's wear coefficient, defined as

$$K = \frac{\text{wear volume} \times \text{hardness}}{\text{load} \times \text{distance}}.$$

The unambiguous measurement of wear requires that there must be a clear distinction between component wear and systems wear.

For each type of direct wear measurement, appropriate detectors are commercially available. During continuous wear measurements it is important to control the operational parameters such as normal force, sliding velocity and temperatures and to store the measured characteristic wear data. By the use of computer-aided test facilities, an improvement of measurement accuracy is possible (Santner, 1990). For example, the necessary corrections for temperature and vibration variations can be taken into account by using software processing on the

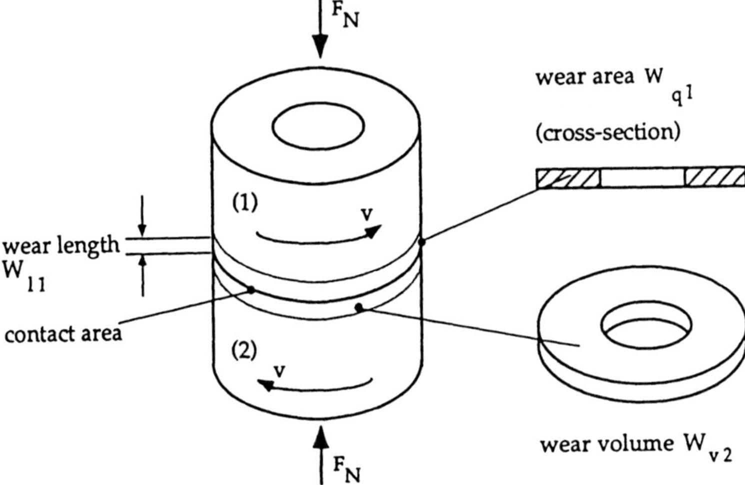

Fig. 11.9 Wear quantities, illustrated for the example of a tribosystem consisting of two sliding cylinders (1), (2): F_N load normal to the contact area, v sliding velocity.

stored data files. The hardware may consist of a personal computer, electronic interfaces such as transducers, signal conditioning modules, analogue to digital converters (ADCs) for transfers of analogue signals to the computer, and digital to analogue converters to transmit digital commands in analogue steering signals for motors etc.

A typical program for computer-aided wear measurements has four main parts:

- input and calibration module;
- running-in wear module;
- main wear module;
- output and storage module.

The input module contains the possibility to define the test parameters such as sliding velocity, normal force, sliding distance for running-in and the main wear test. It asks for system parameters like sample type, counterbody type and surface roughness as well as the calibration factors for the sensors. These calibration factors can also be checked in advance.

The running-in wear module measures and stores the pertinent data of operational parameters and wear at shorter intervals than those in the main wear module.

After reaching a pre-defined maximum wear value the tests are automatically stopped and a printout of all relevant parameters is given. The software program continues to measure elongations and temperatures until the sample temperature has reached room temperature.

In addition to the measurement of wear, the worn material surfaces should be characterized with respect to their chemical composition, microgeometry and microstructure, and the wear debris should be analysed and characterized. For these tasks well-established surface analytical tools, profilometers and microscopes (including the newly developed scanning tunnel and atomic force microscopes) as well as particle detectors are commercially available (Godfrey, 1980), which cannot be discussed within the scope of this chapter.

11.5 METHODOLOGY OF WEAR TESTS

Because of the manifold structural, operational and interaction parameters that influence the metrology of wear, a systematic methodology for wear tests is needed. Laboratory wear tests should be carried out in the following steps:

1. Choose a suitable test configuration for the testpieces of triboelement 1 and triboelement 2, specifying
   (a) geometry of test configuration;
   (b) materials characteristics and properties;
   (c) surface characteristics (clean surfaces before test).
2. Characterize the interfacial element 3 (e.g. lubricant) and the environmental medium or atmosphere (including humidity) 4 in terms of
   (a) chemical nature;
   (b) composition;
   (c) chemical and physical properties.
3. Choose a suitable set of the operational parameters, such as
   (a) type of motion;
   (b) load F_N;
   (c) velocity v;
   (d) temperature T;
   (e) test duration t.
4. Perform the tests as a function of varied
   (a) structural parameters of the triboelements, e.g. hardness, roughness; and/or
   (b) operational parameters, e.g. load cycles, velocity variations.
   The conditions of the tests (e.g. the time dependence of operational parameters) should be controlled by appropriate detectors or sensors and supported by on-line computer techniques.
5. Measure interesting tribometric characteristics, e.g.
   (a) friction quantities;
   (b) wear quantities;
   (c) tribo-induced acoustic quantities, e.g. noise, vibrations;
   (d) tribo-induced thermal quantities, e.g. friction-induced temperature rise.p

6. Characterize the worn surfaces of triboelement 1 and triboelement 2 with respect to
    (a) surface topography, e.g. profilometry, scanning electron microscopy (SEM);
    (b) surface composition and structure, e.g. microprobe, Auger electron spectroscopy (AES).
7. Collect and characterize the wear debris.

The results of tribotests lead to friction and wear data, or generally tribodata, which are system-dependent characteristics and may be presented in the following forms:

1. tribographs, i.e. presentations of a measured friction or wear quantity as a function of
    (a) operational parameters, i.e. kinematics, load F_N, velocity v, temperature T or test duration t;
    (b) structural parameters, e.g. materials pairings, hardness, roughness, environmental humidity etc.;
    (c) interaction parameters, e.g. contact stresses, film thickness to roughness ratio, etc.;
2. transition diagrams (Lossie, Mens and de Gee, 1989), characterizing critical conditions of operational parameters which separate regimes of the efficient performance of a given tribosystem from regimes of inefficient performance (failure);
3. tribomaps (Lim and Ashby, 1987), characterizing operational conditions which separate regimes of different dominating wear mechanisms or different ranges of wear data.

It should be emphasized that for an unambiguous presentation of wear data, system wear and component wear must be distinguished. As can be seen from the example of Fig. 11.10 for an Si_3N_4 ceramic, the values for component wear (ball or disc testpieces) differ from those of the system wear of both partners of a ball-on-disc configuration (Czichos, Becker and Lexow, 1989).

11.6 PRECISION AND REPRODUCIBILITY OF WEAR DATA

One of the primary problems in the metrology of wear is coping with variability. Variability is inherent in wear because of the great number of influencing structural, operational and interaction parameters, and their potential fluctuations and time dependencies (Suh, 1986). Therefore, the metrology of wear requires – in addition to a systematic methodology for the (statistically based) planning, performance and evaluation of wear tests – an estimation of the reproducibility of wear data.

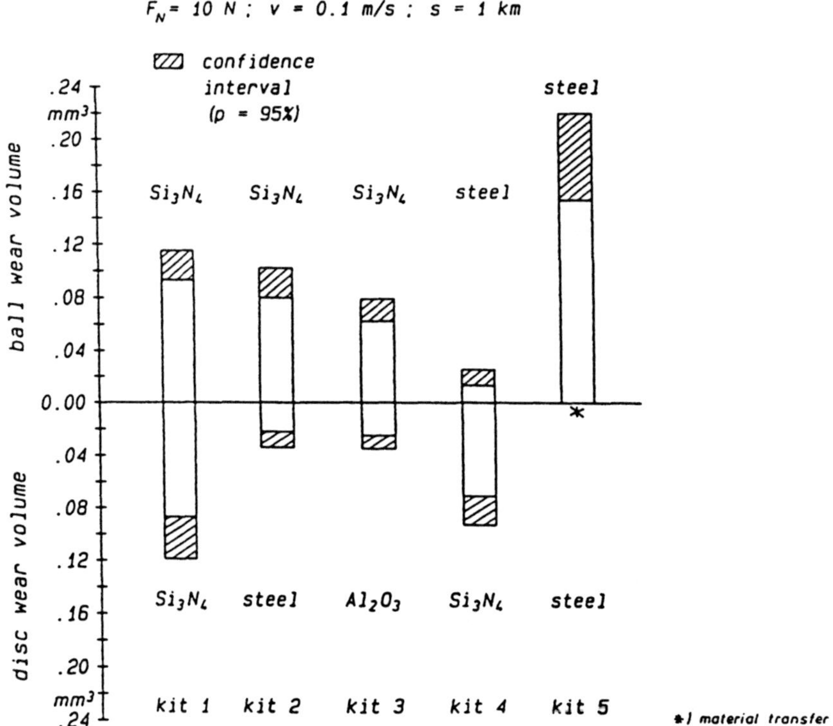

Fig. 11.10 Presentation of wear data for the components of a laboratory tribotesting system (ball-on-disc configuration, VAMAS reference conditions: see Fig. 11.11).

An international round robin comparison on the comparability, repeatability and reproducibility of friction and wear tests has been performed within the framework of the Versailles Project on Advanced Materials and Standards (VAMAS). Under the guidance of an international committee on wear test methods, round robin studies with testpieces of α–Al_2O_3 ceramic and steel were performed with 31 participating institutions from seven countries (Czichos, Becker and Lexow, 1987).

For the sake of simplicity a ball-on-disc specimen configuration was selected as the test system (Fig. 11.11). The testpieces were uniformly manufactured with respect to their geometry, dimensions and surface finish. A detailed instruction sheet containing all parameters and factors relevant to the performance of repeatable tests and specimen pairings ready to be tested were sent from a central source to all participants.

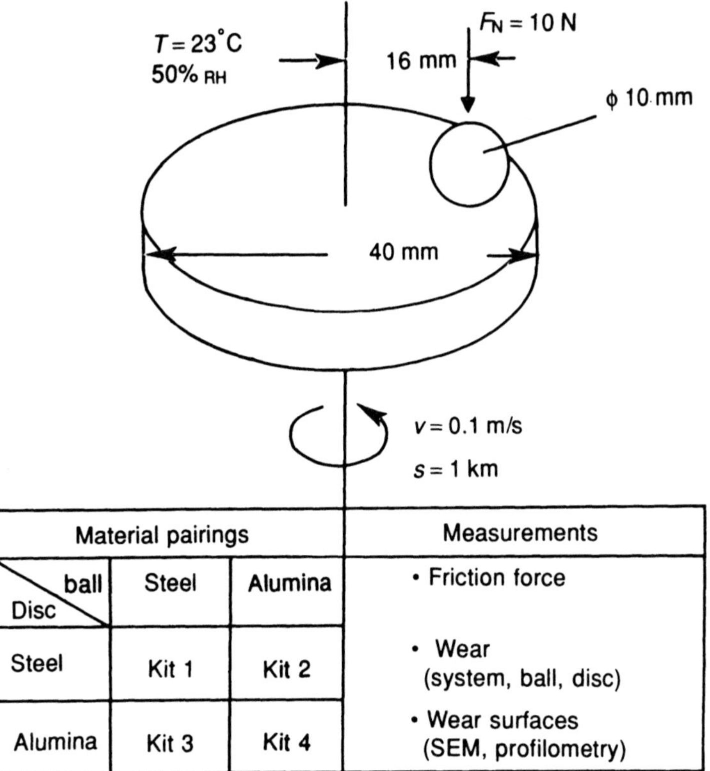

Material pairings			Measurements
ball Disc	Steel	Alumina	• Friction force
Steel	Kit 1	Kit 2	• Wear (system, ball, disc)
Alumina	Kit 3	Kit 4	• Wear surfaces (SEM, profilometry)

Fig. 11.11 Conditions of VAMAS round robin sliding wear tests: F_N normal load; v sliding velocity; s sliding distance; T ambient temperature; RH relative humidity of the atmosphere.

The results of the VAMAS round robin comparison show that a good reproducibility for the numeric friction and wear data has been obtained which can be summarized in terms of the relative standard deviations (s_r and s_R according to ASTM Standard E691 divided by the mean value) as follows:

- repeatability of friction data within laboratories, ±9% to ±3%; inter-laboratory reproducibility, ±18% to ±20%;
- repeatability of testpieces wear data within laboratories, ±5% to ±7%; interlaboratory reproducibility, ±15% to ±20%;
- repeatability of systems wear data within laboratories, ±14%; inter-laboratory reproducibility, ±29% to ±38%.

These results indicate that the repeatability and reproducibility of friction and wear measurements are comparable with those of other engineering quantities provided that the tests are performed under well-controlled conditions, applying standard round robin test procedures.

REFERENCES

Czichos, H. (1978) *Tribology – A Systems Approach to the Science and Technology of Friction, Lubrication and Wear*, Elsevier, Amsterdam.

Czichos, H., Becker, S. and Lexow, J. (1987) Multilaboratory tribotesting: results from the Versailles Advanced Materials and Standards (VAMAS) programme on wear test methods. *Wear*, **114**, 109.

Czichos, H., Becker, S. and Lexow, J. (1989) International multilaboratory sliding wear tests with ceramic and steel. *Wear*, **135**, 171.

DIN 50 321 : 1979 Wear Measuring Quantities (in German). Beuth-Verlag, Berlin.

DIN 50 322 : 1984 Wear – Specification of the Categories of Wear Testing (in German). Beuth-Verlag, Berlin.

Godfrey, D. (1980) Diagnosis of wear mechanisms, in *Wear Control Handbook* (eds W. O. Winer and M. B. Peterson), American Society of Mechanical Engineers, New York, p. 283.

Lim, S. C. and Ashby, M. F. (1987) Wear-mechanisms maps. *Acta Metallurgica*, **35**, 1.

Lossie, C. M., Mens, J. W. M. and de Gee, A. W. J. (1989) Practical applications of the IRG transition diagram technique. *Wear*, **129**, 173.

OECD (1989) OECD glossary on terms and definitions of tribology, in *Wear Control Handbook* (eds W. O. Winer and M. B. Peterson), American Society of Mechanical Engineers, New York.

Santner, E. (1990) Computer-aided testing in tribology (in German). *Materialprüfung*, **32**, 18.

Suh, N. P. (1986) *Tribophysics*, Prentice-Hall Englewood Cliffs, NJ.

Winer, W. O. and Cheng, H. (1980) Film thickness, contact stress and surface temperatures, in *Wear Control Handbook* (eds W. O. Winer and M. B. Peterson), American Society of Mechanical Engineers, New York, p. 81.

CHAPTER 12

Metrology in the science of corrosion

P. E. Francis and S. R. J. Saunders

12.1 INTRODUCTION

Corrosion in aqueous and gaseous environments is a complex topic and depends on a wide range of parameters which need to be measured and controlled. Measurements are not always required to a great accuracy, but the degree of precision and traceability of the measurements must be documented. Since corrosion rates are affected by the composition, surface state and geometry of the testpiece and by the detailed environmental conditions, dimensional metrology, gravimetry, electrical and temperature measurements are all involved in characterizing corrosion behaviour. While most of these measurements can be made to an acceptable degree of accuracy, and traceability of calibrations can be readily demonstrated, corrosion processes frequently involve highly localized conditions which can be difficult to define.

There is a general requirement to develop databases for corrosion, but while various attempts are being made there is concern that much of the information available in the literature has been produced under test conditions that have not been well characterized. As will also be discussed in this chapter, particularly in the case of high-temperature gaseous corrosion, there is a need to define standard test procedures and guidelines so that data may be readily validated and thus provide input for databanks and life prediction models.

In this chapter some general factors governing the accuracy of corrosion measurements are outlined and specific examples of metrological difficulties associated with critical parameters are considered. In aqueous corrosion, critical measurements discussed are the measurement of potential and temperature, and the measurement and control of the test environment. For high-temperature corrosion, continued protection depends, as it does for aqueous corrosion, upon the diffusion barrier characteristics of the corrosion product and its chemical stability. In addition, the mechanical properties of the reaction product are also of major importance since thermal expansion coefficient mismatch induces stresses during growth of the oxide. Thus, in the section on high-temperature corrosion measurement, problems associated with gas composition and determination of the mechanical properties of the protective layers are considered.

12.2 GENERAL CONSIDERATIONS

A brief discussion follows of the various routine measurements used in corrosion testing and of the special precautions that are needed to ensure that valid values are obtained.

Dimensional measurements are readily traceable, but in many cases, since only changes in dimensions are needed, absolute accuracy may not be essential where the same equipment is used. There are difficulties, however, where attack is localized, and in this case statistical methods are required to describe behaviour adequately. Previously the difficulty of obtaining relevant data has inhibited use of the stochastic approach. However, with the advent of advanced image analysis and automated measuring procedures, it is believed that more widespread use of the stochastic approach can now be anticipated. In the case of high-temperature corrosion, significant errors (5–10%) can occur if corrections to the surface area measured at room temperature are not made when calculating the rate of attack per unit area from mass changes derived at high temperatures. Similarly when material recession rates are high, the surface area of relatively small laboratory testpieces will decrease and suitable corrections must be applied to take account of this effect.

Measurements of mass are also readily traceable. However, in aqueous corrosion the testpiece will normally retain insoluble corrosion products and these must be removed by some means which does not cause further damage to allow measurements of mass loss to be made. Chemicals which dissolve the corrosion products whilst causing minimal dissolution of the testpiece are employed, but it is essential to make corrections for any mass loss during the cleaning procedure and there are simple methods for doing this. At high temperatures, care in interpretation of mass change data is required since scale spallation and volatilization and deposition processes can occur.

Methods used to analyse the composition of the bulk electrolyte involve measurements that are readily traceable and an accuracy of 0.5% is usually acceptable. Some difficulties are encountered in defining the concentration of reactants in cracks and crevices and under corrosion product precipitates: these concentrations can differ from the bulk (often by a factor of 10 or more) and are the more important values relating to rates of the corrosion reaction. Diffusion in and out of cavities and under corrosion products prevents bulk composition being maintained. Consequently local conditions must be defined and this frequently involves modelling mass transport processes (Turnbull, 1982).

Specification of the testpiece in terms of dimensions, mass and average chemical composition is straightforward, but other parameters are more difficult to quantify and are often neglected. These include physical defects and chemical inhomogeneity. Examples of chemical inhomogeneity are inclusions and precipitate particles which can have a pronounced effect on localized corrosion. Characterization of the surface chemical and mechanical state is an area where considerable effort is required in order to describe adequately the amount of cold work and inclusion content.

In the case of high-temperature corrosion, a further complication arises due to the effects of specimen geometry on residual stresses in the protective layer. In many laboratory tests, small rectangular samples are used, and the presence of corners and edges can markedly alter behaviour, whereas the application could involve use of the material in tubular form where corners and edges do not exist. Similarly where the corrosion process is controlled by diffusion of anions or cations through the protective scale, the presence of corners etc. alters the diffusion flux so that again the behaviour is very dependent upon specimen shape.

Mass flow rates are readily measured and traceable, but care is required in selecting conditions so that experiments are carried out under the appropriate hydrodynamic conditions at the metal/electrolyte interface (e.g. lamellar or turbulent flow). Defining these conditions can be difficult unless special equipment such as a rotating disc electrode is used, and many measurements are made without sufficient regard to the hydrodynamic conditions. Although errors of 5% in the measurement of flow rate will usually be acceptable, greater accuracy is required near critical velocities where erosion/corrosion is likely to occur. When working with dilute gas mixtures it is clearly important to ensure that flow rates are sufficiently high to eliminate gas diffusion effects where they are inappropriate. Likewise, thermal diffusion can cause demixing of complex gas mixtures.

Design of the tests should allow start-up and shut-down periods to be kept to a minimum to avoid the effects of transients unless for the purpose of simulating plant conditions. At present, testing procedures are often not sufficiently well defined, particularly in high-temperature corrosion, so that a large variety of start-up options are used. The exact choice can have a significant effect upon the final rates of attack observed. For example, with a fecralloy steel exposed to a sulphidizing atmosphere, a thin pre-formed oxide layer delayed the onset of sulphidation, whereas in the absence of the pre-formed oxide layer, sulphides formed from the start of the test (Coley and Saunders, 1987).

Accurate measurement and control of the temperature are important in high-temperature processes where typically activation energies are about 200 kJ mol^{-1}. For example, with alumina- and chromia-forming alloys the rates of attack increase by about one to one and a half orders of magnitude between 800 and 1000 °C, respectively; thus errors of between 25 and 35% would result for a 5 °C change in temperature. Aqueous corrosion processes are also dependent on temperature. For example the corrosion rate of mild steel over a 25-day period increased by a factor of two between 40 and 60 °C (Mercer and Brook, 1978), and the pitting potential of stainless steel was found to change by about 2.5 mV °C^{-1} at temperatures between 30 and 100 °C (Szklarska-Smialowska, 1974). Thus measurement and control of temperature to one degree while introducing errors of about 5% will normally be acceptable, and this is readily achieved in most experimental situations.

12.3 AQUEOUS CORROSION

12.3.1 Potential measurements

Aqueous corrosion occurs by an electrochemical process with anodic ($M \rightarrow M^+ + e$) and cathodic (e.g. $2H^+ + 2e \rightarrow H_2$) processes occurring simultaneously at equal rates which are dependent on the electrochemical potential at the surface of the metal. Thus surface potential is a key parameter in determining the reaction rate of the corrosion process. If the surface potential results from a polarizing current to the metal surface, particular care needs to be taken in the design and positioning of the reference electrode probe. If the probe is too far from the surface of the specimen, electrochemical measurements will include a noticeable *IR* potential drop developed by the polarizing current. The effect of this on measurements of the passivation potential, for example, is illustrated schematically in Fig. 12.1, which shows how *IR* potential drop can introduce an error $E_3 - E_2$ in the measurement of the passivation potential. If the probe is too close to the surface the *IR* drop will be minimized but, because of the shielding effect, current will be reduced at the surface in the vicinity of the probe and the indicated potential will not be representative of the surface as a whole. The influence of shielding on the measurement of the passivation potential is also demonstrated in Fig. 12.1. The current to the specimen will rise more rapidly since the area being monitored by the reference electrode is receiving less polarizing current than the remainder. When the current density reaches the critical value for passivation I_1, passivity will occur so that the current density will decrease and the measured passivation potential will be shifted to a more negative value, from E_2 to E_1. Thus the positioning of the probe is critical to the accurate measurement of surface potential and involves the

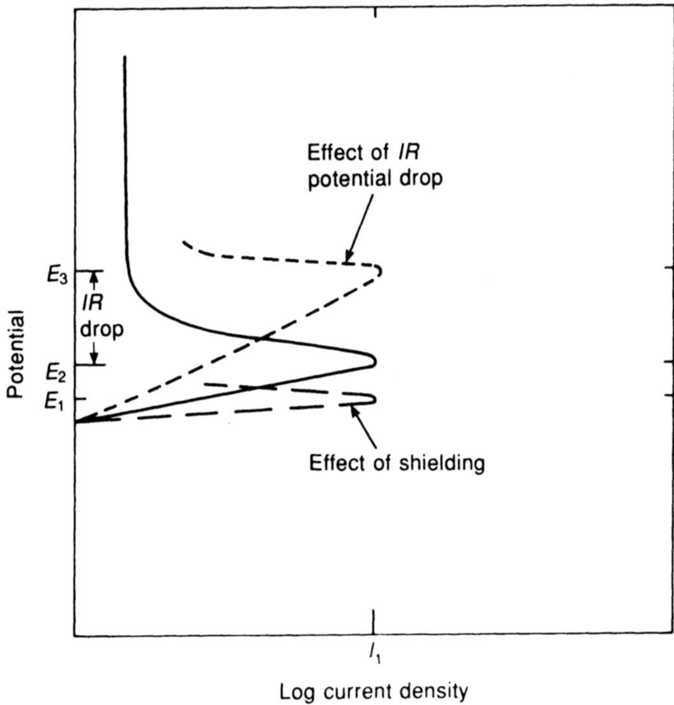

Fig. 12.1 The influence of *IR* potential drop and specimen shielding on the measurement of passivation potential using a polarization diagram.

minimizing of errors caused by *IR* potential drop and by the shielding effect of the probe. These errors are dependent on the resistivity of the test solution: with seawater, which is very conducting, the errors are small, but for less conducting solutions the errors can become significant. For example, in a solution of 0.01 mol l^{-1} potassium chloride the error with a Luggin probe positioned 5 mm from a 10 mm diameter testpiece polarized by a cylindrical counter electrode is about 90 mV when the polarizing current is 5 A m^{-2}, but in the same conditions will be about 2 mV in seawater. ASTM G5 advises a separation between the Luggin probe and the testpiece of twice the diameter of the tip of the probe to minimize the errors.

Errors due to *IR* potential drop can be eliminated or minimized by techniques which include interruption of the polarization current momentarily so that measurement of potential may be obtained free from *IR* drop. Such techniques are very necessary, for example, when the potential of a cathodically protected pipeline buried in soil is being

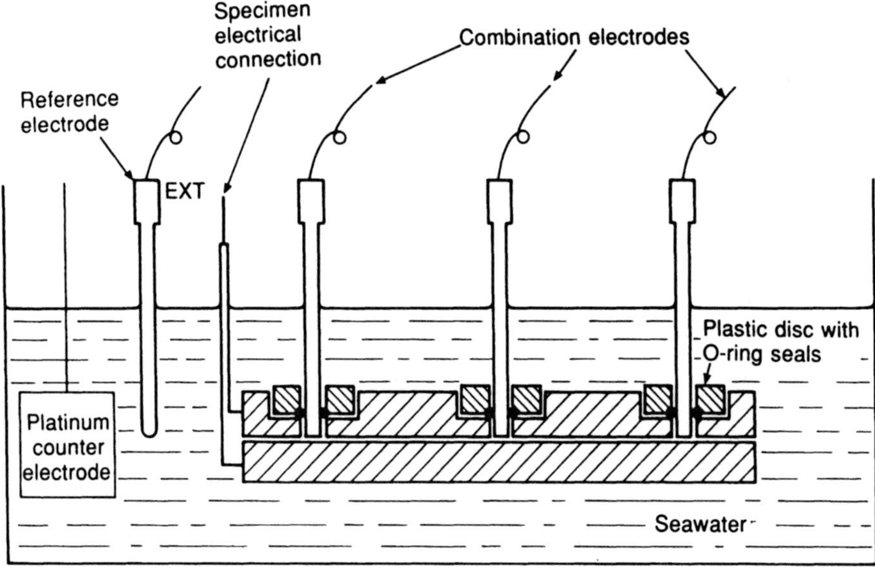

Fig. 12.2 Crevice corrosion cell.

monitored, since accurate measurement of the potential is vital to ensure integrity of the pipeline (BS 7361) and an error of about 5% in the measured potential could have serious consequences.

If the corrosion reactions are largely occurring within cavities (cracks and pits), surface potentials will not necessarily reflect those associated with the corrosion processes occurring within the cavities, and special techniques (Turnbull, 1982) are required to obtain the relevant data. A method used for measuring the potential along a crevice is shown in Fig. 12.2. If very fine probes are employed to measure potentials in pits, errors can occur for example by blocking diffusion processes and electrical shielding, so that the measurement obtained does not represent the situation in an undisturbed pit.

Further errors in potential measurement can occur where metallic material, other than the testpiece, is used in the construction of the test cell and exposed to the solution; an example is when cells are used for tests at elevated temperatures and pressures. If the test cell has an electrical connection to earth and the measuring equipment is not fully floating from earth on *both* measuring inputs, then the metallic test cell will influence the reference electrode (Fig. 12.3) and cause errors in the measured potentials which will be somewhere between that of the testpiece and that of the test cell. Precautions against such errors are especially important when measurements are being carried out on operating plant.

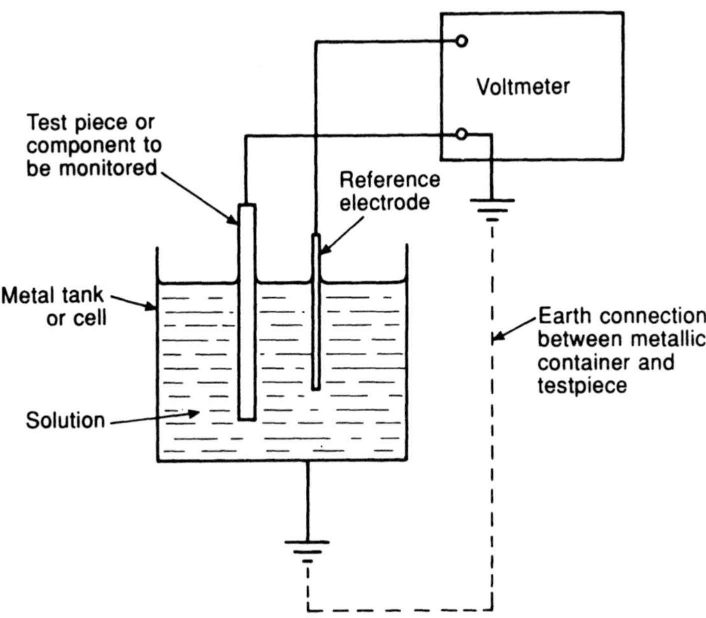

Fig. 12.3 Error in potential measurement caused by low resistance to earth on one input of the voltmeter.

12.3.2 Temperature measurements in heat transfer systems

In situations where heat transfer occurs from the testpiece to the test solution it is important to establish whether a convective or a nucleate boiling process is involved, since nucleation of bubbles can disturb surface hydrodynamics and concentrate electrolyte at the heat transfer surface and within cavities relative to the bulk. This phenomenon is especially applicable to steam power plant when high levels of contaminants can occur where heat flux is enhanced, e.g. at weld imperfections and surface deposits where bubble nucleation is encouraged, even with pure feed water (Gemmill, 1976). The change in heat transfer mechanism occurs at a critical surface temperature as illustrated in Fig. 12.4, and since the temperature gradient in the testpiece is usually small, accurately spaced and calibrated thermocouples are required to avoid large errors in measured surface temperatures. Clearly near T_c it will be essential to ensure that temperature is measured reliably in order to control the experimental conditions.

12.3.3 Measurement and control of the environment

Important environmental parameters include pH, oxygen concentration, and concentrations of aggressive and inhibiting species.

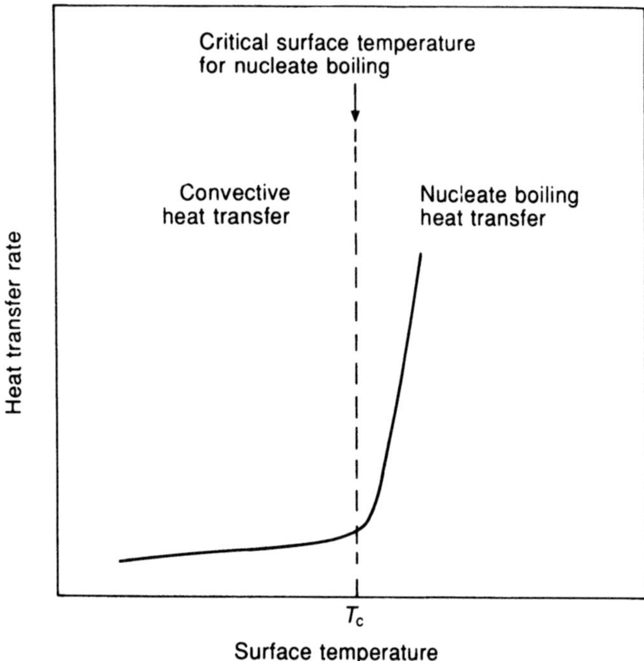

Fig. 12.4 The influence of surface temperature of the testpiece on the heat transfer mechanism.

It is important to ensure that test cells are constructed from materials which will be inert in the experimental environment to avoid contamination by impurity ions. The use of Pyrex glass at temperatures above 60 °C, for example, has been shown to lead to erroneous corrosion data for steel in distilled water because of dissolution of silica from the glass (Mercer and Brook, 1978): silica is known to have corrosion inhibiting properties for steel. In 100-day tests, the corrosion rate of steel was reduced by 50% at 70 °C and by a factor of 10 at 90 °C when Pyrex apparatus was used, as shown in Fig. 12.5. The form of corrosion also changed, with general corrosion giving way to localized attack in the presence of dissolved silica.

Unless the test solution is refreshed during the period of exposure or the cell has a large volume in relation to the surface area of the testpiece, concentration changes resulting from corrosion reactions will occur in the bulk environment during the test. For example, the build-up of copper in a test loop measuring erosion/corrosion data has been shown (Cheung and Thomas, 1988) to increase the critical velocity for film breakdown, as shown in Fig. 12.6, where it is evident that the presence of copper ions halved the mass loss for a given flow velocity.

Measurement of pH is usually considered to be straightforward, but repeatable values are surprisingly difficult to obtain, especially between

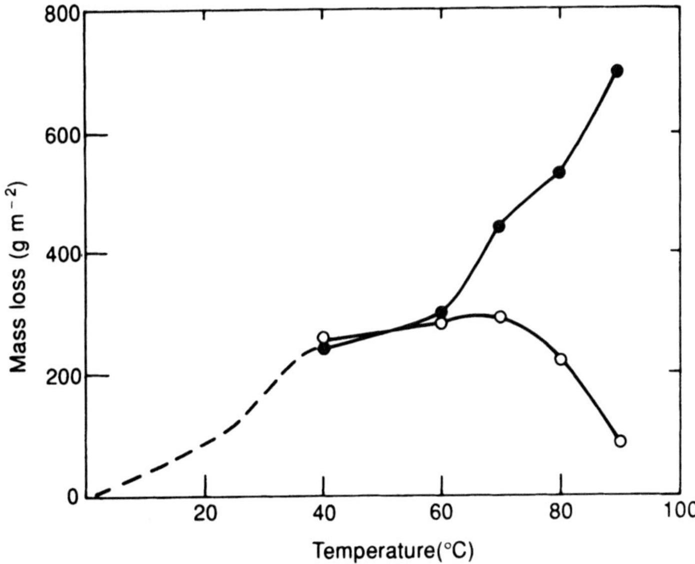

Fig. 12.5 The effect of temperature on corrosion of mild steel in distilled water in new glass vessels -O- and in quartz vessels -●-.

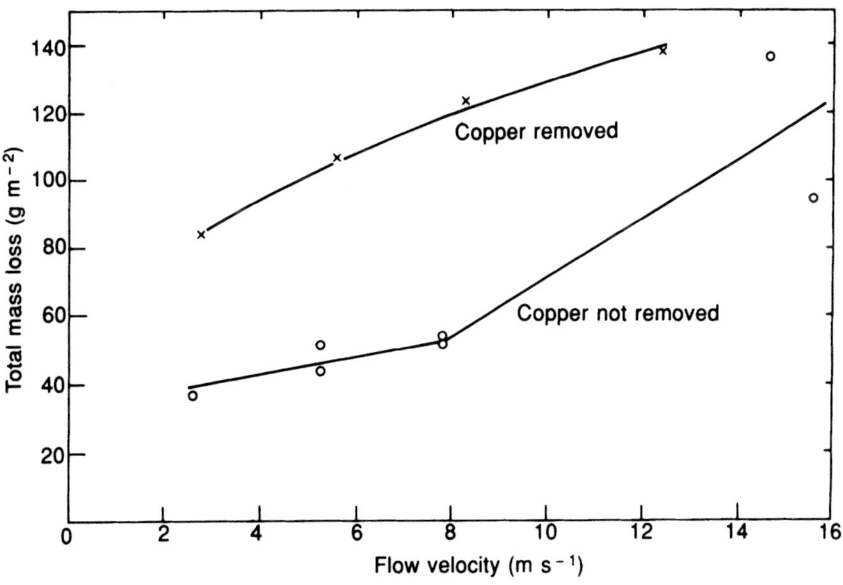

Fig. 12.6 The influence of dissolved copper on the mass loss of 90/10 copper nickel in artificial seawater.

different operators, and a rigorous procedure must be adopted. Often the simple precaution of calibrating the pH meter at two points close to the anticipated test condition is not followed and reliance is placed on a single calibration.

The influence of pH on the corrosion process is demonstrated by potential–pH diagrams (Pourbaix diagrams), which show zones of passivity where solid oxides, hydroxides etc. are thermodynamically stable and can limit corrosion processes, and other zones where corrosion products are soluble and non-protective. The pH value at the boundary between the two zones is critical and in these conditions measurements to better than 2% might be required. Additionally, pH values can influence kinetics. For example in the acidic region, control of pH to ± 0.2 units (i.e. around 5%) could lead to a 95% variation in hydrogen ion concentration in the test solution with equivalent changes in cathodic kinetics in the diffusion-controlled region when first-order kinetics apply, as illustrated in Fig. 12.7. The corrosion rate increases from I_1 to I_2 as the pH value decreases, resulting in an increase in the rate of diffusion of hydrogen ions to the surface of the metal. Thus measurement and control of pH to better than 0.1 units should be the aim, but this

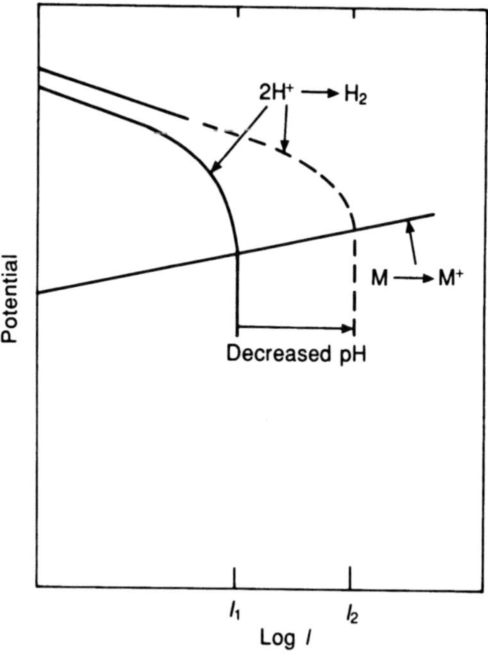

Fig. 12.7 The influence of pH on the corrosion rate with cathodic kinetics under diffusion control.

is difficult to achieve in practice because of the limitations of the glass electrode used.

Analysis of the environment at the end of the test is often necessary. Errors in the measured pH can occur if carbon dioxide from the atmosphere is allowed to dissolve in the solution, which may not be at equilibrium with respect to carbon dioxide, and the pH may therefore change with time. This is especially true when the tests are carried out at elevated temperature.

Solutions are frequently purged with nitrogen or argon to remove oxygen, but difficulties can arise since other dissolved gases will also be removed. The most common problem is encountered in solutions containing carbon dioxide and hydrocarbonate ions since purging will cause the pH of the solution to change as carbon dioxide is removed from the system. Changes in pH of 25% can easily occur, and special gas mixtures containing the equilibrium concentration of carbon dioxide should be obtained to avoid this.

In aerated test solutions, the corrosion reaction will normally be diffusion controlled where first-order kinetics will apply, as illustrated for hydrogen ion reduction in Fig. 12.7; a change in the oxygen concentration of 0.2 ppm, for example, will change the cathodic kinetics by 5%. In deaerated solutions the measurement of oxygen concentration is often more critical since small amounts (ppb) can influence oxidation/reduction equilibria, e.g. in cracks and crevices, and cause large changes in the corrosion potential of steel at elevated temperature, e.g. in steam raising plant, as illustrated in Fig. 12.8 (Indig, 1979). Electrochemical cells isolated from the test solution by an oxygen permeable membrane are used to

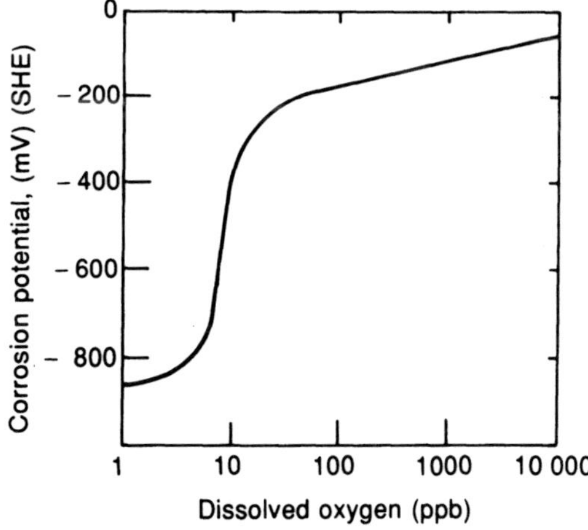

Fig. 12.8 The influence of dissolved oxygen on the corrosion potential of steel at 274 °C.

measure oxygen concentration. These cells are calibrated in air-saturated solutions, but at low oxygen concentrations the membrane can cause measurement errors.

12.4 HIGH-TEMPERATURE CORROSION

12.4.1 Gas composition

In conducting high-temperature corrosion experiments, control of the gas composition is clearly of paramount importance, but this can be readily accomplished only where simple equilibrated mixtures are used. However, in reactions involving complex gas mixtures, the equilibrium composition at the furnace inlet may not be the same as that at temperature near the sample. The question then arises as to how quickly the composition changes to the equilibrium value. In many cases, where the shift reactions are known to be slow, catalysts are used, and indeed the sample itself may act in this capacity. The shifts that can occur in gas composition are illustrated by the data in Table 12.1 which shows the composition of a mixture at the furnace inlet and the equilibrated composition at two temperatures. It can be seen that in this case a very large shift in the CO/CO_2 ratio occurs and significant amounts of CH_4 are formed when this gas mixture is used at the lower temperature of 450 °C. At this low temperature, gas equilibration is very slow and a catalyst should be used, but since the gas also contains hydrogen sulphide which would effectively poison most known catalysts, the experiment is undertaken in non-equilibrium conditions. Thus the gas composition cannot be specified under the conditions of the test and it is therefore essential to determine this at the sample.

Table 12.1 Equilibrium gas compositions (partial pressures, atm)

Gas species	Input composition	Equilibrated composition at 450 °C	Equilibrated composition at 750 °C
CO_2	3.9×10^{-2}	2.0×10^{-1}	5.8×10^{-2}
CO	2.4×10^{-1}	3.2×10^{-2}	2.2×10^{-1}
H_2	2.0×10^{-1}	5.0×10^{-2}	2.1×10^{-1}
H_2O	6.3×10^{-2}	4.0×10^{-2}	4.3×10^{-2}
H_2S	1.2×10^{-3}	1.2×10^{-3}	1.2×10^{-3}
N_2	4.8×10^{-1}	5.7×10^{-1}	4.6×10^{-1}
CH_4		1.1×10^{-1}	1.1×10^{-3}
O_2		6.3×10^{-31}	1.1×10^{-21}
S_2		1.1×10^{-11}	2.4×10^{-9}
$O_2(H_2/H_2O)$		1.0×10^{-31}	
$O_2(CO/CO_2)$		4.3×10^{-34}	

Generally, the chemical stability of the reaction product can be conveniently described by considering the activities of the anion-forming species in the reaction, e.g. oxygen, sulphur, chlorine and carbon. Measurement of these parameters at temperature, therefore, would be effective in describing the gas composition. Zirconia solid electrolyte probes for the measurement of the activity of oxygen, $p(O_2)$, are well established and have been used successfully in the measurement of gas composition in many combustion atmospheres. The probe relies upon conduction of oxygen ions from the reference electrode with a known $p(O_2)$, usually air, across the zirconia to the other electrode in contact with a gas of unknown oxygen content, and the EMF generated in the cell is directly proportional to the difference in oxygen concentration between the two electrodes. There have been various attempts (Taniguchi *et al.*, 1988) to devise similar probes to measure the other important species of gas composition, but none has found widespread acceptance.

An approximate solution to the problem of defining non-equilibrium gas mixtures has been proposed by Norton and Guttman (personal communication, JCR Petten) in which they suggest placing pure materials in the test environment that would define the stable phase. By using several such materials it should be possible to place limits on the uncertainty in the composition of the gas mixture in question. Alternatively, using the example of the gas mixture listed in Table 12.1, Coley and Saunders (1987) applied mass spectrometry to measure the gas composition exiting the furnace and showed that there was little shift in the gas composition upon heating to 450 °C. Thus the $p(O_2)$ in the non-equilibrium gas mixture could be defined by reference to the H_2/H_2O or the CO/CO_2 ratio; the latter produces a lower $p(O_2)$ value. Whether equilibrium gas compositions, H_2/H_2O or CO/CO_2 values are used to define the $p(O_2)$ can determine whether oxides or sulphides are the stable phase and thus greatly influence the rates of reactions (Table 12.1, Fig. 12.9). It can be seen that while both the equilibrium and the H_2/H_2O $p(O_2)$ values fall within the Cr_2O_3 phase field, that calculated using the CO/CO_2 ratio falls on the Cr_2S_3/Cr_2O_3 boundary. A further complication is that kinetic effects shift the phase boundary to higher $p(O_2)$ values, so that the small difference between the $p(O_2)$ values calculated from the H_2/H_2O ratio or assuming equilibrium may be sufficient to allow sulphide formation to occur in the former case while oxide formation occurs in the latter.

12.4.2 Mechanical properties

Measurement of the mechanical properties of thin protective layers is essential to the complete characterization of corrosion reactions so that scale adhesion and spallation can be correctly determined. There have been many attempts to develop models of scale fracture (Evans, 1989; Evans and Hutchinson, 1984; Robestson and Manning, 1990; Schuetze, 1990), but the validation has often had to rely upon estimations of

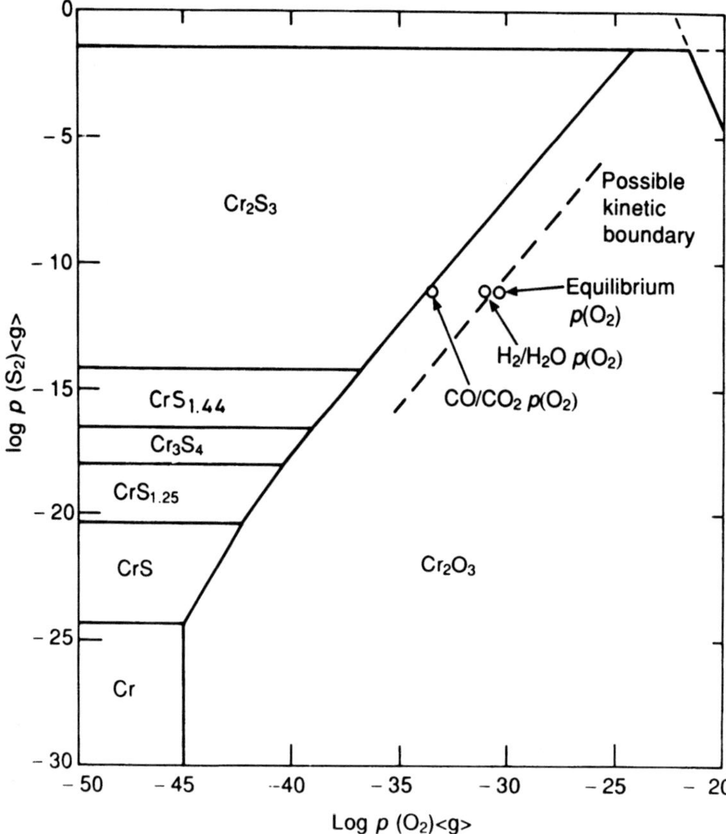

Fig. 12.9 The phase stability diagram for the Cr-O-S system at 450 °C, showing the effect of non-equilibrium gas mixtures on the stability of oxide or sulphide phases.

relevant mechanical properties of the thin layer from the bulk properties of material of related composition. Clearly this approximation is unsatisfactory because it neglects the important role of microstructure in determining mechanical properties of solids. The microelectronics industry has done much to stimulate the development of novel measurement techniques in this area because of its obvious interest in the properties and behaviour of thin films, and one of the measurement methods currently receiving a great deal of attention is depth sensing indentation in the nanometre scale (Nix, 1989; Doerner and Nix, 1986; Oliver and Pharr, 1992) (mechanical property microprobe). Such instruments can give information on the elastic and plastic properties at ambient temperature of the thermally grown layers, and, for example, Young's modulus can be determined from the slope of the unloading curve at peak load (Fig. 12.10). Hardness is obtained in the usual way

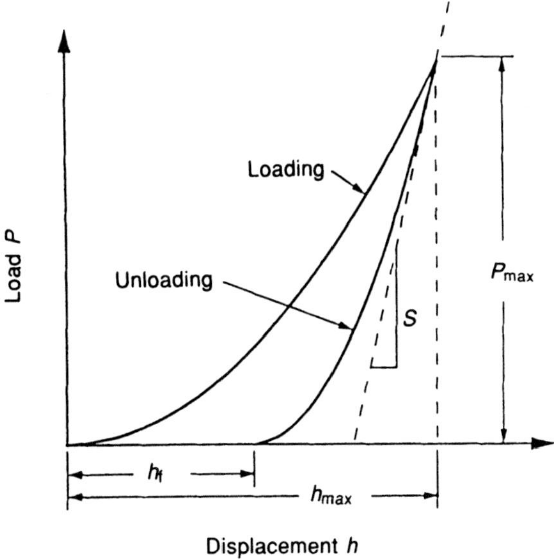

Fig. 12.10 Load–displacement curves obtained from nanoindentation measurements (Oliver and Pharr, 1992, p. 1565).

from the relationship between load and projected area of the indentation, while yield stress can be measured directly using microbeam methods (Nix, 1989). The indenter may also be used to obtain information on adhesion of the scale to the underlying alloy by monitoring acoustic emission during indentation so that onset of cracking can be detected at the oxide/substrate interface.

There are numerous metrological problems concerning this technique which relate to the exact definition of the indenter shape, the determination of the modulus of the indenter itself, the compliance and thermal drift of the instrument and the calibration of the load and displacement measurements. For example, analysis of the slope of the linear portion of the unloading curve is used to give the stiffness of the sample, from which its modulus can be derived if the projected contact area at maximum load can also be measured. Oliver and Pharr (1992) have recently shown that the unloading curves are not linear and are best described by a power law. The slope is then determined by differentiating and evaluating the derivative at peak load and maximum displacement. Furthermore it was found that repeated loading and unloading experiments at the same point gave different shapes to the unloading curve. Figure 12.11 shows the results for stiffness of tungsten using these different procedures, where it can be seen that respective use of a linear or power law fit could alter values by up to 50% and that use of repeated loading,

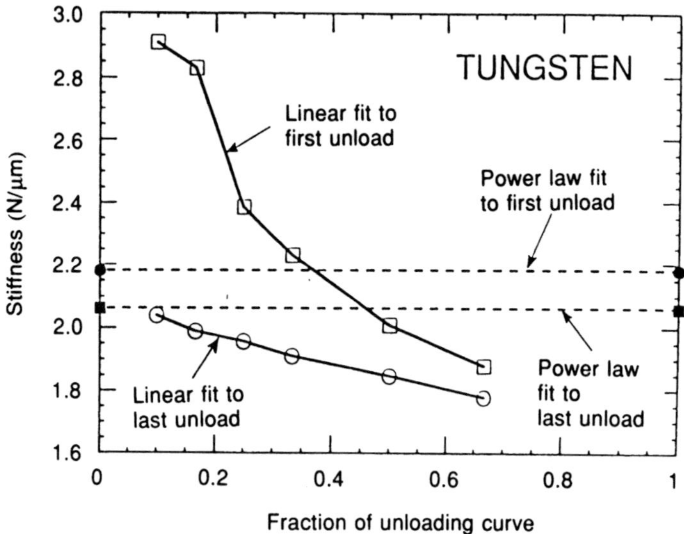

Fig. 12.11 The variation in stiffness of tungsten as a function of the experimental and analytical procedures: see text (after Oliver and Pharr, 1992, p. 1575).

even with the power law analysis of the unloading curves, could introduce differences in stiffness values of almost 10%. This is a clear case where expert metrological input would be beneficial before the technique can gain widespread usage.

It is also assumed that the surface has the same mechanical properties as the bulk, but this will only be the case if extreme care is taken during preparation to avoid introducing mechanical deformation. For the case of thermally formed oxide scale the surface will be rough and, without some procedure to obtain a smooth surface that does not alter properties, the scatter in data will be large or the data may even be misleading. Despite these uncertainties the information obtained by this technique is believed by its proponents to be superior to that derived from bulk materials.

12.5 CONCLUSIONS

The required accuracy and traceability of measurements usually carried out on corroding systems are readily achieved. But there are difficulties associated with measurement of potential and concentration in cracks, crevices and pits and under corrosion products, and large errors occur if bulk values are assumed. Accurate measurement of electrical potential is difficult in poorly conducting solutions because of *IR* potential drop and screening by the reference probe. Measurements of pH are made routinely in most laboratories, but repeatability and the required accuracy are difficult to attain.

A major concern in high-temperature corrosion is the lack of well-defined standards or guidelines for conducting tests and evaluating the testpiece after exposure. This difficulty is now being recognized and some initial attempts have been made to develop improved methods. Frequently data reported in the literature are relatively short term (some hundreds of hours) while component lifetimes are often expected to be many thousands of hours, so that inappropriate extrapolations are sometimes made. In general there are few difficulties with measurement techniques *per se* that are unique to high-temperature corrosion processes, although special care in the application of the measurement method is required. The major problem facing the high-temperature corrosion community is in the definition of realistic environments that adequately simulate relevant corrosion processes and in adequately characterizing the testpiece before exposure. Corrosion reactions are often uniquely sensitive to trace amounts of material in the environment owing to selective adsorption and doping effects, and rates of attack can be markedly altered by apparently minor changes in conditions. Thus while accuracy of measurement is, of course, important it is often a second-order effect where maximum cumulative errors may amount to about 20–50%, whereas lack of control or the wrong choice of the environment can result in measured rates of corrosion which are many orders of magnitude different from true rates. While the present situation with regard to measurement of high-temperature corrosion is very unsatisfactory, there is some cause for optimism that progress can now be made towards the development and specification of improved test procedures.

For both high-temperature and aqueous corrosion processes, intrinsic factors that are related to material properties (composition, surface state and microstructure, for example) and extrinsic factors that determine the environment (start-up procedures, temperature and composition of the media) are both of prime importance in determining corrosion rates. It is clear therefore that future progress in deriving quantitative data on rates of corrosion depends on the development of improved measurement procedures for both intrinsic and extrinsic properties, while correctly identifying critical parameters that require careful control during the test.

REFERENCES

ASTM G5 Standard Reference Test Method for Making Potentiostatic and Potentiodynamic Anodic Polarization Measurements.

British Standard BS 7361 : Part 1 : 1991 Cathodic Protection. Part 1: Code of practice for land and marine applications.

Cheung, W. K. and Thomas J. G. N. (1988) ASTM Special Publication 970 (eds P. E. Francis and T. S. Lee), ASTM, Philadelphia, pp. 190–204.

Coley, K. S. and Saunders, S. R. J. (1987) In *Proceedings of the Conference on Materials for Coal Gasification* (eds W. T. Bakker, S. Dapkunas and V. Hill), ASM Int., Metals Park, Ohio, pp. 105–12.

Doerner, M. F. and Nix, W. D. (1986) *Journal of Materials Research*, **4**, 601–9.

Evans, A. G. and Hutchinson, J. W. (1984) *International Journal of Solid Structures*, **20**, 455–66.

Evans, H. E. (1989) *Materials Science and Engineering*, **A120**, 139–46.

Gemmill, M. G. (1976) Corrosion problems in power plant. High temperature high pressure electrochemistry in aqueous solutions, in *NACE-4* (eds D. de G. Jones, J. Slate and W. R. Staehle), Surrey, 1973, NACE.

Indig, M. E. (1979) EPRI CAC Workshop presentation, Palo Alto.

Mercer, A. D. and Brook, G. M. (1978) *La Tribune du CEBEDEAU*, **31**, 299–306.

Nix, W. D. (1989) *Metallurgical Transactions*, **20A**, 2217–45.

Oliver, W. E. and Pharr, G. M. (1992) *Journal of Materials Research*, **7**, 1564–83.

Robertson, J. and Manning, M. I. (1990) *Materials Science and Technology*, **6**, 81–91.

Schuetze, M. (1990) *Materials Science and Technology*, **6**, 32–8.

Szklarska-Smialowska, Z. (1974) The pitting of iron-chromium-nickel alloys. Localized corrosion, in *NACE-3* (eds W. R. Staehle, B. Brown, J. Kruger and A. Agrawal), National Association of Corrosion Engineers, Houston, Texas, pp. 312–37.

Taniguchi, M., Wakihara, M., Uchida, T., Hirakawa, K. and Nii, J. (1988) *Journal of the Electrochemical Society*, **135**, 217–21.

Turnbull, A. (1982) *Reviews on Coatings and Corrosion*, **5**, 43–171.

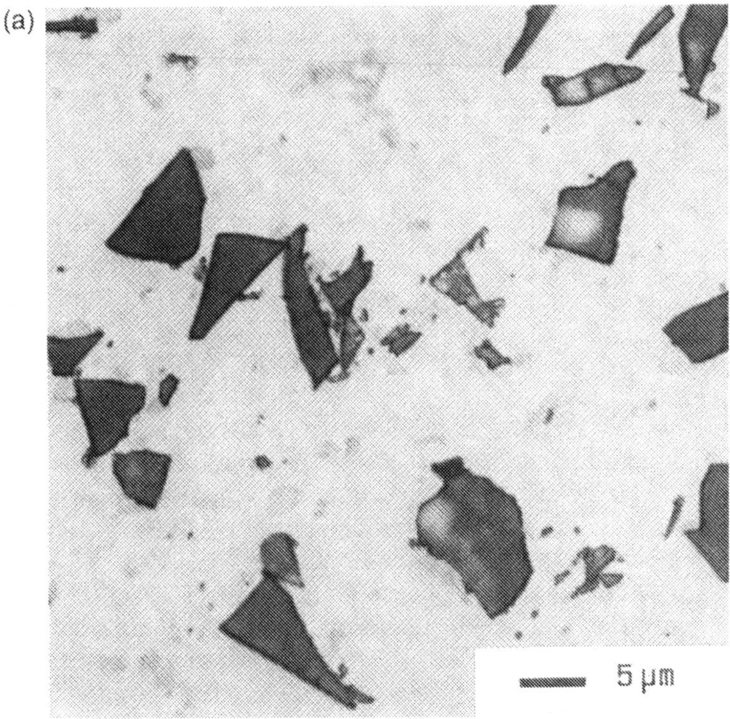

Fig. 13.1 (a) Microstructure of silicon carbide reinforced metal matrix composite: optical micrograph.

replace 'metallography': 'microscopy' and 'micrography' are not suffi-ciently specific to materials, while 'materiography', 'materialography', 'matrography' and 'matography' seem too ugly. Daniel (1991) has suggested 'hyleology' for materials science, from which we could derive, totally from the Greek, 'hyleography'. This is a personal favourite, but it might be simpler, and more pragmatic, to extend the usage of **metallography** to non-metallic materials. This is the practice adopted in this chapter.

Measured parameters which are widely used in assessing materials include grain size, shape and distribution; texture; inclusion volume fraction and uniformity; presence, absence and amount of phases and components; and local and overall composition (for example, Fig. 13.1). These parameters are important because of their effect on engineering properties, but in this chapter we will restrict attention to the microstructu-ral parameters. Two most important considerations are the repeatability

CHAPTER 13

Microstructural metrology

P. J. Goodhew

13.1 INTRODUCTION

Quantitative descriptions of microstructure are beginning to be used routinely for the characterization of engineering materials. There is great concern, especially in Europe, the USA and Japan, that such quantitative descriptions should be standardized, as recognized by the ASTM Symposium on Metallography in Atlantic City in June 1991. Metrological applications of immediate relevance to the engineering use of materials include cleanness assessment, inclusion counting, phase analysis and grain size determination in alloys, volume fractions and size and distribution of particles and fibres in composites, and porosity in ceramics. It is the purpose of this chapter to assess the extent to which metrological reproducibility and traceability can be achieved in these fields, and to point out some of the areas in which procedures will need to be improved or modified if, for example, NAMAS (1989) accreditation is to be sought. It is the intention not to present a fully comprehensive review of what is an enormous subject area, but to give a snapshot of some important problem areas and set the scene for decisions on priorities for future work.

The relatively new phrase **microstructural metrology** raises the problem of terminology in this field. The techniques used to measure and describe the microstructure of engineering materials have a great deal in common, whether the particular material be an alloy, a ceramic or a composite. It is clearly desirable that a uniform set of terms is used to describe the study and classification of microstructures. The well-established terms 'petrology' and 'metallography' refer to the qualitative study of specific classes of materials (rocks and metals respectively), but we lack an all-embracing term covering metals, ceramics, polymers and their composites. The implication of measurement is usually achieved by adding the word 'quantitative', as in 'quantitative metallography'. This latter phrase can sensibly be supplanted by the more general 'microstructural metrology'. It is less obvious what qualitative word should

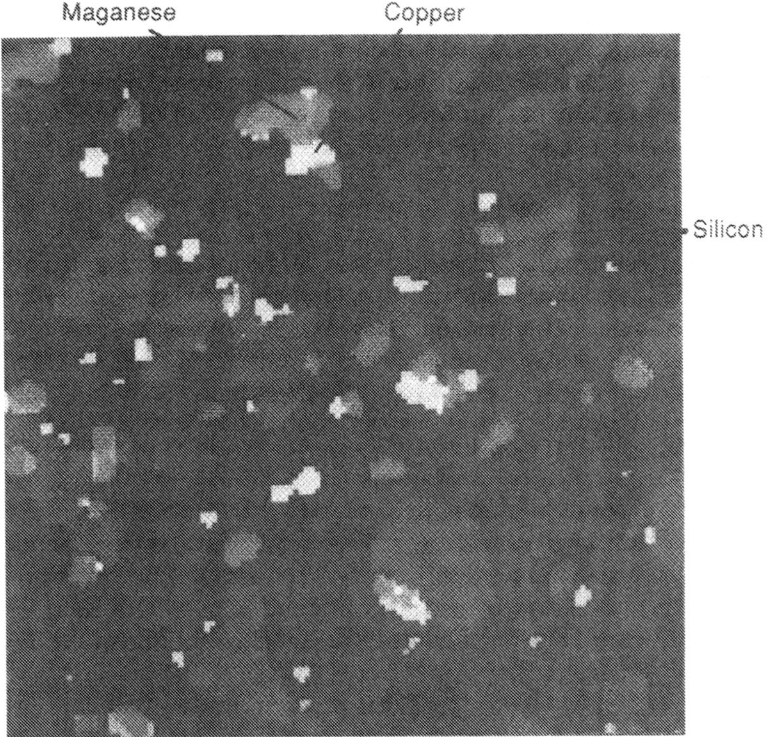

Fig. 13.1 (b) X-ray compositional map.

and reproducibility of microstructural measurements and their trace-ability to primary standards. The measurements must be accurately repeatable by any trained technician and must not depend on the specific skills of any single individual, or at least there should be some quantitative information available on the uncertainties associated with the use of different operators. This is likely to become increasingly important as microstructural data are used in legal and patent proceedings. For the same reasons the dating, timing and archiving of all micro-structural images and data will become widespread. This database requirement will impose the need for standards concerned with data storage formats, access and validation. All these considerations provide a driving force for the full automation of microstructural metrology, from sample selection through specimen preparation to metallographic measurement. These topics will be discussed in greater detail in the following sections.

13.2 SAMPLING STATISTICS

Let us first assume that unlimited areas of perfect artefact-free specimen can be produced, viewed at the appropriate magnification in a perfect microscope and analysed using a perfect image analyser. 'Perfect' in these contexts must mean 'providing the required (defined) accuracy'. Later sections will deal with the problems associated with the attainment of this ideal.

The metrological considerations are:

- the calibration of the measurements and hence their potential accuracy;
- the validity (and necessity) of extending two-dimensional measurements to three-dimensional parameters;
- the statistical significance of the results, as a consequence of the (explicit or implicit) sampling procedure.

Sampling will be considered first. If there is a large volume fraction of the phase to be measured, or if it is easily possible to get a large number of uniform grains within a field of view, then the sampling problem reduces to a straightforward reliance on the measurement scientist spotting any unexpected non-uniformity. Problems are more serious if it is necessary to make measurements in conditions where there is a low volume fraction and/or an inhomogeneous distribution of the measured features.

Chone (1978) has considered this problem in the context of inclusion content assessment in continuously cast steels. His analysis was on the basis that the volume fraction of inclusions could be safely extrapolated from the area fraction found in a section on the assumption of a dilute concentration of spherical inclusions. This may not be entirely satisfactory, but the general conclusions of his analysis are worthy of further study. Chone produced charts showing the surface area which it is necessary to examine in order to make a measurement of inclusion volume fraction with a precision of $\pm\,20\%$. An example is shown in Fig. 13.2. The three bold diagonal lines show the surface area to be examined for populations of particles of three different sizes. In calculating these lines the particles were assumed to be all the same size, which is clearly unrealistic. However, the general conclusions are clear. For instance, in order to measure a volume fraction of 10^{-3} (0.1%) of 100 μm particles it is necessary to analyse $10^{-3}\,m^2$ of specimen (the point marked X on the figure). However, to measure a volume fraction of only 10^{-6} (0.0001%), such as might easily be required in a clean sheet product, requires the assessment of almost 1 m^2 of specimen (point Y on the figure).

A more easily interpreted yardstick for the effort involved in making these measurements is the number of fields of view (or screens) which need to be analysed in each case. If we assume that each field is examined at the appropriate magnification (that is the minimum magnifica-

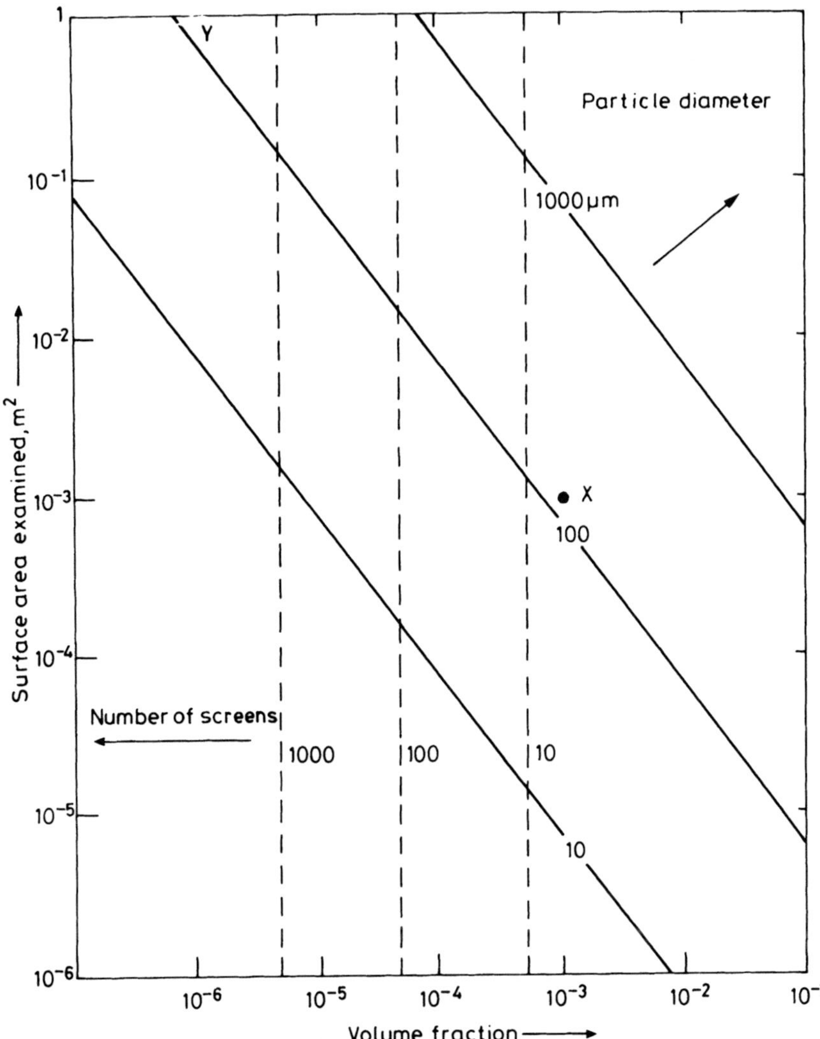

Fig. 13.2 Chone chart showing relationship between surface area to be examined, volume fraction, particle size and number of screens.

tion which will bring the features of interest to measurable size) then we can plot vertical lines which show the number of screens needed. Three of these are shown dashed on Fig. 13.2. We can now see that the measurement at point X needs fewer than 10 screens, whilst that at point Y needs several thousand. The message is clear: the measurement of small volume fractions requires many screens and therefore a long time. It is probably not feasible to analyse 1000 screens without a fully automated image analysis system and even then, in the example given at point Y, many metallographic specimens would be needed to give a total area of 1 m². An assessment similar to this should be made before any quantitative image analysis is started.

There is currently no British or US Standard which sets out the approach just outlined. (A list of relevant British Standards (with ISO equivalents) and ASTM standards will be found at the end of the chapter.) Inclusion levels in steels are commonly measured using charts which tend to be inadequate at low volume fractions. Vander Voort (1988) has discussed the use of image analysis to measure extremely low inclusion contents. He used as an example the assessment of a population of inclusions with a mean size of about 1 µm and a volume fraction of about 10^{-4}. From Fig. 13.2 it can be seen that about 100 screens would need to be analysed, and Vander Voort commented that he used '5 to 500 fields' to achieve his quoted precision of ± 60%. Problems identified as limiting accuracy included the failure to measure small inclusions (below the resolving power of the microscope) and the presence of cracks or holes at the particle/matrix interfaces. It is not possible to do anything about these problems after the measurements have been made!

An ASTM Subcommittee (E0.04) is considering publishing a standard embodying some of these points. Some detail of the stereological parameters which it is possible to deduce by automated image analysis is included in ASTM E1245 : 1989.

13.3 SPECIMEN PREPARATION

The ideal specimen preparation procedure for optical microscopy produces a surface for etching, imaging and analysis which is:

- typical (representative);
- flat and bright;
- free from inclusions picked up during polishing; and
- free from deformation introduced at any earlier stage.

Specimens which conform to this ideal have been described by Bousfield (1988) as having high *integrity*. However, this is not an easily quantifiable measure and therefore it has limited use as a descriptive parameter. It is nevertheless clear that each step of a metallographic specimen preparation technique is prone to many variables. Among these are:

- choice of type of abrasive and its mount;
- choice of abrasive size;
- choice of lubricant;

- load applied to specimen; and
- speed.

It might appear that success could be achieved in specimen polishing by continuing the procedure until the optimum surface is produced, and that this ideal surface would survive further polishing. However, there is increasing evidence that continuing each polishing stage beyond the optimum actually degrades the specimen surface (Bousfield and Bousfield, 1990). There are thus powerful reasons for establishing well-documented procedures which reproducibly achieve the optimum specimen surface. The target of any laboratory aiming to use microstructural metrology must be first to define these procedures for the materials of interest, and then to make them independent of the particular person who is applying them.

One way to improve reproducibility is to use automated preparation equipment (e.g. Bittance, 1989). Currently this seems to be an expensive option, but the initial investment should be seen in the context of the cost of other sophisticated laboratory equipment. For instance many laboratories spend equivalent sums on preparation equipment for (much less reproducible) electron microscope specimens. With such equipment it is possible to reproduce sequences of polishing steps which have been found to be effective. It does not of course remove the requirement that an experienced metallographer must assess the integrity of specimens produced during development of the procedure.

There are some standards relating to the preparation of metallographic specimens. ASTM E3 : 1980 (1986) describes the preparation of metallographic specimens including cutting, mounting and polishing, but is not prescriptive, comprehensive or quantitative. ASTM E768 : 1980 outlines good practice in the particular case of inclusion assessment in steels, but is again only semi-quantitative. It recommends that differential interference contrast (DIC) should be used to monitor residual scratches and topography but admits to its subjectivity. Dark field illumination is also good for revealing scratches.

Etching is obviously of major importance in image analysis and therefore in microstructural metrology. There is more guidance in standards in this area: four of the entries at the end of the chapter make direct reference to both macro- and micro-etching, while it is implicit in several other standards. It remains a problem, however, to define the etch time for grain boundary etching for grain size measurement (see section 13.5). Interference film microscopy (e.g. Buhler and Houghardy, 1979) is in principle superior to conventional etching since it is quantifiably controllable and does not significantly disturb the surface (Fig. 13.3). No standards yet exist in this area.

There may be a future role for laser methods (which would be non-contacting) in assessing surface finish and thus integrity. It is also

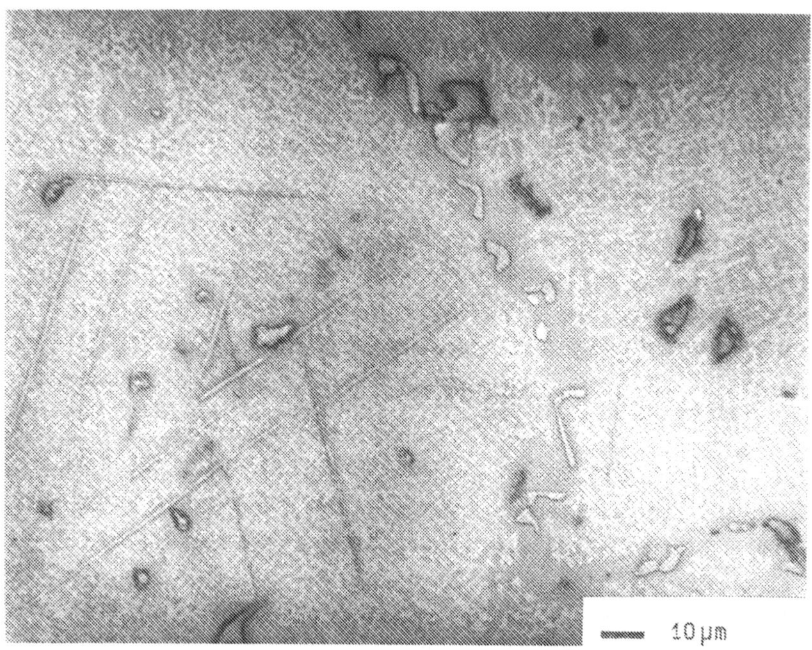

Fig. 13.3 Superalloy specimen coated with Fe_2O_3 coating to give interference film contrast between phases in microstructure. Note that no chemical etching was used in the preparation of the micrograph.

possible that, on an even finer scale, the use of scanning tunnelling microscopy (STM) or atomic force microscopy (AFM) could prove useful. However, the problem with the use of such techniques will be that it is difficult and time consuming to sample a large enough area of surface for statistical significance. Such techniques also find it difficult to cope with a variety of scales of surface roughness and would currently be confused (and possibly damaged) for example by the presence of both fine and coarse scratches.

13.4 STANDARDS AND CALIBRATION

The most obvious candidate for calibration is the magnification of the microscope which is being used to study the specimen. There are a number of subtleties here. It is straightforward to image a ruler which has been calibrated against a primary or secondary standard, although this is rarely useful beyond about 100×. NPL is preparing a line width

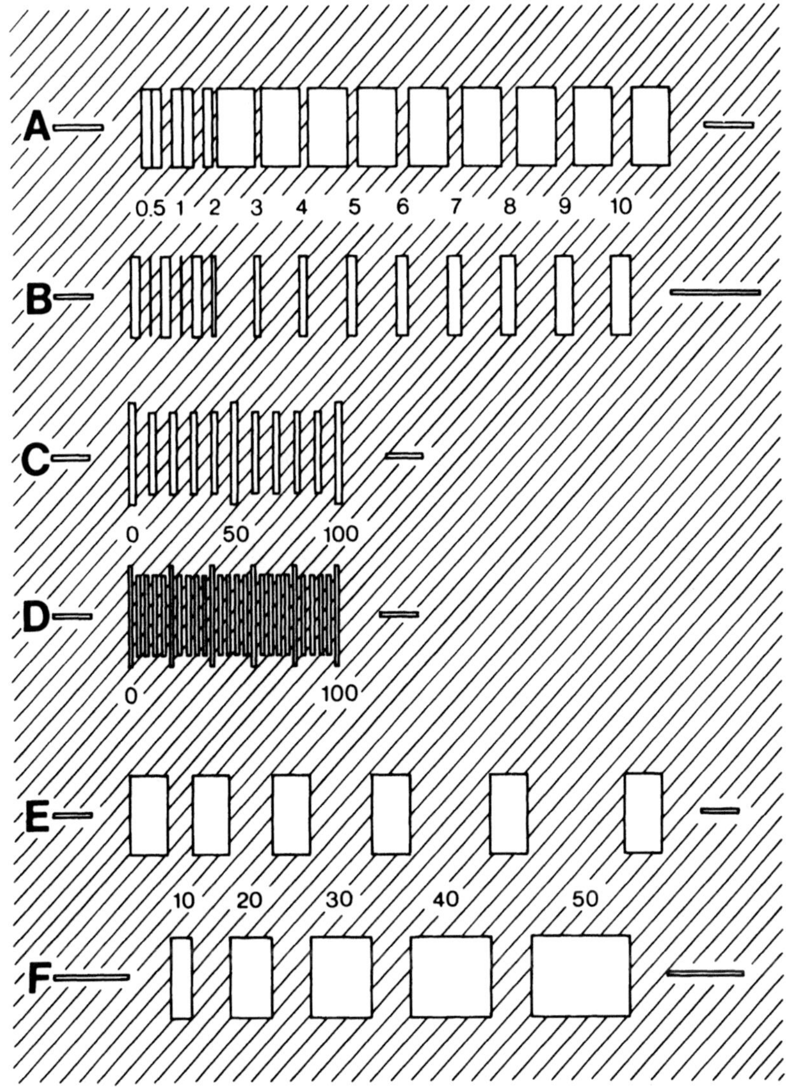

Fig. 13.4 NPL photomask line width standard test pattern (© Crown copyright).

standard with details in the size range 0.5 to 10 μm (Fig. 13.4) which will serve well as a magnification standard for higher (optical microscope) magnifications. This standard has to be calibrated using a travelling microscope fitted with a laser interferometer, and is thus traceable back to the standard metre. It is necessary, however, to calibrate each lens combination which might be used, and to check two orthogonal directions.

Further calibration of the imaging chain of the image analyser must then be undertaken. Variations with supply voltage and temperature should be negligible with digital imaging systems, but will not be for older analogue systems. The essential principle is that each final image magnification must be traceable to the primary standard metre.

The same considerations apply to all types of microscope. For the most frequently used alternative to optical microscopy – scanning electron microscopy (SEM) – quite a range of calibration standards are now available. Several of these, such as metal meshes and populations of polystyrene spheres, are not conveniently traceable to the primary standard. However, traceable standards are beginning to become widely available. For example, electron beam lithographically ruled silicon wafers and scribed metal standards are now available from SIRA in the UK and NIST in the USA (via commercial retailers) with certification guaranteeing 1% accuracy. NPL is working on a 0.2 μm line width standard suitable for the SEM. ASTM E766 : 1986 and E986 : 1986 also outline SEM calibration procedures.

If a stage micrometer is to be used, as will normally be the case for fully automated analyses, then it too should be traceably calibrated in both x and y directions.

More serious calibration problems arise in the use of the image analyser. In addition to calibration of the magnification it is necessary to consider the possibility of image distortion, and the effectiveness of the edge detection algorithms. There is a serious problem in relying on calibration via a reference graticule such as that available from NPL. The setting of thresholds on binary test objects (with sharp boundaries between black and white regions) is not necessarily the same as that appropriate to interfaces between different shades of grey. There is no substitute here for good specimen preparation which provides clear sharp contrast between features which are to be distinguished, together with frequent blind cross-checking between operators using the same specimens. The setting of detection thresholds is still an area where human intervention is frequently necessary. There is potential for the production of standard reference images (not necessarily with sharp edges) by computer. Clearly the computer code would need to be validated, as would the output device, but there is scope for development in this direction.

There is a further aspect of microstructural metrology which is not strictly related to the accuracy or precision of measurement but which is important to the interpretation of direct measurements. Stereological procedures are implicit or explicit in the assessment of three-dimensional characteristics (e.g. grain size) from two-dimensional data (e.g. a micrograph). This is a complex subject, on which several good texts have been written (e.g. Underwood, 1970; Russ, 1986). In this chapter it is sufficient to point out that simple extrapolation from two dimensions to three involves sweeping assumptions about the shape of the particles or grains.

13.5 TWO APPLICATIONS

In this section the principles described above are illustrated by the use of two examples of the application of microstructural metrology.

13.5.1 The measurement of grain size

Grain size is a frequently used and apparently quantitative parameter, and a natural candidate for standardized measurement. Manually this can be achieved either by counting grain boundary intersections with a set of straight lines, or by visual comparison with standard (usually ASTM) idealized grain population images (BS 4490 : 1989; ASTM B390 : 1986, E112 : 1988, E930 : 1989, E1180 : 1987). Automatic image analysis offers a third method, which relies on measuring the area of each grain. Crucial to the success of this method is the continuity of the line indicating the grain boundary: each small gap may lead the analyser to detect a single grain in place of two adjacent grains. Scratches may also be particularly confusing to the analyser. If such methods are to be used then either the polishing, etching and threshold detection must be near perfect, or sophisticated gap-filling routines must be used during the image analysis.

ASTM E112 : 1988 specifies that twins should be ignored in assessing grain size (i.e. that even a multiply twinned crystal should be considered as a single grain). This poses no significant problem for the manual operator, but means that an automatic method must be capable of recognizing and eliminating twin boundaries. This is normally done, by both computer and manual operator, in terms of their straightness (e.g. Friel and Prestridge, 1991). However, the efficacy of the computer routines must be checked periodically before the measurement can be considered to be standard.

A number of other issues arise when automation of grain size measurement is considered. Ito (1990) has considered the effect of image magnification on the measurement and concludes that optimum reproducibility occurs when the number of grains in a field of 400×400 pixels is about 100. Presumably, although the author does not state this explicitly, this number scales up with the total pixel number and a field of 800×800 pixels should be capable of holding 400 grains for measurement.

Duplex grain populations present a further problem. Okada, Katsulai and Oka (1990) have derived expressions for the fraction of volume occupied by the (two) grain populations in a duplex structure, and have

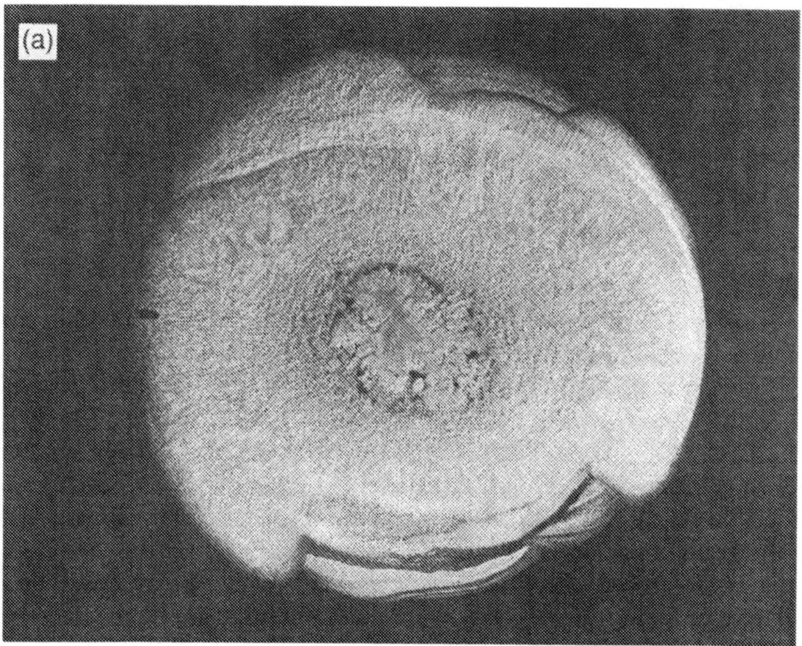

Fig. 13.5 (a) Electron beam button melt specimen showing concentration of impurities at centre of surface of melt: top view.

tested these against computer-generated duplex structures. So far only one ASTM Standard, E1181 : 1987, addresses this problem and gives details of ways of quantifying the deviation of a particular (not necessarily bimodal) distribution of grain sizes from the usual single lognormal distribution. This is one of the more useful standards in that it gives examples of both microstructures and measurements.

13.5.2 Cleanness assessment by electron beam button melting (EBBM)

Cleanness assessment of 'clean alloys' requires that very low inclusion contents are assessed. Because of the practical difficulty in measuring low volume fractions (see section 13.2) an entirely different approach has been devised, in which the solid particles are swept out of a large volume of the alloy by electron beam button melting (EBBM) (e.g. Quested and Chakravorty, 1989) (Fig. 13.5).

Fig. 13.5 (b) Electron beam button melt specimen showing concentration of impurities at centre of surface of melt: side view. Specimen is approximately 75 mm in diameter.

EBBM is obviously only useful for concentrating dilute populations of particles which are stable above the melting point of the alloy. However, it does have the huge statistical advantage that a large volume of material can be sampled quite quickly. At present the measurement of the swept-out particles is only semi-quantitative. Optical and SEM assessment of the particle debris has only been used down to sizes of 50 μm. However, the volume fraction detected can be as low as 10^{-4}. Alloys have successfully been ranked in cleanness but as yet quantitative particle population data, particularly size distributions, have not been reliably produced. There is some evidence that not all particles with sizes below 100 μm are recovered after melting, and this would necessitate correction to the raw data (Shamblen, Culp and Lobek, 1983).

13.6 ELECTRON METALLOGRAPHY

The statistical considerations outlined in section 13.2 apply equally to electron metallography, but are often harder to satisfy because of the smaller area and volume studied.

13.6.1 Scanning electron microscopy (SEM)

The use of SEM is relatively straightforward, although it should be recognized that, in normal usage, SEM only samples the surface and not much 'typical' volume. Clearly only when used in back-scattered mode on polished sections does SEM approach the conditions of use of the optical microscope. It then provides an extension of optical techniques to smaller feature sizes.

13.6.2 Transmission electron microscopy (TEM)

The problems of reliable sampling are a major stumbling block to quantitative use of TEM. Thin areas are difficult to achieve, are rarely extensive, and may tend to represent special regions of the specimen (e.g. those which thin fastest). Very little progress has been made towards standardization of specimen preparation techniques for TEM.

There are many reasons for wishing to use TEM for microstructural metrology. Improved resolution is in principle combined with the ability to identify phases by electron diffraction and/or X-ray analysis (EDX) or electron energy loss spectrometry (EELS). These techniques each need calibration, which is not usually a routine matter for most electron microscopists, and may indeed pose problems which have not yet been addressed in electron metallography. Among the items which are easy to calibrate are magnification and diffraction camera length. Standards in this area would require not only the specification of standard objects for magnification calibration in all magnification ranges from $100\times$ to $1\,000\,000\times$ but recommendations for frequent recalibration because of possible variations in lens currents and temperature.

Analytical techniques such as EDX and EELS are much more difficult to calibrate, since chemically homogeneous standards are not available for many elements. It is relatively easy to calibrate the detection efficiency of the analyser but much more difficult to calibrate and standardize the computer correction routines which are used to convert raw data into composition. This is an area into which metrology has not really yet been introduced, although ASTM Committee E4 has done some work on EDX analysis and will be producing a draft standard shortly.

For both quantitative imaging (e.g. particle counts, phase volumes) and quantitative analysis (e.g. absolute elemental analysis by EELS), local specimen thickness is important. There are many ways of measuring thickness, but calibration of any of them would be quite difficult. This is another area where work is needed.

13.6.3 Electron microprobe analysis (EPMA)

Electron probe micro-analysis (EPMA) or electron microprobe analysis is a fairly mature technology and some thought has been given to accuracy and precision. It should prove possible to set up metrological standards without too much further work. The items which will require attention are the specification of standards and operating procedures and the calibration of routines which are used to correct for atomic number, absorption and fluorescence effects (ZAF corrections).

As with all the techniques discussed in this chapter, there is a problem in the determination of phases present in low volume fractions. A particular aspect of this problem relates to finding a region of potential interest. It is extremely time consuming, in EPMA and also other surface-sensitive analysis techniques such as scanning Auger microscopy (SAM), to locate small regions of interesting composition. The problem is associated with the contrast exhibited by these regions. Most imaging techniques (light microscopy, SEM etc.) produce an image virtually instantaneously, and if features of interest give rise to distinct contrast (e.g. appear dark) it is easy to detect them by eye or by image analyser. However, phases of interesting composition may not be easy to reveal in the image of an EPMA or SAM instrument. They must then be sought on the basis of their composition. Compositional mapping is generally very time consuming, typically taking a factor 10^4 or 10^5 longer than conventional imaging of the same region. Search algorithms are therefore needed to speed up identification of interesting regions, before quantitative assessment can start. Baker and Castle (1988) have made a start on this, in the context of surface analysis, but much more work is needed before a good metrological procedure is reached.

13.7 CONCLUSIONS

Microstructural metrology is in its infancy. In a small number of areas, based on light microscopy, standards are emerging which will help to define sound metrological procedures. At present these are largely confined to grain size determination and particle size analysis. The automation of specimen preparation should help to eliminate operator-specific factors.

Specific topics which need attention and on which standards would be welcome are:

- magnification calibration;
- image analysis threshold setting;
- sampling statistics.

The extension of metrology into electron metallography is possible but difficult and will require considerable effort. In particular, standard procedures will be needed for specimen preparation, magnification and camera length calibration and, for analytical microscopy, correction routines. It might be helpful to consider the experience of laboratories in quite different fields, such as public health and forensic science, where protocols which define technical procedures are rather better established.

ACKNOWLEDGEMENTS

The author would like to acknowledge the help, advice and encouragement of Drs Peter Quested, Mark Gee and Bryan Roebuck of NPL during the preparation of this chapter.

REFERENCES

Baker, M. A. and Castle, J. E. (1988) Location of elementally-rich micro-areas using scanning Auger microscopy, in *Proceedings of EUREM 88*, Institute of Physics Conference Series 93, pp. 267–70.

Bittance, J. C. (1989) Greater precision for materials analysis. *Advanced Materials and Processes*, no. 11, 11–18.

Bousfield, B. (1988) A systematic approach to sample preparation. *Metals and Materials*, 758–61.

Bousfield, B. and Bousfield, T. (1990) Progress towards a metallography standard. *Metals and Materials*, 146–8.

Buhler, H. E. and Hougardy, H. P. (1979) *Atlas der Interferenzschichten Metallographie*, Deutsche Gesellschaft für Metallkunde.

Chone, J. (1978) Échantillonage de billettes de coulée continue en vue de la description de la structure interne et de la propreté inclusionnaire, in *International Symposium ou Quantitative Metallography*, November 1978, Florence, Association Italiana di Metallurgia, pp. 209–24.

Daniel, E. R. (1991) Letter. *Metals and Materials*, 7(8), 522.

Friel, J. J. and Prestridge, E. B. (1991) Grain sizing by image analysis. *Advanced Materials and Processes*, no. 2, 33–7.

Ito, K. (1990) Effect of magnifications of images on automatically measured grain size. *ISIJ International*, **30**, 490–5.

NAMAS (1989) *General Criteria* (M10). *Regulations* (M11). NAMAS Accreditation Standards.

Okada, A., Katsulai, H. and Oka, M. (1990) Three-dimensional distribution of grain structure and image analysis. *ISIJ International*, **30**, 503–10.

Quested, P. N. and Chakravorty, S. (1989) Electron beam button melting as a cleanness assessment technique for superalloys, in *Conference on Electron Beam Melting and Refining State of the Art*, October 1989, Reno, Nevada.

Russ, J. C. (1986) *Practical Stereology*, Plenum Press.

Shamblen, C. E., Culp, S. L. and Lobek, R. W. (1983) Superalloy cleanliness evaluation using the EB button melt test, in *Electron Beam Melting and Refining – State of the Art* (ed. R. Bakish), Bakish Materials Corp., Englewood, NJ, pp. 61–94.

Underwood, E. E. (1970) *Quantitative Stereology*, Addison-Wesley.

Vander Voort, G. F. (1988) Measurement of extremely low inclusion contents by image analysis, in *Effect of Steel Manufacturing Processes on the Quality of Bearing Steels* (ed. J. J. C. Hoo), ASTM STP987, pp. 226–49.

STANDARDS

British Standards

BS 3406 : 1963 (1985)	Methods for Determination of Particle Size Distribution. Part 4: Optical Microscope Method.
BS 3625 : 1963 (1986)	Specification for Eye-Piece and Screen Graticules for the Determination of the Particle Size of Powders.
BS 4490 : 1989	Methods for Micrographic Determination of the Grain Size of Steel (equivalent to ISO 643, EN 103).
BS 5166 : 1974	Method for Metallographic Replica Techniques of Surface Examination (equivalent to ISO 3057).
BS 5411 : 1984	Methods of Test for Metallic and Related Coatings. Part 5: Measurement of Local Thickness of Metal and Oxide Coatings by the Microscopical Examination of Cross-Sections (equivalent to ISO 1463).
BS 5411 : 1989	Methods of Test for Metallic and Related Coatings. Part 16: Scanning Electron Method of Measurement of Local Thickness of Coatings by the Examination of Cross-Sections (equivalent to ISO 9220).
BS 5600 : 1980 : Part 4	Methods of Testing and Chemical Analysis of Hard Metals. Section 4.13: Metallographic Determination of Microstructure (equivalent to ISO 4499). Section 4.14: Metallographic Determination of Porosity and Uncombined Carbon (equivalent to ISO 4505).
BS 6285 : 1982 (1989)	Method for the Macrographic Examination of Steel by Sulphur Print (Baumann Method) (equivalent to ISO 4968).
BS 6533 : 1984	Guide to Macroscopic Examination of Steel by Etching by Strong Mineral Acids (related to ISO 4969).
BS 6617 : 1985 : Part 1	Methods for Determining Decarburization by Microscopic and Microhardness Techniques (related to ISO 3887, EN 104).
BS 6938 : 1988	Recommendation for Documentation of the Preparation of Certified Reference Materials for Metallurgical Analysis of Non-Ferrous Metals.

ASTM Standards

ASTM A247 : 1967 (1984)	Evaluating the Microstructure of Graphite in Iron Castings.
ASTM B276 : 1986	Test Method for Apparent Porosity in Cemented Carbides.
ASTM B390 : 1986	Practice for Evaluating Apparent Grain Size and Distribution of Cemented Tungsten Carbides.
ASTM B487 : 1985	Method for Measurement of Metal and Oxide Coating Thicknesses by Microscopic Examination of a Cross-Section.
ASTM B588 : 1988	Test Method for the Measurement of Thickness of Transparent or Opaque Coatings by Double Beam Interference Microscope.

ASTM B657 : 1987	Method for Metallographic Determination of Microstructure in Cemented Carbides.
ASTM B659 : 1985	Practice for Measuring Thickness of Electrodeposited and Related Coatings.
ASTM B748 : 1985	Measurement of Thickness of Metallic Coatings by Measurement of Cross-Section with a Scanning Electron Microscope.
ASTM C664 : 1987	Test Method for Thickness of Diffusion Coating.
ASTM D2663 : 1987	Test Method for Carbon Black Dispersion in Rubber.
ASTM D3849 : 1987	Test Method for Carbon Black Primary Aggregate Dimension from Electron Microscope Image Analysis.
ASTM E3 : 1980 (1986)	Preparation of Metallographic Specimens.
ASTM E7 : 1989	Terminology Relating to Metallography.
ASTM E20 : 1985	Practice for Particle Size Analysis of Particulate Substances in the Range 0.2 to 75 μm.
ASTM E45 : 1987	Practice for Determining the Inclusion Content of Steel.
ASTM E81 : 1977 (1982) and 1989	Preparing Quantitative Pole Figures of Metals.
ASTM E82 : 1963 (1984)	Determining Orientation of Metal Crystals.
ASTM E112 : 1988	Methods for Determining Average Grain Size.
ASTM E157 : 1988	Assigning Crystallographic Phase Designations in Metallic Systems.
ASTM E175 : 1982	Definitions of Terms Relating to Microscopy.
ASTM E340 : 1987	Macroetching Metal and Alloys.
ASTM E381 : 1979 (1984)	Macroetch Testing, Inspection and Rating of Steel Products.
ASTM E407 : 1970 (1989)	Microetching Metals and Alloys.
ASTM E562 : 1989	Determining Volume Fraction by Systematic Manual Point Counting.
ASTM E766 : 1986	Calibrating the Magnification of SEM using NBS-SRM-484.
ASTM E768 : 1980 (1985)	Practice for Preparing and Evaluating Specimens for Automatic Inclusion Assessment of Steel.
ASTM E807 : 1981 (1986)	Metallographic Laboratory Evaluation.
ASTM E883 : 1986	Guide for Metallographic Photomicrography.
ASTM E930 : 1983 (1989)	Methods for Estimating the Largest Grain Size Observed in a Metallographic Section (ALA Grain Size).
ASTM E963 : 1983 (1989)	Electrolytic Extraction of Phases for Ni and Ni-Fe Superalloys.
ASTM E986 : 1986	Scanning Electron Microscope Performance Characterization.
ASTM E975 : 1984 (1989)	X-Ray Determination of Retained Austenite in Steel.
ASTM E1077 : 1985	Test Method for Estimating the Depth of Decarburization of Steel Specimens.
ASTM E1122 : 1986	Practice of Obtaining JK Inclusion Ratings using Automatic Image Analysis.
ASTM E1180 : 1987	Preparing Sulphur Prints for Macrostructural Examination.
ASTM E1181 : 1987	Test Methods for Characterizing Duplex Grain Sizes.
ASTM E1182 : 1987	Measurement of Surface Layer Thickness by Radial Sectioning.

ASTM E1245 : 1989	Practice for Determining the Inclusion Content of Steel.
ASTM E1268 : 1988	Practice for Assessing the Degree of Bonding or Orientation of Microstructure.
ASTM E1351 : 1990	Production and Evaluation of Field Metallographic Replicas.
ASTM F728 : 1981 (1987)	Method for the Preparation of an Optical Microscope for Dimensional Measurements.

CHAPTER 14

Statistical analysis of data

L. C. Wolstenholme and M. J. Crowder

14.1 INTRODUCTION

The major problem to be tackled in the statistical analysis of data is assessment of the variability in results. No two sets of data are ever likely to be the same even when apparently collected under identical conditions. The quantity of interest is called the response variable. There may be more than one of these, and there may be other recorded factors which may or may not influence the response(s). Often the problem to be examined is how much of the variability in data is due to natural variation and how much is due to these other factors. Quantifying the variability and identifying the source are the principal objectives.

It has to be accepted at the outset that no amount of statistical analysis will result in a 100% definitive conclusion. The result will be a statement of the type 'with 95% confidence, the new technique produces higher readings', or 'with 95% confidence, temperature influences fatigue'. It is then up to the experimenter to decide how to react to this information. Without proper statistical analysis, though, the experimenter has little to go on.

There are many statistical methods available, with a variety of properties, some considered more desirable than others. The purpose of this chapter is to assess methods likely to be appropriate in the context of materials metrology.

14.2 THE IMPORTANCE OF LOOKING AT DATA PLOTS

Some exploratory data analysis based on graphical methods is essential before proceeding to generate summary statistics. It is the case that many statistical tests rely on assumptions about the nature of the random variability in the data, and, whilst some formal tests exist for this purpose, perfectly adequate assessment can often be made by eye using a suitable plot. The process of constructing various plots and graphs is no

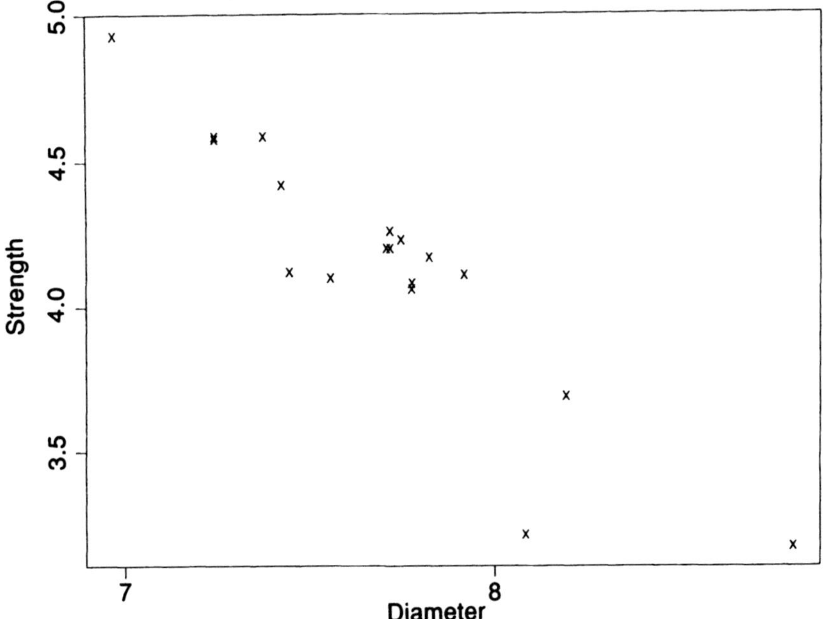

Fig. 14.1 Carbon fibre strength versus log diameter.

longer prohibitively time consuming, as the majority of statistical computer packages will perform the task. In fact, facilities usually exist to extract particular portions of the data to examine it in every way possible.

Figure 14.1 shows some carbon fibre strengths plotted against the log of fibre diameter. The plot shows a strong linear relationship. Figure 14.2 shows a set of results from a low-cycle fatigue experiment where the same material was used in experiments conducted at 14 different laboratories (Thomas and Varma, 1992). A nominal strain level of 1.2% was set (log 1.2 = 0.182) but the actual level was recorded in each case. Between one and four similar experiments were conducted at each laboratory, and each laboratory is indicated by a different letter on the plot. There is some evidence of a relationship between cycles to fatigue life and strain level, but of more striking significance is the clustering of results from individual laboratories. There is a clear indication that a laboratory effect needs to be investigated. It could be that there were differences in material used at each site, though in this particular study it was found that similar laboratory effects were observed across different materials (Thomas and Varma, 1992).

In summary, plotting procedures are a way of revealing data structure and should be used before proceeding to formal analysis, and also applied after analysis to check any assumed or fitted models. A useful reference on exploratory data analysis is Tukey (1977).

Figure 2. Fatigue data: log cycles v. strain

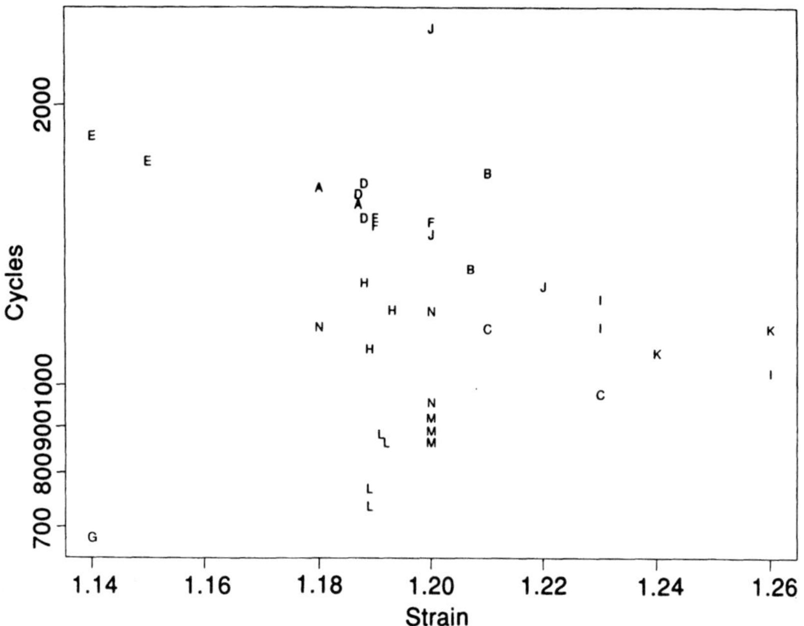

Fig. 14.2 Fatigue data: log cycles versus strain.

14.3 BASIC STATISTICS

The two principal features of any data set are the location of the data on the scale, and the degree to which the data are spread out. The most common measures are respectively the arithmetic mean, and the standard deviation (SD) or its square the variance.

Let a sample of values $x_1, x_2, x_3, \ldots, x_n$, be drawn independently from a population of x-values with mean μ and standard deviation σ. For instance, the x_i might be the weights of n washers taken as a handful from the production line. This scheme is known as simple random sampling. The sample mean $\bar{x}$ is given by

$$\bar{x} = \sum_{i=1}^{n} x_i / n \tag{14.1}$$

The sample standard deviation s is given by

$$s^2 = \frac{\sum (x_i - \bar{x})^2}{(n-1)} \tag{14.2}$$

These sample measures are often used as estimates of the true population parameters. This is written as

$$\bar{x} = \hat{\mu}, \quad s = \hat{\sigma}$$

where ^ denotes 'estimate of'. These are point estimates, and as such carry no indication of how close the estimate might be to the true value. More informative are interval estimates which have an attached degree of certainty. For example, we may be able to say something like 'with 95% confidence, the true value of μ lies between 50.3 and 61.7'. However, such statements often rely on knowledge, assumed or otherwise, about the distribution of values in the population. In many cases this is assumed to be a normal distribution, an assumption which can be assessed to some extent by how nearly the data seem to have a symmetric, bell-shaped distribution. Obviously, with very small samples this assessment may be unreliable, and often past experience is called upon. There is a useful plot which produces an approximate straight line for normal samples: for each data point x_i a normal score z_i is calculated and the $\{x_i\}$ are plotted against the $\{z_i\}$. A facility to do this quickly and painlessly is available on packages such as Minitab (1989). The method of producing an equivalent plot using normal probability paper is found in Chatfield (1983), for example.

More formal tests of normality are available. For example, the Shapiro and Wilk test (1965) is based on a measure of the linearity of the above plot. It is however known that the data analysis techniques considered here are fairly robust (i.e. insensitive) to some departure from normality, so some form of visual impression is generally adequate and the most appealing.

When the experimenter has the luxury of being able to collect a large amount of data, deliberations about distributional form can be bypassed because certain limiting conditions usually ensure the validity of normal-based methods. It is only possible to define 'large' within the observational context, but tens rather than hundreds of observations usually prove adequate.

Interval estimates are called confidence intervals; the end points of the intervals are confidence limits; and the associated probability is the confidence level. A confidence interval for μ in simple random sampling, as defined above, takes the form

$$\bar{x} \pm cs / \sqrt{n}$$

where the value of c is a percentage point from the appropriate sampling distribution of $(\bar{x} - \mu)(\sqrt{n})/s$. For normal populations this will be the t-distribution with ν degrees of freedom (DF), where ν is $n - 1$. As ν becomes larger, the t-values approach the equivalent points of the normal distribution.

Confidence intervals may be the basis for comparing two or more samples. Given m samples, the usual approach is to assume that the m populations have equal variances and that these populations are normal. An estimate of the common variance σ^2 may be constructed as

$$s^2 = \frac{(n_1 - 1)s_1^2 + (n_2 - 1)s_2^2 + \ldots + (n_m - 1)s_m^2}{(n_1 - 1) + (n_2 - 1) + \ldots + (n_m - 1)} \qquad (14.3)$$

where n_i, s_i^2 are the sample size and variance for the ith sample. s^2 is a weighted average of the individual sample variances and is thus a more reliable estimate of σ^2. This estimate may be used to derive confidence intervals for the means of the populations, as above, and for other purposes.

14.3.1 Example 1

The data in Table 14.1 concern the yield from a chemical reaction using three different catalysts C_1, C_2, C_3. It is of interest to know whether there is any difference in the performance of the catalysts.

Table 14.1 Yield of a chemical reaction using three different catalysts

	C_1	C_2	C_3
1	2.5	2.6	2.4
2	3.6	3.1	2.9
3	3.2	3.0	2.8
4	2.7	2.5	2.3

If the samples come from populations with approximately equal means, then the confidence intervals for the three population means will overlap to some degree. If some intervals do not overlap then there is evidence, at some level of uncertainty, that the populations do not all have the same mean. This is roughly how the method known as one-way analysis of variance works: it combines the degree to which the confidence intervals overlap into a single measure which may be compared with the F-distribution. The level of uncertainty is often expressed as a p-value. This represents the probability that the degree of overlap, or rather lack of it, could have arisen by chance under the null hypothesis of equality of population means. If the p-value is very small, say < 0.05, then the hypothesis of equal means looks very unlikely. In this case, to discover where the differences might lie in particular, we have to go back to the confidence intervals. One-way analysis of variance on just two samples is equivalent to the two-sample t-test, using $n_1 + n_2 - 2$ degrees of freedom for the pooled s^2 in equation 14.3.

Table 14.2 shows for the data of Table 14.1 the confidence intervals for each sample based on (a) the sample standard deviation and (b) the pooled standard deviation. In the analysis of variance the p-value at 0.361 is quite large and indicates no significant evidence that the mean yields differ amongst catalysts. This reflects the overlap of the confidence intervals. Whilst the sample standard deviations differ they are not different enough for the assumption of equal variances to be put in serious doubt. Where the assumption of equal population variances is in doubt – and this can be tested formally using the Bartlett test (see, for example, Kennedy and Neville, 1986), not in fact available on many statistical packages – samples may be compared in pairs using the standardized difference

Table 14.2 Confidence intervals and analysis of variance

(a) Sample of standard deviations

	N	Mean	SD	SE	95% CI
C_1	4	3.000	0.497	0.248	(2.210, 3.790)
C_2	4	2.800	0.294	0.147	(2.332, 3.268)
C_3	4	2.600	0.294	0.147	(2.132, 3.068)

(b) Analysis of variance

Source	DF	SS	MS	F	p
Factor	2	0.320	0.160	1.14	0.361
Error	9	1.260	0.140		
Total	11	1.580			

(c) Pooled standard deviation

Level	N	Mean	SD	Individual 95% CIs for mean based on pooled SD
C_1	4	3.0000	0.4967	(———————+———————)
C_2	4	2.8000	0.2944	(———————+———————)
C_3	4	2.6000	0.2944	(———————+———————)

Pooled SD = 0.3742

2.45   2.80   3.15

$$\frac{\bar{x}_1 - \bar{x}_2}{\sqrt{\left(\dfrac{s_1^2}{n_1} + \dfrac{s_2^2}{n_2}\right)}} \tag{14.4}$$

Under the null hypothesis of equal population means this statistic has approximately a *t*-distribution with degrees of freedom given by

$$v = \frac{(s_1^2/n_1 + s_2^2/n_2)^2}{s_1^4/[n_1^2(n_1 - 1)] + s_2^4/[n_2^2(n_2 - 1)]}$$

Statistical packages such as Minitab include this facility, although many statistics books do not.

14.4 ANALYSIS OF VARIANCE (ANOVA)

Inherent in the basic approach outlined above is the assumption that all data in any given sample have been collected under the same conditions. In practice this is not often the case. Identifying the conditions which are likely to influence results is not easy, and unfortunately may be alighted upon too late, i.e. after the experiment has been run, and it is discovered that potentially critical information has not been recorded. It is always better to acquire information which turns out to be redundant, than miss data which could be vital.

For example, it may be thought that measurement of a material property may vary between testing laboratories, so an experiment is set up whereby several pieces from a large batch of a certain material are sent to each laboratory. Prior to the experiment it may have been assumed that all pieces of material have the same properties, but suppose this is not the case. A one-way ANOVA will not be able to separate material differences from laboratory differences. If, however, the sample material is 'matched' across laboratories, then the different effects may be separated using two-way ANOVA.

14.4.1 Example 1 continued

Suppose the experiments were in fact conducted at four different laboratories: the first measurement for each catalyst comes from laboratory 1, and so on. We will equate 'catalyst' here with 'material' and again test for differences between materials, but in the light of new knowledge about the testing environment.

Table 14.3 Two-way analysis of variance

Source of variation	Degrees of freedom	Sum of squares
Laboratories	$l - 1 = 3$	$SSB = m \sum_i (\bar{y}_i - \bar{y})^2 = 1.14$
Materials	$m - 1 = 2$	$SSW = l \sum_j (\bar{y}_j - \bar{y})^2 = 0.32$
Residual	$(l - 1)(m - 1) = 6$	$SSR = \sum_i \sum_j (y_{ij} - \bar{y}_i - \bar{y}_j + \bar{y})^2 = 0.12$
Total	$n - 1 = 11$	$SST = \sum_i \sum_j (y_{ij} - \bar{y})^2 = 1.58$

$\bar{y}$ = overall mean.
$\bar{y}_i$ = mean for laboratory i.
$\bar{y}_j$ = mean for material j.

If there are l laboratories and m materials and each laboratory makes a single measurement on each material, there will be $n = m \times l$ measurements in total. Let y_{ij} denote the measurement made at laboratory i on material j, and $\bar{y}$ the mean of all n observations. A measure of the total variation among all observations is the total sum of squares $SST = \sum_i \sum_j (y_{ij} - \bar{y})^2$. ANOVA divides this total variation into between-samples, within-samples and residual variation. In the present context these contributions represent variation between laboratories SSB, variation between materials SSW and 'left-over' variation SSR which cannot be attributed to either of the main sources. Table 14.3 shows the breakdown for the data of Example 1.

The model underlying the analysis is one where the expected response $E(y_{ij})$ (e.g. material property measurement) is the sum of effects consisting of an overall mean value μ and departures from this average due to the different effects:

$$E(y_{ij}) = \mu + \alpha_i + \beta_j$$

where α_i is the effect due to the ith laboratory and β_j is the effect due to the jth material. By definition $\sum \alpha_i = 0$ and $\sum \beta_j = 0$.

The measured responses y_{ij} are assumed to vary around the expected value by some quantity e_{ij}:

$$y_{ij} = \mu + \alpha_i + \beta_j + e_{ij} \tag{14.5}$$

It is perhaps potentially misleading that e is often referred to as the 'error'. The term 'random departure' is a better description for this natural variation around the expected response. It will be assumed here that α_i and β_j are fixed effects, i.e. they correspond to laboratories and materials which do not vary from one occasion to another. In many cases this is unrealistic and an alternative approach, whereby α_i and/or β_j are also subject to random variation, may be preferred. The incorporation of what are termed 'random effects' changes the analysis in certain respects and this is covered in detail in Crowder and Wolstenholme (1994).

Fundamental to the model is that the $\{e_{ij}\}$ are independent with zero mean. Further, if we assume that the $\{e_{ij}\}$ have constant variance and are normally distributed, then, under the null hypothesis that there are no significant material or laboratory effects, the ratios of mean squares $SSW(l-1)/SSR$ and $SSB(m-1)/SSR$ may be compared with the F-distribution and p-values calculated. For the results in Table 14.3 these F-ratios are respectively 19.0 and 8.0, indicating that there is a significant difference both between laboratories (p-value < 0.01) and between catalysts (p-value < 0.025). The earlier one-way ANOVA failed to detect these differences because laboratory differences were masking material differences. We can note also the mean square $SSR/[(l-1)(m-1)]$, which provides an estimate of the underlying variability, is 0.02 compared with the earlier pooled variance 0.3742. The model has in fact explained over 92% of the variability in the results, the residual 8% being SSR/SST.

Possible interaction effects between laboratories and materials may be built into the model and assessed via replication of measurements, say each laboratory/material combination observed r times. The resulting ANOVA is given in Table 14.4 and it can be seen that for the case $r = 1$ it reduces to the form of Table 14.3.

Table 14.4 Two-way analysis of variance with interaction

Source of variation	DF	SS	MS	F-ratio
Laboratories	$l-1$	$mr \sum_i (\bar{y}_i - \bar{y})^2$	$SS/(l-1)$	
Materials	$m-1$	$lr \sum_j (\bar{y}_j - \bar{y})^2$	$SS/(m-1)$	
Lab. x Mat.	$(l-1)(m-1)$	$r \sum_i \sum_j (\bar{y}_{ij} - \bar{y}_i - \bar{y}_j + \bar{y})^2$	$SS/[(l-1)(m-1)]$	
Residual	$n-lm$	$\sum_i \sum_j \sum_k (y_{ijk} - \bar{y}_{ij})^2$	$SS/(n-lm)$	
Total	$n-1$	$\sum_i \sum_j \sum_k (y_{ij} - \bar{y})^2$		

$y_{ijk} = k$th observation for laboratory i and material j.
$\bar{y}_{ij} =$ mean of r replications for laboratory i and material j, $n = lmr$.
Other terms as for Table 14.3.

14.5 INCOMPLETE DATA

14.5.1 Missing data

As we all know, in real life things do not always work out as planned. In experiments destined for statistical analysis, two situations may arise leading to 'incomplete' data, and these usually require important adjustments to procedures. The first case is that of missing values. Standard multi-way analysis of variance assumes a perfectly balanced set of results, that is the same number of results for each set of conditions. What if one experiment goes wrong? It may be possible to repeat it, but suppose the right kind of experimental material is not available, or the experiment is too long to repeat. Some computer packages will cope with this by substituting some sort of estimated reading, maybe based on average results. The user should be aware of such an approach and be able to judge whether or not it is reasonable.

Alternatively, analysis of unbalanced data may be carried out by multiple regression using binary variables. This approach is not in essence different to analysis of variance. Multiple regression on balanced data yields exactly the same results as ANOVA, but ANOVA is easier to use.

The regression model is given by:

$$y_i = \mu + \alpha_1 x_1 + \alpha_2 x_2 + \ldots + \alpha_l x_l + \beta_1 z_1 + \beta_2 z_2 + \ldots + \beta_m z_m + e_i$$

where y_i is the ith observation, and $\{x_j\}$ and $\{z_k\}$ are binary or indicator variables taking the value 0 or 1 dependent on whether or not the jth treatment (e.g. laboratory) or kth block (e.g. material) is present. The regression coefficients $\{\alpha_j\}$ and $\{\beta_k\}$ are the treatment and block effects respectively, but it is not possible to determine all α_j and β_k uniquely without using some constraints such as $\sum \alpha_j$ and $\sum \beta_k$ both equal zero. Differences between treatments will be determined by values of some α_j which are different from zero, and similarly differences between blocks will be determined by values of some β_k which are different from zero. The judgement as to whether a coefficient is significantly different from zero depends on the assumptions made about the error structure. As before, the assumption is made that departures $\{e_i\}$ have zero mean and constant variance and are normally distributed. Under these criteria, the estimated regression coefficients are normally distributed and p-values for testing zero effects may be assigned via an estimate of the residual variance and the t-distribution. Very small p-values suggest that the corresponding effects are non-zero.

Table 14.5 Creep rate measurements at time 400 hours

	Bar				
Laboratory	1	2	3	4	5
1	62.3	58.4	56.5	77.9	62.5
2	64	61.5	66.5	72	64
3	89	72	–	74	82
4	71.4	71.2	75.8 74.3	–	73
5	62.5	68.4	70.5	70.9	67
6	64	67	71	75	73
7	70.4	70.6	77.8	82.2	76.5
8	66.2 67.6	73	82.8	81.3	66
9	69	64	76	89	84

Consider the data in Table 14.5 concerning creep rate measurements of a material made by nine different laboratories. The data have been extracted from part of a study by Gould and Loveday (1992). Whilst it was felt that the material was homogeneous and would display similar properties throughout, measurements were made on portions of material taken from five different bars. A one-way ANOVA may be performed on either the rows (laboratories) or the columns (bars), but this will ignore the effect of the other variable. Such comparisons are of limited value unless it can be established that the other variable has no effect. With the regression model, individual effects may be assessed for both materials and laboratories. In the practical application of the regression analysis, one laboratory and one material must be taken as a reference point and only $l-1$ binary variables used for laboratories and $m-1$ binary variables used for materials. The estimated constant term then includes the effect of the reference laboratory and the effect of the reference material, and the coefficients for laboratories and materials are relative to the reference point. It is entirely arbitrary which laboratory and material are used as the reference. Table 14.6 shows the results obtained for the data of Table 14.5 using laboratory 1 and bar 1 as the reference point.

Table 14.6 Regression analysis of unbalanced data using binary variables

Predictor	Coefficient	SD	t-ratio	p
Constant	60.241	2.845	21.17	0.000
c_{22}	2.080	3.378	0.62	0.543
c_{23}	16.413	3.606	4.55	0.000
c_{24}	10.204	3.420	2.98	0.006
c_{25}	4.340	3.378	1.28	0.208
c_{26}	6.480	3.378	1.92	0.064
c_{27}	11.980	3.378	3.55	0.001
c_{28}	9.843	3.244	3.03	0.005
c_{29}	12.808	3.607	3.55	0.001
c_{32}	-1.135	2.463	-0.46	0.648
c_{33}	6.010	2.496	2.41	0.022
c_{34}	8.929	2.556	3.49	0.001
c_{35}	2.592	2.558	1.01	0.319

$s = 5.340$ $R^2 = 64.6\%$ $R^2 \text{ (adj.)} = 51.0\%$

The partial regression coefficients c_{XX} represent estimates of $\{\alpha_i - \alpha_1\}$ and $\{\beta_j - \beta_1\}$ and are shown with their calculated p-values. The predictor constant refers to $\mu + \alpha_1 + \beta_1$ in the model. The significance of the regression coefficients is relative to the reference point of laboratory 1 and bar 1. Predictors c_{22} to c_{29} refer to laboratories 2 to 9, and predictors c_{32} to c_{35} refer to bars 2 to 5. Very small p-values indicate that the laboratory or bar has a significantly different result to that of the reference point. It can be seen therefore that bars 3 and 4 are very different to bar 1 and, even more significantly, laboratories 3, 4, 7, 8 and 9 give markedly different results compared with laboratory 1. Comparisons between other pairs of laboratories or pairs of materials may be made via differences between the respective regression coefficients, their standard errors and the t-distribution. The difference $c_{23} - c_{24}$, for example, gives $\hat{\alpha}_3 - \hat{\alpha}_4$.

As with multi-way ANOVA, interaction terms may be built into a regression model, but in the example considered there are too few replications to estimate these effects. To put it another way, there are too few residual degrees of freedom available to spread over the possible laboratory/material interactions.

It is possible to have a combination of quantitative and binary variables. For example, Fig. 14.2 shows that a model for cycles to fatigue failure might include a variable for log strain as well as binary variables for laboratories. Further reading may be found in for example the classic reference for regression analysis by Draper and Smith (1981).

14.5.2 Censored data

The second type of incomplete data is where we have censored observations. This is where the observation is not known exactly, but is known only to lie in some interval. Frequently the value is of the right-censored variety, i.e. the response is known only to be larger than some value. This often arises in cases where an experiment is terminated after a certain length of time, or once a certain amount of uncensored data has been collected. It is very important that censored data are not simply discarded, though it has to be admitted that censored data complicate the analysis quite considerably. As a simple example, consider the following observations from a fatigue experiment which was terminated after 673 cycles. Some experimental units did not fail, giving rise to lifetimes known only to be greater than 673, i.e. right-censored observations, denoted 673*

$$31, 58, 157, 185, 300, 470, 497, 673^*, 673^*, 673^*, 673^*$$

Precise calculation of the sample mean is not possible since some of the sample values are unknown. If only the uncensored values were used the mean would be unrealistically low. However, the sample median (the point where 50% of data lie above and 50% of data lie below) may be calculated (470), and provides a reasonable estimate of central tendency as long as the population is not too skew. This of course only works in the example shown because the right-censored observations are all to the right of the median. Similarly, any left-censored observations would need to lie to the left of the median.

In general terms, censored observations require us to deal more directly with the distributional form of the population. This often involves the use of likelihood functions, and to date this is generally outside the scope of the commonly used non-specialist packages. A likelihood function is a measure of the probability, under the assumed statistical model, of obtaining the data that were actually observed. Likelihood functions are able to deal with all types of censored observations, provide an efficient method of estimating unknown parameters, and enable different models for the behaviour of the response variable to be compared.

14.6 CONCLUDING REMARKS

The key to successful data analysis is identification: identification of variables, identification of data structure and identification of appropriate methods. There is considerable benefit to be derived from more advanced modelling methods, and it is important that the experimenter should be able to recognize the kind of analysis possible, though not necessarily be personally able to perform it. Whilst most of us can effectively treat ourselves with aspirin, antibiotic treatment requires medical advice! Similarly, the scientist, having identified the scope of a data analysis problem, can gain much from the assistance of a professional statistician.

The purpose of statistical analysis of data may be twofold:

- to make statements about the results of today;
- to make statements about the results of tomorrow, i.e. to make predictions.

This chapter is largely confined to the first purpose but, having used the methods described to estimate the parameters for today's results, we can make predictions about replications under the same conditions.

REFERENCES

Chatfield, C. (1983) *Statistics for Technology*, 3rd edn, Chapman and Hall, London.

Crowder, M. J. and Wolstenholme, L. C. (1994) Statistical methodology for inter-comparison studies, in this book, Chapter 15.

Draper, N. R. and Smith, H. (1981) *Applied Regression Analysis*, 2nd edn. Wiley, New York.

Gould, D. and Loveday, M. S (1992) *The Certification of Nimonic 75 Alloy as a Creep Reference Material*.

Kennedy, J. B. and Neville, A. M. (1986) *Basic Statistical Methods for Engineers and Scientists*, 3rd edn, Harper and Row, New York.

Minitab (1989) *Minitab Release 7*, Minitab Inc., USA.

Shapiro, S. S. and Wilk, M. B. (1965) An analysis of variance test for normality (complete samples). *Biometrika*, **52**, 591–611.

Thomas, G. B. and Varma, R. K. (1992) In *Harmonisation of Testing Practice for High Temperature Materials* (eds M. S. Loveday and T. B. Gibbons), Elsevier.

Tukey, J. W. (1977) *Exploratory Data Analysis*, Addison-Wesley, Reading, MA.

CHAPTER 15

Statistical methodology for intercomparison studies

M. J. Crowder and L. C. Wolstenholme

15.1 INTRODUCTION

Intercomparison studies are a necessary part of the movement towards greater standardization in scientific and industrial practice. The studies might be set up to compare different companies within an industry or different industries using similar processes. They might also be on a national or international scale; for example, the European Community Bureau of Reference has initiated a programme of such studies.

The area of main concern in this book is materials metrology and the problems which arise when international companies compare data on the same material property that has been measured in different laboratories. The analysis of such intercomparison studies poses problems which fall outside routine, day-to-day statistical methodology. The intention here is solely to discuss the statistical issues, without consideration of the many other important aspects such as those examined by Marchandise and Colinet (1983).

The situation to be examined is as follows. A reservoir of material has been prepared under conditions carefully controlled to achieve homogeneity of the relevant properties. Samples of the material are sent out to a number of laboratories where measurements are made of the physical, mechanical or chemical property under examination. In a perfect outcome, all the figures returned by the participants would be precisely equal. Departure from this ideal arises from a variety of sources including heterogeneity of the material, systematic variation between laboratories, and random variation within laboratories.

All methods of making statistical inferences from statistical data are based on models which determine a probability framework for the observations. Such probability models range from the highly structured – as in normal linear models which give rise to *t*-tests, for instance – to the weakly structured, as used for so-called distribution-free or non-parametric methods. In any case, in order to make probability statements resulting from statistical analysis of data, some initial probability model must be prescribed.

For the present application – intercomparison studies – the tasks are (a) to identify appropriate probability models for the resulting data, and (b) to determine efficient statistical methods associated with these models. Part (a) is dependent on informed discussion to focus upon the relevant factors, and how they are likely to affect the observed data. Part (b) falls more within the realm of mathematics. As always, the process may require iteration. In the first cycle a model is adopted, the chosen methods are applied, and then the results are scrutinized. If the last operation reveals unsatisfactory aspects then these will be used to modify the model in an appropriate way and a second cycle will be performed. Of course, statistical judgement is necessary to prevent over- modification of the framework to suit any one particular set of data unduly.

A different question, which is not addressed in this chapter, is that of identifying and accommodating anomalies in the data. These may take the form of outliers, that is odd observations which stand out from the rest in some way, or more consistent patterns, such as that arising from a non-conforming laboratory. The literature on this aspect includes Rocke (1983), Olkin and Sobel (1987) and Davies (1988).

In the following sections, an approach currently advocated will first be discussed. Then some suggestions will be made regarding general methodology. Finally, a possible programme for development along these lines will be set out.

15.2 ONE-WAY ANOVA

Marchandise and Colinet (1983) gave a method for analysing data from several laboratories. Suppose that there are l laboratories and that the ith laboratory makes n_i measurements (for notation see Appendix 15.A). The basis for their proposed method was not stated in their paper but appears to be a one-way analysis of variance (ANOVA) with random effects (e.g. Searle, 1971, Chapter 9). In this framework, the underlying model (representation) for the jth measurement y_{ij} made in the ith laboratory is

$$y_{ij} = \mu + \alpha_i^{\text{L}} + e_{ij}, \quad i = 1, \ldots, l; \ j = 1, \ldots, n_i \tag{15.1}$$

Here μ is the overall mean level of the type of measurement under consideration; α_i^{L} is the ith laboratory effect, representing a consistent upward or downward adjustment to μ associated with laboratory i; and e_{ij} is a random error term. It is assumed in this model that the α_i^{L} are independent random variables with mean 0 and variance σ_{L}^2; this is to say that over the population of similar laboratories the α_i^{L} have arithmetic average (mean) zero, and a definable variation expressed as the variance (whose square root is called a standard deviation). The e_{ij} are assumed to be independent random variables with mean 0 and variance σ^2; they are regarded as unpredictable, unexplained, non-repeatable departures of y_{ij} from its expected value. Lastly, the α_i^{L} and e_{ij} are all assumed to be independent of one other.

Models are at the root of any statistical analysis, though in some accounts they are not made explicit. Such concealment can give the mistaken impression that the statistical methodology is automatic in the sense that no input is required other than the data. In the present chapter all model assumptions will be set out for appraisal. Even with good data a statistical analysis based on an inappropriate model is unlikely to yield the most useful results.

The use of random effects α_i^{L} is appropriate when the l laboratories are regarded as a random sample from a large pool of similar laboratories and inferences are to be made about such laboratories in general. The more familiar analysis using fixed effects (where the α_i^{L} are not random variables) would be appropriate when conclusions are to be restricted to the l participating laboratories and no others. This distinction, which may at first sight appear to be a mere logical nicety, is important in practice because it governs the conduct of the statistical analysis and what may be reported in the conclusions.

Table 15.1 Tensile property of alloy samples measured at three laboratories

Laboratory	Measurements				No.	Mean	SD
1	3.05	3.01	2.85		3	2.97	0.11
2	3.02	3.08	3.21	2.97	4	3.07	0.10
3	2.97	3.16	3.02		3	3.05	0.10

Table 15.2 ANOVA for tensile property data

Source of variation	DF	SS	E (MS)
Between labs	$l - 1 = 2$	SSB = 0.01824	$\sigma^2 + a\sigma_{\text{L}}^2$
Within labs	$n - l = 7$	SSW = 0.07400	σ^2
Total	$n - 1 = 9$	SST = 0.09224	—

The data in Table 15.1 are concocted to provide a small illustration of one-way ANOVA with random effects. Here, $l = 3$ (number of laboratories) and $n = \sum_{i=1}^{l} n_i = 10$ (total number of measurements). In the notation described above $y_{11} = 3.05$, $y_{23} = 3.21$ etc. The table also gives the laboratory means and standard deviations. The purpose of ANOVA is to partition the overall variation in the data into components identifiable with the different sources, here between laboratories and within laboratories. This partition is displayed in Table 15.2, where

$$a = \left(n - \sum_{i=1}^{l} n_i^2 / n \right) \div (l - 1) = 3.3$$

and DF and SS denote respectively degrees of freedom and sum of squares. Formulas for the latter are given in Appendix 15.B. The algebraic identity $SST = SSB + SSW$ is interpreted as a partition of the overall variation in the data into between- and within-laboratory components. Mean squares (MS) are defined as $MS = SS/DF$, and their expected values $E(MS)$ in Table 15.2 can be calculated as described in e.g. Crowder and Hand (1990, section 3.3).

To assess the relative magnitude of between- and within-laboratory variation the sums of squares SSB and SSW may be compared via an F-statistic:

$$F = \{SSB/(l-1)\} \div \{SSW/(n-l)\}$$

A large value of F in excess of 1 suggests a correspondingly large value of $a\sigma_L^2/\sigma^2$, i.e. that the between-laboratory variation is not negligible. In the balanced case (when the n_i are all equal), and under the usual normality assumptions (that the α_i^k and e_{ij} are all normally distributed), F has a variance ratio F-distribution with DF $l - 1$ and $n - l$. In the unbalanced case (n_i not all equal) the precise F-distribution of F fails (Searle, 1971, section 9.9a). For the (unbalanced) data of Table 15.1,

$$F = \{0.01824/2\} \div \{0.07400/7\} = 0.86$$

This, being smaller than 1, suggests that there are not strong differences between laboratories.

An estimate of σ_L^2, which is a measure of the between-laboratory variation, may be obtained by inspection of the $E(MS)$ column in Table 15.2. Evidently, the combination

$$a^{-1}\{SSB/(l-1) - SSW/(n-l)\}$$

has an expected value σ_L^2 as required for an unbiased estimator. This is what Marchandise and Colinet quote as their S_b^2 in slightly different notation; their S_w^2 is $SSR/(n-l)$ here. An unfortunate property of such ANOVA-based estimators of variance components is that they can turn out to be negative. In the present case σ_L^2 will be negative whenever F is less than 1. Searle (1971, section 9.8b) discusses this problem at length but there is no completely satisfying resolution of it. Such a failure occurs for the data of Table 15.1, for which the estimate comes out as -0.00044.

For estimation of the overall mean level μ, weighted combinations of laboratory means of the form $\tilde{\mu}_w = \sum_{i=1}^{l} w_i \bar{y}_i / w_+$, where $w_+ = \sum_{i=1}^{l} w_i$, are obvious suggestions. Marchandise and Colinet consider two such estimators: the overall mean $\tilde{\mu}_1 = \bar{y}$, corresponding to weights $w_i = n_i$, and a straight average of laboratory means $\tilde{\mu}_2 = l^{-1} \sum_{i=1}^{l} \bar{y}_i$, with equal weights $w_i = 1$. For the data of Table 15.1, $\tilde{\mu}_1 = 3.034$ and $\tilde{\mu}_2 = 3.030$. Marchandise and Colinet recommend, without explanation, the use of $\tilde{\mu}_1$ when F is small and of $\tilde{\mu}_2$ when F is large. However, it is possible that the two estimates might differ widely and, at the same time, small perturbations of the data might make the F-value cross any chosen significance boundary. Such discontinuities are disturbing. To give an extreme numerical example, suppose that $l = 5$, the n_is are (10, 10, 10, 10, 1) and the $\bar{y}_i$ s are (10.0, 10.0, 10.0, 10.0, 100.0). Then one finds that $\tilde{\mu}_1 = 28.0$, $\tilde{\mu}_2 = 12.195$ and $F = 1975.61/s^2$, where $s^2 = \mathrm{SSW}/(n - l)$ is the within-laboratories mean square. Thus, adopting Marchandise and Colinet's rule, we should estimate μ as 28.0 if $s^2 > 1975.61F_0$ and as 12.195 if $s^2 < 1975.61F_0$, where F_0 is the adopted significance value for F. Such sudden jumps are counter-intuitive. It would seem much more sensible to discard the single result from laboratory 5 and estimate μ as 10.0.

A traditional criterion for the performance of an estimator is its variance over repeated samples. Thus, an estimator which varies little about the target value μ over the different sample values which can arise is to be preferred over one which can stray far from μ. For the weighted combination $\tilde{\mu}_w$ the variance is

$$\mathrm{var}(\tilde{\mu}_w) = w_+^{-2} \sum_{i=1}^{l} w_i^2 \left(\sigma_L^2 + \sigma^2/n_i \right)$$

yielding variances

$$v_1 = n^{-1} \left(\sigma^2 + n^{-1} \sigma_L^2 \sum_{i=1}^{l} n_i^2 \right)$$

$$v_2 = l^{-1} \left(\sigma_L^2 + l^{-1} \sigma^2 \sum_{i=1}^{l} n_i^{-1} \right)$$

for $\tilde{\mu}_1$ and $\tilde{\mu}_2$ respectively. It can be shown (see Appendix 15.C) that $v_1 \leqslant v_2$ when σ_L^2/σ^2 is small, and that $v_1 \geqslant v_2$ when σ_L^2/σ^2 is large. This is possibly the justification underlying Marchandise and Colinet's rule for choosing between $\tilde{\mu}_1$ and $\tilde{\mu}_2$. In fact, the (unconditional) variance of the estimation procedure is actually $p_1 v_1 + (1 - p_1) v_2$, where p_1 is the probability of selecting $\tilde{\mu}_1$.

The optimal choice of weights w_i, which yields a minimum of $\mathrm{var}(\tilde{\mu}_w)$, is $w_i = (\sigma_L^2 + \sigma^2/n_i)^{-1}$. Unfortunately, this choice depends on the unknown values of σ^2 and σ_L^2 and so it is not possible to employ this estimator in practice. Note that these weights are adaptive in the sense that they yield $\tilde{\mu}_1$ as $\sigma_L^2/\sigma^2 \rightarrow 0$ and $\tilde{\mu}_2$ as $\sigma_L^2/\sigma^2 \rightarrow \infty$. The corresponding estimator is the maximum likelihood estimator (see Appendix 15.D).

15.3 MORE GENERAL ANOVA METHODS

In the current context, one-way ANOVA as described above is often unrealistic in that no allowance is made for heterogeneity of the experimental material. The situation is illustrated in the following example.

A quantity of an alloy was prepared and rolled into bars, and lengths cut from four of these bars were sent out to six laboratories. The values in Table 15.3 are measurements of a tensile property, and form a balanced layout in which each laboratory has conducted a test on one length from each of the four bars. The data can be subjected to routine two-way ANOVA, and Table 15.4 gives the result. As in the one-way ANOVA, between-laboratories variation may be assessed via an F-test:

$$F = (14.444/5) \div (11.101/15) = 3.90$$

for which the significance is $p = 0.018$. Similarly, for between-bars variation,

$$F = (8.833/3) \div (11.101/15) = 3.98$$

yielding $p = 0.029$. Both p-values are small and so there appear to be statistically substantial differences both between laboratories and between bars.

Table 15.3 Tensile property of four alloy bars measured at six laboratories

Bar	Laboratory						
	1	2	3	4	5	6	*Mean*
1	4.61	2.72	5.24	4.28	3.74	4.82	4.23
2	4.25	3.05	3.77	5.39	3.74	6.74	4.49
3	4.79	6.65	5.96	5.81	4.31	7.40	5.82
4	5.03	5.72	5.18	4.52	3.26	6.44	5.02
Mean	4.67	4.53	5.04	5.00	3.76	6.35	4.89

The model underlying the two-way ANOVA with random effects is

$$y_{ij} = \mu + \alpha_i^L + \alpha_j^B + e_{ij}, \quad i = 1, \ldots, l; j = 1, \ldots, b \tag{15.2}$$

where μ is the overall mean level; the laboratory effects α_i^L are independent with mean 0 and variance σ_L^2; and the bar effects α_j^B are independent with mean 0 and variance σ_B^2. The α_i^L, α_j^B and e_{ij} are all assumed to be independent. The adoption of random effects for laboratories follows the practice in section 15.2; regarding that for bars, it seems natural to treat them as a random sample from a population of such bars.

Table 15.4 Two-way ANOVA for tensile property data

Source	DF	SS	E (MS)
Laboratories	$l - 1 = 5$	14.444	$\sigma^2 + b\sigma_L^2$
Bars	$b - 1 = 3$	8.833	$\sigma^2 + l\sigma_B^2$
Residual	$(l - 1)(b - 1)$	11.101	σ^2

The expected mean squares given in Table 15.4 lead to estimates of σ_L^2 and σ_B^2, the so-called variance components. Thus the random variation between laboratories can be estimated as

$$\tilde{\sigma}_L^2 = (14.444/5 - 11.101/15)/3 = 0.716$$

and that between bars as

$$\tilde{\sigma}_B^2 = (8.833/3 - 11.101/15)/5 = 0.441$$

The overall mean μ is estimable as $\tilde{\mu} = \bar{y} = 2.96$ from the data as before.

Often, the measurement of a material property – particularly a mechanical one – is performed in slightly different ways within the participating laboratories. An appropriate statistical analysis should therefore take these differences into account. With l laboratories, b bars and m methods a complete $l \times b \times m$ factorial layout would correspond to each laboratory using each method on one length cut from each bar. For such a situation the previous model would now be augmented to

$$y_{ijk} = \mu + \alpha_i^L + \alpha_j^B + \beta_k^M + \alpha_{ik}^{LM} + e_{ijk}, \quad i = 1,...,l; \ j = 1, \ldots, b; \ k = 1, \ldots, m \quad (15.3)$$

for y_{ijk}, the measurement at laboratory i on bar j using method k. The new terms in the model are β_k^M, the (fixed) effect for method k, and α_{ik}^{LM}, the (random) effect for the laboratory $\times$ method interaction. The use of fixed effects for methods implies that interest is directed at the current k methods *per se*, and not at a large collection of variations of method from which these k have been selected at random. Let $\sigma_k^2 = \text{var}(\alpha_{ik}^{LM})$; the smaller σ_k^2 is, the more consistent over laboratories is method k. In the usual treatment of this model σ_k^2 is taken to be homogeneous over k. However, for the present application it is important to allow the possibility of differing values. Given data $\{y_{ijk}\}$, a three-way ANOVA can be performed and inferences made along the same lines as for the one- and two-way ANOVAs described above.

ANOVA has been the most widely advocated approach to intercomparison studies in the literature. Youden and Steiner (1975) demonstrate two-way ANOVA for data like that of Table 15.1 but with possibly more than one replicate per cell. They take the laboratories effect as random but ignore the between-samples variation, corresponding to between-bars here, as not being of interest in their context (of chemical analyses). Mandel (1959) presents a modified version treating both laboratories and bars as fixed effects, but incorporating Tukey's (1949) extra term for interaction. Mandel and Lashof (1959) expound this method at greater length.

ANOVA-based procedures can give rise to certain difficulties, not all of which are easy to resolve. For example, it is possible to obtain negative estimates for positive quantities, the variance components (e.g. Searle, 1971, section 9.8b). In some cases it might not even be possible to find such estimators at all (e.g. Crowder and Hand, 1990, Examples 3.1 and 3.2). In addition, when the data contain missing values, or when the layout is unbalanced, the ANOVA approach is generally more awkward to apply. There might be numerical problems with increased computational complexity, and problems of interpretation since the various effects become entangled. Expected mean squares become even more laborious to calculate (Crowder and Hand, 1990, sections 3.3 and 3.4). There can be many alternative estimators for the variance components but no very definite rule for choosing between them (Searle, 1971, section 10.7). Finally, the F-ratios fail in general to have F-distributions (Searle, 1971, section 9.9).

15.4 LIKELIHOOD APPROACH

The ANOVA methods illustrated above are not the only possible approach. A basis for much statistical theory and practice is the likelihood function. In general terms one uses the assumed model (e.g. equations 15.1, 15.2 and 15.3) to calculate the probability of observing the actual recorded data. This probability involves the parameters of the model (e.g. σ^2 and σ_L^2) and, when regarded as a function of the parameters with the data fixed at the observed values, it is called the likelihood function. It can be used for inference in a number of ways. For instance, the maximum likelihood estimators of the parameters are the values which maximize the likelihood function, i.e. give maximum probability to the data actually observed.

Most modern statistical methodology involves the likelihood, or closely related functions, in some way. Standard statistical theory shows that likelihood-based methods are usually optimal in an asymptotic sense, i.e. as the number of observations increases. This asymptotic optimality often continues to hold down to smaller sample sizes. What this means in effect is that the likelihood function extracts as much information from the data as it is possible to extract. Other methods may perform as well in some cases, but they cannot do better.

Just as ANOVA suffers from certain problems so does maximum likelihood, but they are different in kind. Generally speaking, the firm theoretical foundation of likelihood-based methods enables problems to be solved in a structured rather than an *ad hoc* manner. Practical aspects, such as the often heavier computational effort required for maximum likelihood than for ANOVA, are now easier to handle with modern computing equipment. General accounts of likelihood methodology for the types of model considered here can be found in Laird and Ware (1982) and Crowder and Hand (1990, Chapter 6). A more specialized discussion is given by Crowder (1993).

15.5 DISCUSSION

As indicated in section 15.1, the discussion of formal methodology can be separated into aspects concerning models and methods. It is argued below that appropriate use should be made of fixed and random effects in the models, and that appropriate ANOVA methods should be employed in the first instance. In the longer term, serious consideration should be given to more formal likelihood-based methods.

15.5.1 Models

The main dichotomy here involves the adoption of fixed or random effects, in particular for laboratory differences. In certain circumstances either may be appropriate. In the examples treated in sections 15.2 and 15.3, the laboratories were regarded as a random selection from a larger pool of such establishments. However, if the inferences are to be drawn solely for the current fixed set of laboratories, and not broadened to laboratories of the type in general, then fixed laboratory effects are appropriate. The effect on the analysis of the data of this change of attitude is to compute the F-statistics and estimate the parameters differently in general. Marchandise and Colinet (1983) appear to have based their analysis on random laboratory effects.

For the other main factors – materials and methods of measurement – it would seem uncontroversial to adopt respectively random and fixed effects in most situations. In the case of materials, the effect may be properly omitted altogether from the model if either complete homogeneity obtains or all the measurements are made on different batches, bars or samples. This is implicit in the approach adopted by Marchandise and Colinet.

15.5.2 Methods

The main choice here is between ANOVA and likelihood methods. The former are more familiar to practitioners in the current context, and easier to implement because of the present availability of computer packages. However, even for the basic case of balanced layouts, the calculation and presentation of expected mean squares is not routinely undertaken in most packages. Fewer packages are designed to cope with unbalanced layouts, and there the additional calculation of expected mean squares is an even greater problem. Likelihood methods have certain theoretical advantages over other methods, and are more flexible in that a wider range of different data types can be handled.

15.5.3 Practical considerations

The above discussion of pure models and methods does not remove the need for good statistical practice. It is always unwise to take on trust the assumption that the adopted model describes perfectly the generation of the data. Hence, a thorough analysis must include a certain amount of less formal data examination. This includes (1) an initial screening to look for outliers, anomalies or unexpected patterns before fitting the model; and (2) critically examining the results after fitting the model, e.g. looking at residuals to identify possible violations of model assumptions. Depending on the outcome of these investigations it may be decided that the model needs to be modified, e.g. to allow for non-additivity of the effects via interaction terms, or to introduce other previously omitted factors, or apply transformations.

Regarding the formal methods employed, there is scope for checking their efficacy in certain respects. For instance, asymptotic approximations are used routinely to derive standard errors for estimates; to derive confidence intervals; and to perform hypothesis tests. It is often feasible to check, and even improve upon, the accuracy of such approximations by simulation, particularly in the case of small data sets.

15.5.4 Development programme

The description and treatment of the subject above has been necessarily condensed. It would be easy to fill a book with a full treatment, and this has been done by many authors. In order to develop the material into a system usable in practice, the following aspects are identified.

(a) Catalogue

Only a few basic situations have been described above and the calculations and results have not been presented in detail. For the purpose of setting up a working system for data processing, however, a much more comprehensive catalogue would be required. A wider range of situations would need to be covered, and the consequences and details would need to be put on record unambiguously.

(b) Computing

The informal examination of data, discussed in section 15.5.3, is provided for in many current statistical packages. Techniques include the search for outliers; testing homogeneity of variances; ANOVA; plotting residuals against normal scores and omitted factors; and testing samples for assumed distributional form. In contrast to the availability of routine tools, likelihood-based methods are not commonly supplied in packages. This is probably due to the greater complexity of the computations, e.g. iterative non-linear optimization is usually necessary. However, there is a substantial body of experience to be found in the literature (e.g. Crowder and Hand, 1990, Chapters 5 and 6), and this could provide a foundation for the development of computer programs for the present purpose.

(c) Design

The study of the design of experiments has a long history in statistics, and the amount of published material on the subject is massive. However, almost all of it is concerned with the case of independent observations, whereas the introduction of random effects induces non-independence of the observations. It appears that the amount of material readily available on design for the present application is rather limited. However, although this area of knowledge is underdeveloped there are some clear directions for progress. First steps would involve the appraisal of standard designs for independent data in the present framework, both theoretically and via simulation.

15.6 CONCLUSIONS

The need for proper statistical approaches to intercomparison studies is becoming more widely accepted. The two main components focused upon in this chapter are models and methods. Models are best specified by a process of careful discussion between the materials metrologist, for instance, and the statistician. Methods belong more to the realm of statistical theory: if the data and the model form a useful combination, the methods should follow naturally. Thus the more important aspect from the point of view of the materials metrologist is modelling.

The useful combination alluded to above also involves the design of studies which, as mentioned in section 15.5.4, has not received an enormous amount of attention in the present context. Put at its most basic level, a badly designed study might yield data incapable of answering some of the questions of interest. More research is needed here.

ACKNOWLEDGEMENTS

The work here results from the identification of certain problems by scientists at the National Physical Laboratory. We are grateful to Maurice Cox, Tom Gibbons, Malcolm Loveday, Brian Dyson and others at NPL for introducing and explaining these problems to us.

APPENDIX 15.A NOTATION

l	number of laboratories
n_i	number of measurements made at laboratory i
n	total number of measurements
y_{ij}	jth measurement made at ith laboratory
b	number of bars

m	number of measurement methods
y_{ijk}	measurement at laboratory i on bar j using method k
$\bar{y}$	overall sample mean
$\bar{y}_i$	sample mean for laboratory i
s^2	residual mean square
μ	overall (population) mean value
e_{ij}	random error
σ^2	variance of e_{ij}
α_i^L	random laboratory effect
α_j^B	random bar effect
σ_L^2	variance of α_i^L
σ_B^2	variance of α_j^B
SST	total sum of squares
SSW	within-laboratories sum of squares
SSB	between-laboratories sum of squares
F	variance ratio statistic
p	significance level of test
$\tilde{\mu}_w$	weighted mean estimate of μ
w_i	weight for $\bar{y}_i$ in weighted mean estimate
$\hat{\mu}$	maximum likelihood estimate of μ
β_k^M	fixed method effect
α_{ik}^{LM}	random laboratory $\times$ method interaction
σ_k^2	variance of α_{ik}^{LM}

APPENDIX 15.B   COMPUTATION OF SUMS OF SQUARES IN ONE-WAY ANOVA

The sums of squares are given by

$$\text{SST} = \sum_{i=1}^{l} \sum_{j=1}^{n_i} (y_{ij} - \bar{y})^2 \quad \text{(total variation about overall mean)}$$

$$\text{SSB} = \sum_{i=1}^{l} n_i (\bar{y}_i - \bar{y})^2 \quad \text{(variation between laboratory means)}$$

$$\text{SSW} = \sum_{i=1}^{l} (n_i - 1)s_i^2 \quad \text{(residual variation within laboratories)}$$

where

$$\bar{y} = n^{-1} \sum_{i=1}^{l} \sum_{j=1}^{n_i} y_{ij}$$

is the overall sample mean,

$$\bar{y}_i = n_i^{-1} \sum_{j=1}^{n_i} y_{ij}$$

is the *i*th laboratory sample mean, and

$$s_i^2 = (n_i - 1)^{-1} \sum_{j=1}^{n_i} (y_{ij} - \bar{y}_i)^2$$

is the *i*th laboratory sample variance.

APPENDIX 15.C COMPARISON OF ESTIMATORS $\tilde{\mu}_1$ AND $\tilde{\mu}_2$

The variances of the estimators $\tilde{\mu}_1$ and $\tilde{\mu}_2$ are

$$v_1 = n^{-1}(\sigma^2 + n^{-1}\sigma_L^2 \sum_{i=1}^{l} n_i^2)$$

$$v_2 = l^{-1}(\sigma_L^2 + l^{-1}\sigma^2 \sum_{i=1}^{l} n_i^{-1})$$

respectively. The ratio of these is

$$r = v_1/v_2 = (l/n)\{1 + n^{-1}\rho \sum_{i=1}^{l} n_i^2\} \div \{\rho + l^{-1} \sum_{i=1}^{l} n_i^{-1}\}$$

where $\rho = \sigma_L^2 / \sigma^2$. As $\rho \to 0$,

$$r \to l^2 / \left(n \sum_{i=1}^{l} n_i^{-1} \right) = H/A \leqslant 1$$

where

$$A = l^{-1}n = l^{-1} \sum_{i=1}^{l} n_i$$

is the arithmetic mean and

$$H = \left(l^{-1} \sum_{i=1}^{l} n_i^{-1} \right)^{-1}$$

is the harmonic mean of the n_i. As $\rho \to \infty$,

$$r \to l \sum_{i=1}^{l} n_i^2 / n^2 \geqslant 1$$

since

$$\sum_{i=1}^{l} n_i^2 \geq \left(\sum_{i=1}^{l} n_i\right)^2 / l = n^2 / l$$

APPENDIX 15.D MAXIMUM LIKELIHOOD ESTIMATOR OF μ

Let y_i be the $n_i \times 1$ vector $(y_{i1}, \ldots, y_{in_i})^T$ of data from the ith laboratory. Then the y_i are independent random variables with means $\mu 1_{n_i}$ and covariance matrices

$$\Sigma_i = \sigma^2 I_{n_i} + \sigma_L^2 J_{n_i}$$

where I_{n_i} denotes the $n_i \times n_i$ unit matrix and J_{n_i} is the $n_i \times n_i$ matrix of 1s. The corresponding normal log-likelihood function is

$$L = -(1/2) \sum_{i=1}^{l} \{n_i \log(2\pi) + \log(\det \Sigma_i) + Q_i\}$$

where

$$Q_i = (y_i - \mu 1_{n_i})^T \Sigma_i^{-1} (y_i - \mu 1_{n_i})$$

The derivative of L with respect of μ is

$$\partial L / \partial \mu = -(1/2) \sum_{i=1}^{l} \partial Q_i / \partial \mu = \sum_{i=1}^{l} 1_{n_i}^T \Sigma_i^{-1} (y_i - \mu 1_{n_i})$$

$$= \sum_{i=1}^{l} \left(\sum_{jk} y_{ij} \Sigma_i^{jk}\right) - \mu \sum_{i=1}^{l} \left(\sum_{jk} \Sigma_i^{jk}\right)$$

where Σ_i^{jk} denotes the (j, k)th element of Σ_i^{-1}. The form of the MLE $\hat{\mu}$ follows from equating $\partial L / \partial \mu$ to zero, using

$$\Sigma_i^{-1} = \sigma^{-2} \{I_{n_i} - \sigma_L^2 J_{n_i} / (\sigma^2 + n_i \sigma_L^2)\}$$

REFERENCES

Crowder, M. J. (1993) Interlaboratory comparisons: round robins with random effects. *Applied Statistics*, **42**, 409–25.

Crowder, M. J. and Hand, D. J. (1990) *Analysis of Repeated Measures*, Chapman and Hall, London.

Davies, P. L. (1988) Statistical evaluation of interlaboratory tests. *Fresenius Zeitschrift fur Analytische Chemie*, **331**, 513–19.

Laird, N. M. and Ware, J. H. (1982) Random effects models for longitudinal data. *Biometrics*, **38**, 963–74.

Mandel, J. (1959) The measuring process. *Technometrics*, **1**, 251–67.

Mandel, J. and Lashof, T. W. (1959) The interlaboratory evaluation of testing methods. *ASTM Bulletin*, **299**, 53–61.

Marchandise, H. and Colinet, E. (1983) Assessment of methods of assigning certified values to reference materials. *Fresenius Zeitschrift fur Analytische Chemie*, **316**, 669–72.

Olkin, I. and Sobel, M. (1987) A model for interlaboratory differences, in *Advances in Multivariate Statistical Analysis* (ed. A .K. Gupta), Reidel, Dordrecht, pp. 303–14.

Rocke, D. M. (1983) Robust statistical analysis of interlaboratory studies. *Biometrika*, **70**, 421–31.

Searle, S. R. (1971) *Linear Models*, Wiley, New York.

Tukey, J. W. (1949) One degree of freedom for non-additivity. *Biometrics*, **5**, 232–42.

Youden, W. J. and Steiner, E. H. (1975) *Statistical Manual of the Association of Official Analytical Chemists*. Association of Official Analytical Chemists, Arlington, VA.

Index